# Bayesian Entrepreneurship

# Bayesian Entrepreneurship

edited by Ajay Agrawal, Arnaldo Camuffo, Alfonso Gambardella,
Joshua Gans, Erin L. Scott, and Scott Stern

The MIT Press
Cambridge, Massachusetts
London, England

The MIT Press
Massachusetts Institute of Technology
77 Massachusetts Avenue
Cambridge, MA 02139
mitpress.mit.edu



The MIT Press would like to thank the anonymous peer reviewers who provided comments on drafts of this book. The generous work of academic experts is essential for establishing the authority and quality of our publications. We acknowledge with gratitude the contributions of these otherwise uncredited readers.

This book was set in Times New Roman by Westchester Publishing Services. Printed and bound in the United States of America.

Library of Congress Cataloging-in-Publication Data is available.

ISBN: 978-0-262-05215-3

EU Authorised Representative: Easy Access System Europe, Mustamäe tee 50, 10621 Tallinn, Estonia | Email: gpsr.requests@easproject.com

# Contents

# Foreword

The greatest myth of entrepreneurship may be that entrepreneurs are born, not made. Those who subscribe to this view argue that it is the innate gifts of the individual—a heroic tolerance of risk, extreme self-confidence, omniscient vision, and ultimate powers of persuasion—that determine whether someone becomes a successful entrepreneur. Accordingly, efforts to teach entrepreneurship are futile and, by extension, so are efforts to conduct scholarly research on the topic beyond hagiographies.

The myth of the entrepreneur has been repeated so often that it can be easy to overlook its flawed logic. All professions benefit from talent—the innate skills and traits that make some more capable than others. But they also benefit from training—from lawyers to engineers, scientists to athletes. Certainly, gifted brain surgeons are endowed with a combination of exceptional intellectual and emotional capacities and manual dexterity. And yet, no one argues that "brain surgeons are born, not made"—nor (thankfully) has anyone been persuaded to stop training surgeons, lawyers, engineers, actors, musicians, dancers, accountants, economists, scientists, or athletes, all of whom benefit from their innate talent when combined with rigorous and systematic training.

No different from any other professional, an entrepreneur should engage in a profession—of entrepreneuring. We intentionally use the word "entrepreneuring" instead of "entrepreneurship." Entrepreneurship, in its literal meaning, suggests a state of being. Someone who is an entrepreneur, however, is not just in a state of being an entrepreneur. They are engaged in a specific set of activities—entrepreneuring—just as engineers are engaged in engineering. The word "entrepreneurship" overemphasizes the qualities of the individual, whereas "entrepreneuring" recognizes the activities, behaviors, and practices that—through training—allow people to become better entrepreneurs.

A systemic body of knowledge and principles is essential to effective professional education. And yet, unlike other fields within management, entrepreneurship has lacked a unifying framework defining the central questions and key skills and offering a conceptually coherent means to address them. It is hampered by the lack of what Michael Porter's Five Forces framework provided to the field of strategy or what the efficient market hypothesis and capital asset pricing model contributed to the field of finance. Entrepreneurship has been a phenomenon without a central theory or even a set of theories.

The collection of papers in this volume of Bayesian entrepreneurship is an attempt to (re)orient the field of entrepreneurship around the focal challenge facing the entrepreneur: decision-making under uncertainty and the processes of experimentation, learning, and adaptation. By structuring the field around a core activity that is relevant to all entrepreneurs, regardless of the market opportunity they are addressing or the technological solutions they are generating, this approach makes a potentially large contribution to the study and practice of entrepreneurship.

Bayesian entrepreneurship is an attempt to orient the field around entrepreneurs' prior beliefs about their opportunity space and how those beliefs evolve over the course of their entrepreneurial journey. If we take a Bayesian perspective, then entrepreneurship is more about establishing hypotheses rather than having the right "vision," and running the right experiments rather than writing one-hundred-page business plans. In fact, a Bayesian view of entrepreneuring accepts that entrepreneurs rarely have their initial vision right, given that their priors are based on incomplete information, and thus are necessarily flawed. What they can do is systematically learn through experimentation.

Whatever their starting point, entrepreneurs will, by definition, face varying degrees of uncertainty and risk. For those who follow the model developed at Flagship Pioneering and who are engaged in completely unprecedented science, technology, or markets, the challenge is one of uncertainty in the sense Frank Knight described it more than a hundred years ago. In this "Knightian" world, entrepreneurs are akin to pioneers who venture into new, uncharted lands in order to discover (and settle) the most attractive places, and they need to constantly experiment in order to chart their territory. For others solving more familiar problems in well-charted terrain, their job will be to experiment to resolve specific risks. Regardless, Bayesian entrepreneurs are experimenters par excellence.

An entrepreneur earns rents—for themselves, their team, and their investors—by discovering among the vast pool of uncertain opportunities those that offer the most attractive risk–return profiles. The process by which this occurs is anything but straightforward. In most cases, the Knightian entrepreneur, like the pioneer, has only a vague notion about what might lie over the horizon. Other entrepreneurs will be focused on which path up the mountain is more likely to yield a rapid ascent, given the risks inherent in both. They may be able to make probability estimates of technical or market risks, but outcomes can vary dramatically. Both probe, both experiment, and both update their plans and visions (their priors) based on what they learn. Bayesian perspectives are useful for understanding the approaches of entrepreneurs along the full spectrum of risk and uncertainty.

If we take a Bayesian perspective, then entrepreneurial strategies emerge through an evolutionary process of exploration (search), selection (experimentation and evaluation), and adaptation (learning). Emergent processes generate outcomes that are not fully predictable at the outset yet might appear perfectly rational in hindsight. There are serious implications of this concept for the study and teaching of entrepreneurship. Too often, case studies of entrepreneurship draw lessons that are distilled based on a particular company's successful strategy. This has a number of flaws for training a more effective Bayesian entrepreneur: Could they repeat such a strategy (e.g., how could you create the next Amazon)? Was it luck or a judgment based on information gathering and experimentation? What were the alternative paths explored but not selected? Although stories of failure often teach us more, they are frequently pushed out of the typical entrepreneurship classroom in favor of heroic, idiosyncratic successes.

An emergent (Bayesian) perspective warns us of the fallacy in teaching only from the simplified, linear success stories. Entrepreneurs change their mind all the time ("update their priors" in Bayesian terms). Steve Jobs was famously adamant that he didn't want third parties developing apps for the iPhone—until he changed his mind. Amazon Web Services was not the product of a grand vision of Jeff Bezos. It emerged from a small experiment to open Amazon's application interfaces to third-party developers in order to drive more traffic to Amazon's retail site from "affiliate" sellers. It was only after seeing the results of this experiment that Amazon leaders began to realize the greater potential of web services. Netflix initially used traditional per-rental pricing of videos (similar to incumbents such as Blockbuster) but later changed its strategy to a monthly subscription pricing model. Moderna did not set out to create a new vaccine technology but rather to enable a patient's body to make any needed biotherapeutic drug. After building its unprecedented mRNA platform, the use of this new therapeutic modality to make a prophylactic vaccine for healthy subjects (not patients) emerged as the preferred application.

When viewed as the result of an emergent process of search and selection, a successful strategy is just one outcome along a path of many alternative evolutionary branches. Teaching entrepreneurship through a Bayesian lens means following these paths and focusing on the (many) choices along the path—and learning lessons about the search and selection processes themselves—rather than simply the final selected outcome that we observe as a success. It also means learning about the paths not taken and the cases where, even when entrepreneurs searched widely and designed, executed, and evaluated their experiments effectively, the outcome was still disappointing. It is also an opportunity to teach about when to stop along a particular branch—in a way that is more systematic and dispassionate—so as to reallocate resources (including the entrepreneur's own time and energy) to other activities.

The papers in this volume help us understand the economics behind this Bayesian process. They give analytical rigor to activities that skillful, reflective entrepreneurs and venture builders do on a daily basis. The economics of Bayesian entrepreneurship are a reminder that experimentation is, after all, not free. Entrepreneurs do not have unlimited resources. They do not have unlimited time, money, or talent. Allocating resources to one experiment means not allocating it to another. It is one thing to believe that entrepreneurs should experiment and update their beliefs based on new data—it is quite another to understand what the best experimental strategies may be under different circumstances.

The Bayesian approach developed in this volume begins to help us understand how to make choices among different experimental strategies and how those choices may impact venture performance—and how those pathways might be more or less successful in different regions and countries, depending upon the resources of the ecosystems in which entrepreneurs build their ventures. It also highlights an important organizational aspect of the entrepreneurial process: that individuals and their teams must be able to adapt their priors (and behavioral economic research suggests this may not be so easy in practice!), but so too must their stakeholders (employees, directors, investors, partners, etc.). The entrepreneurs' burden is not just to run the right experiment, and keep their minds open to updating, but also to convince others as well.

The notion that entrepreneuring revolves around experimentation also invites inquiry into questions of organizational process. Many writings highlight, even celebrate, the seeming chaotic nature of entrepreneurial ventures. They attribute their dynamism and adaptability to their lack of processes and structure (which are, in turn, equated with "bureaucratic" corporate enterprises). Indeed, in this perspective, the words "entrepreneur" and "process" do not belong in the same sentence. A Bayesian perspective—which puts experimentation at the heart of the entrepreneurial process—challenges this perspective. Experimentation is a process that requires careful consideration of design, disciplined execution, rigorous collection and analysis of data, and thoughtful evaluation. Experiments can produce surprises (they often do), but the process of experimentation itself should not be a surprise. And by the same logic, the process of experimentation can be shaped and structured to be more systematic and undertaken in parallel and at scale, as has been the focus of decades of work at Flagship Pioneering, bringing rigor to entrepreneuring.

From a Bayesian perspective, there is nothing ad hoc about the entrepreneurial process. It is a rigorous, careful process of experimentation, iterative learning, and adaptation.

Noubar Afeyan, Flagship Pioneering
Fiona Murray, MIT Sloan School of Management
Gary Pisano, Flagship Pioneering

# Preface

The field of entrepreneurship has long been characterized by debates over whether success is a product of innate talent or whether it can be systematically nurtured and taught. In this book, we introduce the concept of Bayesian entrepreneurship—a novel, synthetic approach that reconceptualizes entrepreneurial decision-making as a dynamic process of forming, testing, and updating subjective beliefs in the face of uncertainty. Drawing on rigorous Bayesian learning principles, the book integrates insights from decision theory, economics, and behavioral studies to provide a coherent framework for understanding how entrepreneurs identify opportunities, experiment with strategies, and ultimately persuade stakeholders. The aim is to offer researchers and graduate students new tools and pathways to analyze and teach entrepreneurial phenomena.

A Bayesian approach to entrepreneurial decision-making offers a framework to assess how entrepreneurs form beliefs about the prospects for a given opportunity, how these beliefs evolve over time through active experimentation and learning, and the consequences of such beliefs for entrepreneurial strategy and performance. The goal of this book is to shape distinctive implications and empirical approaches to the study of entrepreneurship guided by three founding premises. First, the entrepreneur must hold stronger positive beliefs about the opportunity *relative* to others. This involves a distinct theory that translates into a different perspective on the opportunity's prospects. Second, this systematic divergence in beliefs impacts how an entrepreneur will undertake learning about an opportunity. Notably, the demand for "experiments" is fundamentally influenced by beliefs about the opportunity. For example, relative to a disinterested agent, a Bayesian entrepreneur will conduct experiments that are more likely to allow for false positives than false negatives. Finally, entrepreneurs are more likely to convince those who share their idiosyncratically positive beliefs about an opportunity (with implications for homophily and firm culture), yet will also engage in choosing experiments that cater to those with different (more negative) beliefs than they themselves hold.

Chapter 1, by Ajay Agrawal, Arnaldo Camuffo, Alfonso Gambardella, Joshua Gans, Erin L. Scott, and Scott Stern, lays the theoretical groundwork for the Bayesian approach to entrepreneurship. The chapter argues that entrepreneurial decision-making can be rigorously understood as a process in which individuals hold heterogeneous and, often, strongly optimistic prior beliefs about opportunities. These priors are systematically updated through experimentation and learning via Bayes' rule. By synthesizing insights from traditional economic theory and emerging behavioral research, the authors demonstrate how the Bayesian framework provides a structured method for evaluating risk, testing hypotheses, and ultimately guiding strategic choices in uncertain environments.

In chapter 2, Alfonso Gambardella, Luca Gius, and Scott Stern shift the focus to the nature of the entrepreneurial agent—termed *Homo entrepreneuricus*. The contributing scholars examine how individual entrepreneurs differ from other economic agents by virtue of their idiosyncratic, often contrarian, beliefs and cognitive predispositions. By combining empirical observations with theoretical modeling, the chapter illustrates how these unique traits shape opportunity recognition and decision-making. The authors argue that understanding these distinct features is essential to explaining why entrepreneurs consistently pursue ventures that defy conventional wisdom, thereby setting the stage for a Bayesian analysis of their behavior.

Joshua Gans, in chapter 3, critically contrasts the classical notion of Knightian uncertainty with the Bayesian perspective on entrepreneurial decision-making. Through a detailed exploration of uncertainty types, the author demonstrates that although entrepreneurs undoubtedly face profound unknowns, the systematic application of Bayesian updating allows for the quantification and management of such uncertainty. By formalizing the conditions under which probabilistic reasoning remains valid, even in the presence of incomplete information, this chapter provides a robust argument for abandoning fatalistic views of uncertainty in favor of a model that embraces learning and adaptability.

In chapter 4, Ashish Arora brings theory into practice by presenting a user-oriented examination of Bayesian entrepreneurship. The author, drawing from both academic research and real-world case studies, illustrates how entrepreneurs apply Bayesian principles to navigate complex market dynamics and secure necessary resources. The chapter emphasizes the role of experimentation, not only as a means for internal learning but also as a tool for persuading external stakeholders such as investors and partners. This practical orientation bridges the gap between abstract theory and the tangible challenges of entrepreneurial execution.

In chapter 5, Thomas Åstebro, Frank M. Fossen, and Cédric Gutierrez turn to the behavioral aspects of decision-making in entrepreneurship. The contributing authors investigate whether the apparent biases and heuristics observed in entrepreneurial behavior are simply flaws or if they can be reinterpreted as rational strategies under a Bayesian framework. By juxtaposing traditional interpretations of overconfidence and optimism with models of systematic belief updating, the chapter challenges prevailing notions and highlights how even seemingly irrational behavior may have adaptive value when viewed through the lens of Bayesian learning.

Having explored what Bayesian entrepreneurship is, in part II, the book then considers where priors come from. Timo Ehrig, Teppo Felin, and Todd Zenger (chapter 6) examine the foundational role of causal reasoning in shaping entrepreneurial priors. Authored by experts in both entrepreneurship and decision science, the chapter argues that the process of constructing a causal model of opportunity is central to forming robust prior beliefs. Through a blend of theoretical discussion and empirical evidence, the authors show how entrepreneurs use logical reasoning and scientific principles to develop, test, and refine their hypotheses about value creation.

In chapter 7, Elena Novelli presents a detailed exposition of how a scientific approach to decision-making naturally aligns with Bayesian learning. By conceptualizing entrepreneurial action as a series of experiments that test underlying hypotheses, the chapter elucidates the mechanisms by which systematic data collection and analysis improve decision quality. The discussion highlights the importance of rigorous experimentation in reducing uncertainty and accelerating the convergence of beliefs, thereby enhancing overall venture performance.

Chapter 8 from Ramana Nanda explores the intricate interplay between initial beliefs, experimental design, learning, and persuasive communication. The authors demonstrate that the design of experiments not only informs an entrepreneur's self-assessment but also plays a critical role in convincing external stakeholders to invest in the venture. By analyzing how different experimental strategies can yield varied signals and updates, this chapter provides a nuanced understanding of how Bayesian persuasion can be effectively employed to align divergent beliefs among partners, investors, and other decision-makers.

Part III of the book focuses on how priors are updated through experimentation and communication. Hyunjin Kim, in chapter 9, focuses on the centrality of data in the entrepreneurial learning process. Through an in-depth analysis of various data collection and interpretation methodologies, the author argues that accurate, timely, and well-structured information is vital for updating prior beliefs. The chapter outlines practical techniques for minimizing noise and bias in experimental data, thereby ensuring that learning is both meaningful and actionable for entrepreneurial decision-making.

In chapter 10, Joshua L. Krieger deepens the discussion by exploring how continuous Bayesian updating drives adaptive learning in entrepreneurial ventures. By providing formal models and empirical evidence, this chapter illustrates how iterative belief revision—prompted by sequential experiments—leads to improved strategic decisions. The author emphasise that this dynamic process is essential not only for refining current ventures but also for identifying when to pivot or terminate a project in light of new evidence.

Susan Cohen and Rem Koning (chapter 11) delve into the role of advice and external information in shaping entrepreneurial choices. The contributing scholars examine how the dissemination and reception of advice can be understood as a Bayesian process, where the credibility and prior beliefs of both the adviser and the entrepreneur interact to produce a persuasive outcome. The chapter provides insights into how entrepreneurs can leverage external expertise to update their beliefs, thereby enhancing decision quality and strategic alignment with key stakeholders.

The final part of the book considers the pedagogy and practice of entrepreneurs and how these are impacted on by a Bayesian approach. Erin L. Scott (chapter 12) investigates the interrelationship between learning processes, strategic commitment, and decision-making in entrepreneurial ventures. The author articulates how early-stage learning influences long-term commitment decisions and how Bayesian updating can mitigate the risks associated with premature commitment. By examining both theoretical models and case studies, this chapter offers a comprehensive framework for understanding how initial choices set the trajectory for future venture performance.

In chapter 13, Chiara Spina focuses on the determinants of effective learning in entrepreneurial contexts. The author identifies and analyzes a range of factors—from individual cognitive capabilities to external environmental influences—that facilitate or impede the learning process. The chapter emphasizes that systematic experimentation, when properly designed and executed, not only accelerates learning but also enhances the entrepreneur's ability to adapt to evolving market conditions.

The final chapter (chapter 14) from Andrea Coali, Claudia Frosi, Diego Jannace, Saeid Kazemi, and Abhinav Pandey, integrates the theoretical and practical dimensions of the book by advocating for the incorporation of causal reasoning into entrepreneurial education. The authors argue that by teaching future entrepreneurs to construct and critically evaluate causal models, academic programs can better prepare them for the uncertainties of real-world ventures. This chapter outlines practical pedagogical strategies and offers a road map for embedding Bayesian methods into the curriculum, ensuring that the next generation of entrepreneurs is equipped with the tools to make sound, data-driven decisions.

In sum, this book is designed for researchers in entrepreneurship and graduate students seeking innovative analytical tools and theoretical frameworks. By weaving together rigorous Bayesian methods with practical insights from entrepreneurial practice, the volume not only challenges conventional perspectives but also provides a comprehensive guide for systematically exploring and understanding the complex dynamics of entrepreneurial phenomena.

# I

# What Is Bayesian Entrepreneurship?

# 1

## The Foundations of Bayesian Entrepreneurship

Ajay Agrawal, Arnaldo Camuffo, Alfonso Gambardella, Joshua Gans, Erin L. Scott, and Scott Stern

### 1.1 Introduction

The study of entrepreneurship is inherently linked to the study of decision-making. As a phenomenon, potential entrepreneurs choose to pursue some sort of opportunity (or not), undertake activities that allow them to learn about that opportunity (or not), and make choices that allow them to realize the potential of that opportunity (or not). At each stage of this dynamic process, founders face a set of decisions that inevitably shape the future path by which they pursue a particular opportunity and the organization they build in order to take advantage of that opportunity. There is, of course, no requirement that entrepreneurs (or others) make these decisions optimally, but at the same time, there is no requirement that entrepreneurial decision-making is inherently flawed or biased. There is, of course, no requirement that all entrepreneurs make their decisions in the same way, and certain approaches to entrepreneurial decision-making may be far more effective than others. Centering the role of decision-making in entrepreneurship is simply highlighting that entrepreneurs do indeed engage in decision-making (i.e., they are *not* simply following directions) and that their decision-making is consequential for performance. The decisions they make, and possibly how they are made, affect the financial and nonfinancial benefits they accrue.

A central challenge, then, in the field of entrepreneurship is the development of a theoretical framework that is simultaneously broad enough to allow for a wide range of entrepreneurial decision-making approaches yet also specific enough that researchers are able to ask and answer questions of fundamental theoretical and empirical interest and potentially provide guidance for practitioners or policymakers. Striking this balance has proved elusive. Consider the contrast between two broad approaches that characterize a large fraction of existing studies of entrepreneurial decision-making: environmental and behavioral. On the one hand, the environmental approach, grounded in economics and finance, largely abstracts away from how entrepreneurs identify opportunities or how they make decisions and focuses instead on the impact of the microeconomic, strategic, and institutional environment on the choice to enter entrepreneurship or the impact of entrepreneurship on outcomes such as earnings or wealth (Evans and Jovanovic, 1989; Hamilton, 2000; Hurst and Lusardi, 2004; Lazear, 2004; Åstebro et al., 2011). On the other hand, more behavioral approaches have tended to abstract away from the potential for entrepreneurs to optimize and focus more on the role of the background and experiences of entrepreneurs (Shane, 2003), the propensity for entrepreneurs to learn through action and local contingency (Bhide, 1994; Sarasvathy, 2001), and the potential role of biases and heuristics to distort entrepreneurial choice. Both of these broad approaches and their many offshoots have proven valuable in shaping our understanding of the role of the external environment in shaping entrepreneurial choice and the role of entrepreneurial motivation and behavior in enacting those choices.

Over the past decade, a third approach has emerged that focuses squarely on the interplay between entrepreneurial decision-making and the process of entrepreneurial learning (Rosenberg, 1992; Murray and Tripsas, 2004; Sull, 2004). Three related (but distinct) areas are highlighted in this rapidly emerging body of research. First, entrepreneurs are decision-makers making active choices under uncertainty (i.e., entrepreneurship is itself an experiment), including critically the choice to pursue a particular opportunity (Nanda and Rhodes-Kropf, 2016; Manso, 2016). In other words, although entrepreneurs may be subject to biases or constrained by their local environment, the choice to pursue a particular opportunity (and not others) is clearly a choice made by the entrepreneurs themselves, and the construction of the opportunity is itself uncertain (or else there would be no opportunity to pursue). Second, whether and how to pursue an opportunity is shaped by purposeful learning and experimentation as a tool for entrepreneurial decision-making (among many others, Gans et al., 2019, Camuffo et al., 2020, and, importantly, Ries, 2011, in the domain of entrepreneurial practice). Entrepreneurial experimentation allows entrepreneurs to (somehow) reduce the uncertainty associated with the pursuit of an opportunity, and this process of uncertainty reduction shapes the choices that entrepreneurs then make on an ongoing basis. However, as emphasized by Camuffo et al. (2020), we are at an early stage of understanding how entrepreneurs learn from experiments and how different types of experimentation (e.g., more iterative vs. scientific approaches) impact entrepreneurial strategy and performance. Finally, to ultimately realize the potential of the opportunity they are pursuing, entrepreneurs must separately *persuade* others of the value associated with that opportunity. Success in pursuing a given opportunity depends on the ability of the entrepreneur to attract resources and capabilities from others (Stevenson, 1983), and the choices that entrepreneurs make are not in isolation. Instead, the successful realization of an entrepreneurial opportunity depends on choices by other decision-makers—potential investors, employees, and even customers—who will also be making decisions about whether to invest time, effort, or resources to engage with a nascent venture. For each of these domains—how entrepreneurs form their initial beliefs about opportunities, how they learn from proactive experimentation, and how they ultimately persuade others (or not)—this work confronts directly the fact that entrepreneurial decision-making takes place under significant uncertainty and that a primary means by which entrepreneurs realize the opportunity is through the resolution of that uncertainty (in ultimately a public way).

The purpose of this chapter is to synthesize the different streams of work in this third approach under an umbrella that we will refer to as "Bayesian entrepreneurship." This synthesis builds on two distinct but related insights. First, different (and often disparate) studies of the process of entrepreneurial decision-making—the formation of beliefs, entrepreneurial learning and experimentation, and entrepreneurial communication—are each linked to one another. At a broad level, what you believe influences what types of experiments and learning you might undertake, and each of these will then influence what other actors you are able to persuade in order to attract resources and capabilities. Second, a necessary condition on the nature of the pursuit of entrepreneurial opportunity allows us to place additional structure on this linkage: Entrepreneurs *choose* the opportunities they pursue. Although this may seem obvious, the fact that entrepreneurs choose to pursue a *particular* opportunity has far-reaching implications. Simply put, entrepreneurs (almost by construction) hold "contrarian" beliefs about the opportunities they choose to pursue (Masters and Thiel, 2014). Almost by construction, an entrepreneur choosing a particular opportunity must hold a *relatively more positive* belief about the potential for that opportunity. Conversely, if entrepreneurs believed that the value of the opportunity they were pursuing was equally valued by others (who perceived the same chance of success, risk, etc.), then the rationale for pursuing that opportunity would disappear. Bayesian entrepreneurship brings these two insights together: How does the fact that entrepreneurs choose the opportunity they pursue impact entrepreneurial experimentation, strategy, and performance?

Consider the case of Zappos (as memorably recounted in Hsieh, 2010), founded by Nick Swinmurn based on his belief in the potential for online shoe retail. Although there were, of course, many other online retail businesses started during the dot-com "boom," very few other growth-oriented businesses entered the shoe category, as footwear traditionally required customized fittings, bulky shipping, and a high level of returns and exchanges. Despite these observable challenges (which presumably dissuaded others) and his lack of background in the shoe industry (or even retail!), Swinmurn formed a specific and clear hypothesis about why there was such a large opportunity in the footwear domain: Although footwear was a large and growing category ($40 billion), retail was highly fragmented, and so there was an opportunity to create a one-stop online offering that was "a network among all the separate shoe stores." Armed with his contrarian hypothesis, he sought to attract resources (in the form of venture capital) and capabilities (in the form of retail experience) that would help him realize the value of this opportunity. Notably, although he was rejected or dismissed by many, those who joined at an early stage—investors Alfred Lin and Tony Hsieh and former Nordstrom's executive Fred Mossler—were explicitly attracted to join the venture precisely because they shared the same contrarian theory held by Swinmurn (with the understanding that others who did not fund or join disagreed with them). However, this shared belief by the founding team then had consequences for the venture going forward. Specifically, although the founders interpreted their early growth numbers in a positive light (believing it gave them a reason to continue to move forward), larger venture capitalists (most notably Sequoia) looked at the same data and came to the conclusion that the venture was unlikely to succeed (and declined to invest). Finally, despite this feedback from Sequoia, and with cash running perilously low, the team (contrary to prevailing wisdom to reduce expenses to extend the runway) opted to allocate a significant portion of its dwindling resources toward purchasing inventory. This decision was underpinned by the hypothesis that a substantial in-stock inventory, coupled with an exceptional customer service experience, would catalyze e-commerce traction. More precisely, Hsieh characterizes the experiment (which involved investing the remainder of his own personal financial resources) as a test to "either save Zappos or ensure our speedy demise." The experiment itself was a success, and after a record run of observable sales growth and bottom-line performance, more traditional venture capitalists (including Sequoia!) invested in more mature stages of the company prior to its acquisition by Amazon in 2009.

Although most start-ups do not experience the success of Zappos, its history nonetheless captures a number of recurrent themes that are consistent with a Bayesian entrepreneurship approach. First, although Nick Swinmurn may be more optimistic or not than the general population in terms of background effect, there is no doubt that Swinmurn was specifically optimistic about the opportunity associated with online retail. Importantly, he did not have an extensive background or prior career in this domain but instead observed the opportunity as a particularly fruitful (albeit particularly challenging) path to pursue in the broader domain of online commerce. This characteristic feature brings us to our first broad insight into Bayesian entrepreneurship: Since an entrepreneur *chooses* the specific opportunity they are pursuing, a given entrepreneur is *relatively optimistic* about that opportunity relative to other decision-makers. In other words, a Bayesian approach captures the idea that an entrepreneur has a systematically different theory of why an opportunity is valuable relative to others who have chosen *not* to pursue that opportunity. More generally, Bayesian entrepreneurs are endowed with relatively favorable priors about the opportunities they choose to pursue, a state that is also associated with a contrarian hypothesis about how they might create and capture value from that opportunity in a way that others do not foresee.

Second, the Bayesian approach offers distinct and novel insight into the nature of purposeful learning and experimentation on the part of founders. First, at a broad level, the fact that founders have relatively more positive priors does not imply that they are *certain*; outside of degenerate cases, priors are noisy, and the "theory" underlying a given prior may be more or less well developed. As such, the existence of calculable uncertainty about an opportunity leads founders to have a *demand for experimentation*. In particular, the demand for experiments (which are costly in terms of time, effort, and, potentially, opportunity cost) will be shaped by the priors held by the entrepreneur. When founders are more uncertain about their theory of value creation and capture or hold more noisy priors, the demand for experimentation will increase. Just because the founding team at Zappos had more positive beliefs did not mean they were immune to feedback or learning through experimentation; instead, they tailored experiments that allowed them to update their priors in order to both assess whether the venture was likely to be viable and make decisions about their overall entrepreneurial strategy. Moreover, this demand for experiments (i.e., opportunities to learn about the opportunity in order to make better decisions) is fundamentally influenced by beliefs about the opportunity. For example, relative to a disinterested agent, a Bayesian entrepreneur will conduct experiments that are more likely to allow for false positives (a signal to continue despite the opportunity not being valuable) than false negatives (a signal to disband even though the opportunity is worthwhile). This sort of biased experiment (in which entrepreneurs choose a "best foot forward" approach that allows them to only receive negative feedback in the case when the opportunity is, in fact, not worthwhile) can be seen directly in the experiment chosen in Zappos where they undertook a costly experiment that was "optimized to succeed" in order to give the team the confidence that they should continue to pursue the opportunity.

Finally, this approach yields novel insight into who (and how) entrepreneurs are able to attract resources and capabilities by providing information to other agents. Notably, entrepreneurs are more likely to persuade those who share their idiosyncratically more positive beliefs about an opportunity. For example, in the Zappos case, the investors and industry experts who were attracted to the venture were attracted to it precisely because they held the same optimistic beliefs as the founder of the original idea. This Bayesian dynamic suggests that the emergence of homophily within start-up organizations might reflect not simply a preference for association among similar types but attraction to an opportunity for those who share similar beliefs (Van den Steen, 2005, 2011). More subtly, the existence of heterogeneous priors by different agents suggests that, when needing to attract resources and capabilities of those whose priors differ from that of the entrepreneur, the experiments that are conducted (and the way that evidence is interpreted) will cater to those with different (more negative) beliefs than they themselves hold. In other words, the natural distortions that arise in attempts at Bayesian persuasion (Kamenica and Gentzkow, 2011) are a natural consequence of a Bayesian entrepreneurship approach.

The remainder of the chapter proceeds as follows. First, we undertake a bit of a deck-clearing exercise by delineating what Bayesian entrepreneurship is *not*, followed by the development of a formal model of a Bayesian entrepreneur. After illustrating some broad properties of the model, we dig into the role of priors, the nature of learning and experimentation, and the role of informativeness. We then briefly highlight a connection to two prior bodies of work (by subsets of the authors) in line with this approach, including a Bayesian approach to entrepreneurial strategy and the scientific method to entrepreneurial decision-making.

## 1.2 What Bayesian Entrepreneurship Is Not

In order to make useful progress for how a Bayesian entrepreneurship approach might usefully inform theory, empirics, and practice in entrepreneurship, it is useful to start with more definitional questions about what exactly Bayesian entrepreneurship means. At one level, the idea that Bayesian logic is related to entrepreneurial decision-making is intuitive and appealing. Certainly, entrepreneurs pursuing opportunity in the face of uncertainty must have some beliefs (perhaps even something we might call "priors") about the probability of success. A central dogma of entrepreneurship is the informational value of learning and experimentation at the earliest stages of new venture formation. To the extent that Bayesian reasoning involves how priors are updated through learning (and how that updating informs practical decision-making), the utilization of a Bayesian approach toward the study of entrepreneurial decision-making seems like a promising approach.

Despite this appeal, it is useful to note that some of the most central approaches and paradigms for studying entrepreneurship either abstract away from the central logical consequences of a Bayesian approach or (often implicitly) impose assumptions that make a Bayesian approach infeasible (or perhaps meaningless). And many approaches that seem, at first blush, to embrace a Bayesian logic (e.g., the lean start-up movement and its focus on learning) nonetheless do not draw out the consequences of an implicitly Bayesian approach in terms of their practical implementation. As such, before turning to a more precise definition of what we mean by Bayesian entrepreneurship, a useful deck-clearing exercise involves first grappling with what Bayesian entrepreneurship is *not*.

### 1.2.1 Not Knightian Uncertainty

A long tradition in entrepreneurship is premised on the idea that entrepreneurs face a form of fundamental uncertainty—the most extreme type of uncertainty described by Knight (1921)—in which agents are unable to form any meaningful probability distribution over the space of possible outcomes. The essential argument in favor of Knightian uncertainty is that it is impossible for an entrepreneur to forecast all the potential contingencies and circumstances that might arise as they pursue a particular opportunity. If one cannot foresee the many potential contingencies that might arise, the notion of probabilistic reasoning simply does not apply.

In contrast, a Bayesian approach to entrepreneurship operates on a fundamentally different premise. It acknowledges that while absolute certainty is unattainable and the future inherently uncertain, entrepreneurs can still form rational probability estimates based on available information. Bayesian reasoning accepts the premise of uncertainty but posits that uncertainty can be quantified and updated through experience and new information. This perspective would argue against the Knightian view by suggesting that even in the absence of complete information, the perception of entrepreneurial opportunity (which might reflect prior experiences, idiosyncratic market data, or trends) allows entrepreneurs to nonetheless create a subjective probability distribution over the outcomes of interest. These probability distributions are then updated as new evidence becomes available, reflecting a dynamic learning process. Put another way, the Bayesian approach does not deny the existence of uncertainty but rather provides a rigorous framework for managing it. Although Knightian uncertainty underscores the limits of predictability in entrepreneurial ventures, the Bayesian approach embraces these limits as a starting point for systematic improvement in decision-making under uncertainty.

### 1.2.2 Not Shared Priors

Traditional equilibrium models in economics and finance often operate under the assumption that all agents, including entrepreneurs and investors such as venture capitalists, share a common perspective on the underlying distribution of possible outcomes. This assumption facilitates the analysis of financial contracting and the impact of external variables (e.g., wealth levels) on decision-making processes. Within this framework, there's an implicit consensus on the risks and rewards associated with entrepreneurial ventures, leaving little room for divergent beliefs or the notion that parties might agree to disagree.

The Bayesian approach, by contrast, fundamentally diverges from this assumption by recognizing and accommodating the heterogeneity of priors among different agents. It acknowledges that each decision-maker may hold an idiosyncratic set of priors shaping their individual probability distributions regarding the outcomes of entrepreneurial endeavors. This diversity of priors is seen not as a hurdle to be overcome but rather as a realistic and valuable aspect of the entrepreneurial process.

Allowing for the possibility of agreeing to disagree, especially in the early stages of venture formation and investment, the Bayesian approach offers a more nuanced understanding of how decisions are made under uncertainty. It opens the door to exploring how these varied priors influence negotiation, risk assessment, and the eventual structuring of financial contracts between entrepreneurs and investors. Far from requiring consensus, the Bayesian framework permits a dynamic interaction between differing views, recognizing that such differences are inherent to the process of innovation and value creation.

### 1.2.3 Not Effectuation

Entrepreneurial effectuation posits a process where opportunities are not discovered but rather created through iterative actions within the entrepreneur's immediate environment (Bhide, 1994; Sarasvathy, 2001). This contrasts sharply with the Bayesian approach, which emphasizes the identification and evaluation of opportunities based on preexisting theories of value creation and capture. In effectuation, the entrepreneur progresses by leveraging available resources and relationships, adapting to feedback, and iterating on their actions, effectively making the opportunity as they go. This method is more about shaping the future through a series of contingent choices rather than predicting it.

Conversely, the Bayesian perspective is grounded in the notion that entrepreneurs start with a specific hypothesis about an opportunity, informed by their prior experiences and beliefs. They then purposefully engage in experimentation to test the validity of their assumptions, adjusting their strategy based on new evidence. This process is analytical and systematic, relying on probabilistic reasoning to refine the entrepreneur's understanding and approach to the opportunity before them. The key distinction here is not just in the method but in the underlying philosophy of opportunity itself. Whereas effectuation sees opportunity as emergent and cocreated with stakeholders, the Bayesian approach views opportunity as something that can be perceived and assessed through rigorous analysis and experimentation.

It is worthwhile noting, however, that although this chapter draws a clear demarcation between these approaches, the formal modeling methodologies outlined here might well form the basis for a future constructive reconciliation of what is, at this stage, different viewpoints in approaching entrepreneurial studies. That is, our expectation is that a richer set of articulations of what is and is not exogenous to entrepreneurial choice may lay bare the truth in both approaches.

### 1.2.4 Not Simply Biases and Heuristics

A considerable portion of the literature on entrepreneurship has been dedicated to the exploration of biases such as overoptimism and overconfidence, alongside heuristics such as pattern matching, which are thought to significantly influence entrepreneurial decision-making (Åstebro et al., 2014). These biases and heuristics are often portrayed as systematic deviations from rationality that can lead entrepreneurs to make suboptimal decisions. However, the Bayesian approach to entrepreneurship provides a nuanced perspective on these phenomena. It suggests that what is traditionally labeled as biases might, in fact, be the result of varying priors among decision-makers.

Under a Bayesian lens, the apparent overoptimism of entrepreneurs, for instance, may not universally reflect a bias toward unwarranted positivity. Instead, it can be seen as a natural outcome of the unique information sets, or priors, that these individuals possess about their ventures. Entrepreneurs, almost by definition, believe in the potential of their projects more than the average person does. This belief is not merely a bias but a reflection of their unique perspective and information, which informs their probability distributions differently from those of external observers.

Similarly, the use of heuristics, such as pattern matching, often critiqued for its simplicity and potential for error, can be reinterpreted within the Bayesian framework.[1] These heuristics may represent efficient, albeit imperfect, strategies for updating beliefs in the face of new evidence, especially under constraints of time and information.

It is important to clarify that the Bayesian approach does not outright refute the existence or impact of biases and heuristics in entrepreneurship. Rather, it offers an alternative interpretation that these phenomena may sometimes reflect rational updating of beliefs based on individual priors and information. This perspective encourages a deeper investigation into the nature of so-called biases and heuristics, suggesting that they may not always signify errors in judgment but could instead be adaptive responses to the unique challenges faced by entrepreneurs.

By reframing the discussion around biases and heuristics within the context of Bayesian reasoning, we open the door to a more sophisticated understanding of entrepreneurial decision-making. This approach not only acknowledges the complexity of the entrepreneurial environment but also emphasizes the role of subjective experiences and information in shaping entrepreneurs' decisions. Through this lens, we can begin to see these decision-makers not as inherently flawed in their reasoning but rather as individuals navigating uncertainty with the tools and information at their disposal.

## 1.3 What Bayesian Entrepreneurship Is

Of course, simply saying what Bayesian entrepreneurship is *not* does not actually clarify what Bayesian entrepreneurship is. At its core, Bayesian entrepreneurship is a framework that puts the spotlight on the critical interaction between an entrepreneur's prior beliefs and the capacity for learning and experimentation to update those beliefs. This dynamic process allows entrepreneurs to fundamentally formulate and then test theories about value creation and capture and, in so doing, attract resources and capabilities to a venture and allow that venture to choose and implement an overall entrepreneurial strategy.

The essence of Bayesian entrepreneurship lies in its iterative process of belief adjustment. Entrepreneurs set out hypotheses about their business models, market opportunities, and customer preferences and then engage in experimentation to validate or refute these hypotheses. Each piece of evidence collected through these experiments serves to update the entrepreneur's priors, refining their understanding of the venture's potential success and guiding their subsequent decisions.

This dynamic process of updating beliefs in light of new evidence differentiates the Bayesian approach from more static models of decision-making. It recognizes the inherent uncertainty of entrepreneurial ventures and provides a structured method for navigating this uncertainty. By valuing both the initial set of beliefs and the continual learning process, Bayesian entrepreneurship offers a robust framework for understanding how entrepreneurs can more effectively assess and react to the evolving landscape of opportunities and challenges they face.

From a scholarly standpoint, the Bayesian approach blends the sizable streams of research on entrepreneurial strategy as choice (Gans et al., 2019; Agrawal et al., 2021), entrepreneurial theories (Ehrig and Schmidt, 2022; Felin and Zenger, 2009, 2017), entrepreneurship as experimentation (Kerr et al., 2014; Koning et al., 2022; Lindholm-Dahlstrand et al., 2019), and the scientific approach to entrepreneurial decision-making (Camuffo et al., 2020; Camuffo, Gambardella, Messinese et al., 2024; Zellweger and Zenger, 2023).

From a practitioners' standpoint, the Bayesian approach provides an overarching rationale to understand the widespread adoption, often in combination, of methods such as design thinking (Liedtka, 2018), the business model canvas (Osterwalder and Pigneur, 2010), minimum viable products, and A/B testing (the lean start-up method; Eisenmann et al., 2012; Maurya, 2022; Ries, 2011) that explicitly or implicitly advocate for structured processes of formation, testing, and updating of beliefs in developing new ventures.

To clear the field from conflicting interpretations, we clarify that the Bayesian approach does not intend to be a rigid model of entrepreneurial learning. Instead, we see Bayesian entrepreneurship as providing an actionable and possibly teachable "algorithm" (Minniti and Bygrave, 2001) that entrepreneurs can flexibly use to navigate uncertainty; frame, formulate, and solve iterated choice problems; learn by incorporating new information in their beliefs; and ultimately improve decisions concerning their entrepreneurial projects.

Given this end, there are three key tenets, or core claims, that characterize a Bayesian entrepreneurship approach:

1. **Heterogeneous prior beliefs**: Agents do not share a common prior belief about important elements of the state space—that is, they effectively agree to disagree.
2. **Entrepreneurs have stronger prior beliefs than other agents**, which defines the nature of any disagreement with other agents.
3. **Agents update beliefs using Bayes' rule**—that is, as a general statement, uncertainty is not so pervasive that it is impossible to update their beliefs and learn prior to making certain choices. But specifically, updating uses Bayes' rule where possible.

Next, we elaborate on each of these and show that they direct entrepreneurial research toward a set of research questions that are important to both positive and normative theories of entrepreneurship but that, in the absence of an explicit

Bayesian entrepreneurship approach as defined by the three tenets, have been understudied in the discipline.

### 1.3.1 Prior Beliefs Are Heterogeneous

Much of game theory rests on the assumption that it is not possible for rational agents possessing both common knowledge and common priors to disagree. The common prior assumption is that, given the same information, agents would have identical beliefs about the probabilities of states arising. Aumann (1976) proved a famous result that rational agents would not agree to disagree in the sense that they would understand and have common knowledge (or in Geanakoplos and Polemarchakis, 1982, the weaker notion of mutual knowledge) that they had differing priors. If those priors were the result of different information, then rational agents, understanding that the other agent may know something different, will infer each other's information, leading to agreement.

The conclusion of this set of assumptions is not typically borne out in practical reality where persistent disagreement is observed. This includes disagreement in very high-stakes situations, such as those that surround innovation and entrepreneurship. This situation has not been lost on many economic theorists (notably Yildiz, 2000, and Van den Steen, 2001), who have developed ways to move on from the common priors assumption. Indeed, Harsanyi (1967) noted that rational agents with the same information may well assign different subjective probabilities to states. As Van den Steen (2001) notes, what is required is a foundation for the notion that agents might have differing priors while acknowledging and having common knowledge of the fact that they have different priors.

Van den Steen (2001) outlines potential paths forward that we develop below using the notation in table 1.1. His starting point is a game with agents, $i \in \{1, \dots, N\}$ and a finite state space. For our purposes here, it is useful to consider

**Table 1.1**
Key notation definitions

| Symbol | Definition |
|---|---|
| $i$ | Agent index, $i \in \{1, \dots, N\}$ |
| $V$ | Value of a venture conditional on success |
| $C$ | Cost of exploiting the entrepreneurial opportunity |
| $x$ | Random variable where $x \in \{0, V\}$ |
| $\tilde{\mu}_i$ | Posterior probability of $i$ |
| $\mu_i$ | Prior probability that $i$ holds that value will be $V$ |

a state space where the value created by an entrepreneurial idea is either 0 (a failure) or $V(>0)$ (a success). A simple approach is to assume that each agent has a prior, $\mu_i \equiv \Pr_i[V]$. A more elaborate setup involves a common prior $\mu$, but agents observe information that is objectively correct with some probability but over which agents have subjective beliefs regarding that information being correct. The difference in priors comes from differences in the informativeness of the signal. These beliefs, $\tilde{\mu}_i$, can then be used as the foundational, but subjective, different priors in a resulting model.[2]

The second approach is cumbersome relative to the first, but Van den Steen (2001, p. 16) notes that it makes it "easier to judge the 'reasonableness' of priors in the analysis." This allows the researcher (i) to link results to psychological biases and behavioral considerations such as overconfidence, (ii) to create a link between beliefs and some fundamental reality that can assist in evaluating efficiency, and (iii) to tackle the issue of differing information versus differing beliefs and, as a result, could allow interpretations that differences are the result of distinct theories about the world rather than the facts per se. This also allows for a role of experimental communication whereby some agent could produce facts that could convince someone else of their position.

This second approach connects with what we describe below as a theory-driven process. That is, although a Bayesian entrepreneurship approach can take heterogeneous priors as a starting point, it also makes explicit an analysis of how priors are formed in the first place. The theory-driven approach specifies a specific class of processes that elaborates on how entrepreneurs build the future state spaces (and prior beliefs) underlying their ideas. Under that approach, prior heterogeneity stems from both differences in the envisioned state space and how entrepreneurs elicit the prior probability distribution.

As an illustration of how this second approach can be formalized, assume two agents, *i* and *j*, have the same information. Other things equal, they should form the same prior, $\mu$. However, based on their idiosyncratic background, knowledge, and experience, they build future state spaces and elicit prior probability distribution differently (e.g., maximum entropy vs. indifference rule vs. other methods; Jaynes, 1968). Their prior belief distributions are $\mu_i = \mu + \phi_i$ and $\mu_i = \mu + \phi_j$, with $\phi_i$ and $\phi_j$ subjective parameters corresponding to different subjective future state spaces and idiosyncratic ways in which priors are formed. Under this condition, ex ante $\phi_i \neq \phi_j$, and hence priors can be different ("agree to disagree" condition). However, ex post (over time) priors tend to converge through social processes (e.g., communication, persuasion, as described below) so that $\phi_i = \phi_j$ ("common knowledge" condition; Samuelson, 2004).

### 1.3.2 Entrepreneurs Have Relatively Strong Prior Beliefs

The second tenet of Bayesian entrepreneurship concerns the nature of disagreements between agents. Informally, one might claim that if there are two broad states regarding an entrepreneurial idea, success and failure, then the entrepreneur has stronger prior beliefs about the success of the idea than other agents do.

As a term, however, "optimism" has both positive and normative connotations. The normative connotation is that the entrepreneur's beliefs are mistaken relative to a ground truth probability of success that is lower than the entrepreneur's subjective probability. As that ground truth is neither knowable nor observable to the entrepreneur, even ex post, that normative connotation is not a reasonable claim on which to base analysis.

Instead, this tenet of Bayesian entrepreneurship focuses on the positive component of entrepreneurs'optimism—namely, that entrepreneurs have stronger prior beliefs than other agents, which involves a higher likelihood that their project would be successful by whatever criteria one may consider. This is the approach taken by Van den Steen (2005). Thus, we will use the following definition to rank beliefs:

**Definition 1.1. Stronger prior.** An individual, $i$, has a stronger prior belief about an idea than individual $j$, if $\mu_i > \mu_j$.

This definition does not define belief strength per se but rather *relative strength* without reference to the consequences of those beliefs. In applications, an analyst might make additional assumptions or focus on cases where belief differences are consequential. For instance, if by investing $C$ in exploitation an entrepreneur resolves uncertainty about the venture and generates payoffs of $V$ or 0 as the case may be, a case of interest is where $\mu_i > \mu^*$, where $\mu^* \equiv C/V$. In that case, without any additional information, $i$ would be *predisposed* to choose to exploit an opportunity. Belief strength could also be evaluated relative to available experiments—that is, at a level at which even a negative signal resulting in $\tilde{\mu}_i(0)$ would still involve $\tilde{\mu}_i(0)V \geq C$, implying that $i$ has such strong beliefs that no information would dissuade them from pursuing an entrepreneurial opportunity.

The relative definition of belief strength, however, does not require any such thresholds to be met. It only requires that (a) individuals hold heterogeneous priors and (b) they can be ranked using those priors. Thus, it could also be the case that $\mu_i V < C$ but that $\mu_i$ is high enough that it is worth gathering information (i.e., engaging in exploration). Interestingly, a single experiment is insufficient to persuade another to pursue the venture, so another will not explore.

The second tenet of Bayesian entrepreneurship takes this definition of the relative strength of prior beliefs and claims, according to this ordering, that an *entrepreneur has stronger prior beliefs about their ideas than all (or most) others*. The claim is a focused one. Bayesian entrepreneurship simply requires that the entrepreneur has stronger beliefs about the idea they are pursuing, even if they might be a pessimist about realizations of other uncertain variables.

The claim that entrepreneurs have stronger beliefs is made without reference as to why this is so. One possibility is that this is a trait of entrepreneurs (see, e.g., Dushnitsky, 2010; Fraser and Greene, 2006; Hmieleski and Baron, 2009; Landier and Thesmar, 2008) and can be simply taken as a starting point. Alternatively, it is possible to imagine that this claim is justified by a selection mechanism. For instance, one could ground opportunity recognition and initial action in the context of evolutionary game theory.[3] Justifying this claim more carefully, both theoretically and empirically, is, of course, a research direction that works within Bayesian entrepreneurship, which makes it more salient and compelling.

This claim defines a central challenge for the entrepreneur. To the extent that they do not control key resources necessary to exploit an opportunity, an entrepreneur will have to convince and reduce the level of disagreement with agents who hold weaker prior beliefs. Interestingly, reducing agreement is desirable, but only for instrumental reasons. In a competitive context, entrepreneurs may benefit if others do not share their priors. Weaker beliefs on the part of others imply that they are less likely to pursue an opportunity in competition with the entrepreneur, and entrepreneurs may be concerned about signals that positively update those beliefs of others (Gans, 2023). This would change their incentives regarding the disclosure of information. Indeed, it may well be that this could be the basis of another micro-foundation for the optimistic priors of entrepreneurs, as being optimistic in a relative sense is a precondition for the entrepreneur to find it worthwhile to pursue an opportunity (Robson and Samuelson, 2011).

A key implication of this challenge is that the probability of a venture's success will not just be driven by differences in the preferences or incentives of entrepreneurs. It will also depend on the skills of the entrepreneur (and indeed, as we will see, others) in being able to interpret data, signals, and experiments in a Bayesian manner and choose those experiments understanding data-generating processes and biases. That is, research should focus on improvements in the cognitive process that will allow entrepreneurs to interpret information by taking into account their priors and preferences.

### 1.3.3 Agents Update Beliefs Using Bayes' Rule

The final tenet of Bayesian entrepreneurship concerns how prior beliefs are updated—that is, how agents learn from new information. Consistent with the name for this approach, those beliefs are updated using Bayes' rule to generate a posterior belief from a prior belief. This is a process we call "Bayesian learning." As we will emphasize below, Bayesian learning can be simple or it can be deep. Simple Bayesian learning presumes that posterior beliefs depend only on the prior and not on how the prior is formed. Deep Bayesian learning takes into account how priors are formed and how this impacts what is actually learned. In what follows, we will initially assume Bayesian learning is simple, leaving the consideration of the technically more challenging exposition of deep Bayesian learning until section 1.5.

Given its prominence for Bayesian entrepreneurship, it is useful here to recall how (simple) Bayesian learning actually works. Consider an entrepreneur who has subjective beliefs that their business idea will be successful. The entrepreneur can simply decide to pursue their idea based on what they know or consider collecting data to learn more about it—that is, a signal or experiment.

For Bayesians, a prior probability, simply referred to as a prior, is the probability distribution of a random variable *prior* to the consideration of new evidence or information. It is a subjective probability encapsulating the beliefs an individual might have about a random variable before gathering any data. It is this prior that is updated using new information and, for the purposes of this research program, that updating procedure is given by Bayes' rule. Thus, if $\mu$ is a prior probability attached to an event or state, $\tilde{\mu}_i$ is the updated probability that takes into account new information using Bayes' rule. Specifically,

$$\tilde{\mu}_i = \frac{\Pr[\text{Signal}|\mu_i]\mu_i}{\Pr[\text{Signal}]}.$$

The output of this updating procedure, $\tilde{\mu}_i$, is called $i$'s posterior beliefs. The posterior belief is what a decision-maker will use after collecting information to make a decision. The prior beliefs will guide whether it is worth collecting information or, instead, making a decision without such information. For many applications, a full version of Bayesian updating is not necessarily required, and it is robust to some relaxation; see the discussion in Jakobsen (2023).

Bayes' rule summarizes the three main steps of Bayesian learning, which is applicable to entrepreneurs (Zellweger and Zenger, 2023):

1. Forming a prior distribution about an entrepreneurial idea
2. Collecting data (e.g., through an experiment) and determining the likelihood function using the information about the parameters available in the data
3. Combining the prior distribution and the likelihood function—using Bayes' theorem—to form the posterior distribution

The posterior distribution reflects the entrepreneur's updated prior, which balances prior knowledge with observed data and is used to conduct inferences about the idea. Bayesian entrepreneurship can be seen as a way to systematize a series of advances that recently occurred in entrepreneurship theory and practice.

As already noted, entrepreneurs and others understand that entrepreneurs have strong priors, so they agree to disagree along those lines. However, this also molds their expectations regarding the informativeness of experiments and other information received. Importantly, all agents expect that new information will reduce disagreement and hence posteriors will converge with new information (Kartik et al., 2021). However, Bayesians believe that their priors are correct, and given a set of possible signals, their own expected posterior equals their own prior.

### 1.3.4 How Are Priors Formed?

The three tenets of Bayesian entrepreneurship put priors at the centre of analysis. With this focus the question of where those priors come from is raised. This, itself, becomes an important research question, and later in the paper, we will show that different models of prior formation can have different implications for how entrepreneurs gather information and make choices. Given this, here, we note some of the main approaches to the formation of priors that have been undertaken in the literature to date. This is not an exhaustive list but an indicator of where development has taken place and how these give rise to an important research agenda in Bayesian entrepreneurship.

Entrepreneurs' priors form idiosyncratically and are heterogeneous, so that individuals might have different priors about the same idea or opportunity, even starting from the same available information. The idiosyncrasy and heterogeneity of entrepreneurs' prior beliefs stem from a variety of factors.

First, they derive from individual differences rooted in physical and psychological traits, personality, or other individual factors (Baum et al., 2014). These generate differences in cognitive processes (e.g., memory, thought, perception), emotions, and so forth, which result in heterogeneous levels of self-efficacy and achievement drive (Frese and Gielnik, 2014).

Second, they derive from differences in entrepreneurs' knowledge, skills, and abilities. Human capital is unevenly distributed across entrepreneurs in type and quantity (Lazear, 2004; Unger et al., 2011). It stems from unique knowledge of a specific matter (Shane, 2000), education (Martin et al., 2013), experience in a particular knowledge domain, or talent (Eesley and Roberts, 2012), and such differences in endowment shape what entrepreneurs explore (exploration space), why and how they do it, and for how long (Gimeno et al., 1997).

Third, they derive from the context in which they operate, as context also contributes to the definition of the individual–opportunity nexus (Shane, 2003). For example, geographic location (Delgado et al., 2010), the ecosystem in which entrepreneurs operate (Acs et al., 2017), their relevant institutions such as universities (Tartari and Stern, 2021), accelerators (Cohen et al., 2019), as well as the access to investors, shape entrepreneurs' prior beliefs. A similar role is played by the social networks entrepreneurs build and are embedded in (Stuart and Sorenson, 2005).

Fourth, they derive from entrepreneurs' preferences (Roach and Sauermann, 2015), interests (Van de Ven et al., 2007), motivation (Baum and Locke, 2004; Guzman et al., 2020), incentives, and goals.

Finally, entrepreneurs form prior beliefs based on their cognitive processes (Baron, 1998), which include intuition (Blume and Covin, 2011), creativity (Amabile, 1997), imagination (Patvardhan and Ramachandran, 2020), as well as logical reasoning and causal thinking (Camuffo, Gambardella and Pignataro, 2024; Felin and Zenger, 2009, 2017). Logical reasoning and causal thinking (Pearl (2000)) are especially important because, through them, entrepreneurs can intentionally and deliberately form priors through theories (Felin and Zenger, 2017; Ehrig and Schmidt, 2022; Agarwal et al., 2025) and hence *choose* (Agrawal et al., 2021; Gans et al., 2019, 2021) which idea is worth pursuing.

## 1.4 Experiments Under Simple Bayesian Learning

Bayesian entrepreneurship places entrepreneurial processes for decision-making at the center of entrepreneurial research. That focus gives priority to the types of information that are available to entrepreneurs and the purposes to which that information is put. Here, we will follow a convention in information economics that uses the term "experiment" or "signal" as a formal construct for any information that entrepreneurs might gather that itself requires a choice to generate and then utilize. More broadly, information may include information that is passively acquired by entrepreneurs that itself can be used in Bayesian learning. In this section and the next, we focus on the entrepreneur's choices in incurring the costs of generating experiments to understand the underlying value of their idea, as well as strategies that may be employed in its commercialization. This section will assume that priors are given and entrepreneurs use simple Bayesian learning, whereas the next section will consider deep Bayesian learning, which requires the causal structure associated with any experiments or signals to be appropriately understood by the entrepreneur.

### 1.4.1 How Priors Impact Experimental Interpretation

Priors are the root primitive for considering a Bayesian approach to entrepreneurial decision-making. They matter precisely because entrepreneurs will engage in various activities to de-risk and refine their commercialization plans, which will be a key driver of their potential success. Not only that, these activities will be designed to influence or persuade others changing the set of resources that may be brought under an entrepreneur's control. To that end, here, we consider how posterior beliefs are formed through experimentation.

We derive a fundamental and, we believe, underappreciated result that entrepreneurs' demand for and the form of experimentation they choose will depend critically on their priors. In particular, stronger priors will generate a demand for experiments designed to yield clear signals that the venture should not be pursued while at the same time providing less clear signals that the venture should proceed.

The key starting point for this is to explicitly consider the role of experiments in entrepreneurial decision-making and how Bayesian learning can identify their demand characteristics. Specifically, we want to demonstrate how priors—both of entrepreneurs and others they interact with—drive the value of experiments. We will consider other issues such as the type of experiments chosen below.

Let's return to our ongoing simple example. Recall that an entrepreneur ($E$) has a prior, $\mu_E$, that a venture idea will be successful, earning them $V$, versus not successful, earning them 0. That is, the true underlying state space is $x \in \{0, V\}$. If the cost of launching that venture were $C$, then the expected return from that venture given the prior of $\mu_i$ is:

$$R(\mu_i) = \mu_i V - C.$$

If $R(\mu_i)$ exceeds the return of the entrepreneur's alternative opportunity (something we typically normalize to 0 in theoretical models), then, without more information, an entrepreneur might choose to launch the venture. Using our earlier notation, this implies that the entrepreneur will exploit the opportunity, in the absence of further information, if and only if $\mu_i \geq \mu^* \equiv C/V$.

**Table 1.2**
Experiment maps from signal to state, $\Pr[s|\theta]$

| | (Good news) $s=1$ | (Bad news) $s=0$ |
|---|---|---|
| (Success) $V$ | $\lambda_1$ | $1-\lambda_1$ |
| (Failure) 0 | $1-\lambda_0$ | $\lambda_0$ |

What if there is a signal, $s$, that takes on a value of 1 with probability $\lambda_1$ if the venture will be successful and a value of 0 with probability $\lambda_0$ if the venture is unlikely to be successful—that is, $\Pr[s=1|x=V]=\lambda_1$ and $\Pr[s=0|x=0]=\lambda_0$ (see table 1.2). Note here that $\Pr[s=1|\mu_E]=\lambda_1\mu_E+(1-\lambda_0)(1-\mu_E)$ and $\Pr[s=0|\mu_E]=(1-\lambda_1)\mu_E+\lambda_0(1-\mu_E)$. Note also that if the entrepreneur follows the signal in deciding whether to invest, the rate of entrepreneurial implementation (i.e., starts) will be $\Pr[s=1|\mu_E]$, and the observed rate of successful ventures will be $\lambda_1\mu_E$. Thus, observed success will increase with $\mu_E$ if the entrepreneur is in control of whether the venture is launched following an experiment.

It is useful to orient ourselves by translating the information in table 1.2 to some familiar terms.

- Sensitivity or true-positive rate: $\lambda_1\mu_i$
- Specificity or true-negative rate: $\lambda_0(1-\mu_i)$
- False-positive probability: $(1-\lambda_1)(1-\mu_i)$
- False-negative probability: $(1-\lambda_0)\mu_i$

Notice, therefore, that an experiment that involves both a higher $\lambda_1$ and a higher $\lambda_0$ increases both the sensitivity and specificity and reduces the probabilities of false positives and false negatives. Below, we will consider situations where the entrepreneur must choose experiments that trade off $\lambda_1$ and $\lambda_0$ and so face the option of reducing either false positives or false negatives but only at the expense of each other. It can be seen here that the prior, $\mu_i$, will likely factor into such trade-offs.

### 1.4.2 How Priors Impact Experimental Value

The above shows how priors impact the information contained in an experiment. Not surprisingly, for this reason, those priors also impact the experiment's value to a decision-maker. We explore that here.

Suppose that it costs $c << C$ to obtain that signal. To calculate whether it is worthwhile to obtain the signal prior to making a decision to launch the venture, we need to calculate the posterior beliefs contingent on "good news" (i.e., $s=1$)

or "bad news" (i.e., $s = 0$). Those are:

$$\tilde{\mu}_i(1) = \frac{\lambda_1 \mu_i}{\lambda_1 \mu_i + (1 - \lambda_0)(1 - \mu_i)}$$

$$\tilde{\mu}_i(0) = \frac{(1 - \lambda_1)\mu_i}{(1 - \lambda_1)\mu_i + \lambda_0(1 - \mu_i)}.$$

Given this, the expected returns, contingent on the signal realizations, are $R(\tilde{\mu}_i(1)) = \tilde{\mu}_i(1)V - C$ and $R(\tilde{\mu}_i(0)) = \tilde{\mu}_i(0)V - C$.

What level of $\{\lambda_1, \lambda_0\}$ implies that a signal would be useful (even if $c$ is arbitrarily low)? A natural interpretation of usefulness would be that good news should lead to a decision to launch the venture—that is, $R(\tilde{\mu}_i(1)) > 0$—and bad news should lead to a decision to abandon the venture—$R(\tilde{\mu}_i(0)) < 0$. Combining these two conditions implies that:

$$\frac{\lambda_1 \mu_i}{\lambda_1 \mu_i + (1-\lambda_0)(1-\mu_i)} V > C > \frac{(1-\lambda_1)\mu_i}{(1-\lambda_1)\mu_i + \lambda_0(1-\mu_i)} V.$$

A necessary condition for this inequality to hold is:

$$\lambda_1((1 - \lambda_1)\mu_i + \lambda_0(1 - \mu_i)) > (1 - \lambda_1)(\lambda_1 \mu_i + (1 - \lambda_0)(1 - \mu_i))$$

$$\Leftrightarrow \lambda_1 \lambda_0 (1 - \mu_i) > (1 - \lambda_1)(1 - \lambda_0)(1 - \mu_i) \Leftrightarrow \lambda_1 + \lambda_0 > 1.$$

Intuitively, the precision of the signal (i.e., the probability the signal is a correct indicator of the underlying state) needs to be sufficiently high. If these conditions hold, then the signal will determine the decision taken.

The important feature to note regarding the necessary conditions for the informativeness of a signal in terms of the decision to be taken is that it only depends on $\lambda_1 + \lambda_0$ and not on prior beliefs, $\mu_i$, or parameters of the decision context. This means that any agent, regardless of beliefs or preferences, will agree that, following good news, $\tilde{\mu}_i(1) \geq \mu_i$ and, following bad news, $\tilde{\mu}_i(0) \leq \mu_i$. This is a standard finding in information economics that the Blackwell ordering of "informativeness" is independent both of prior beliefs and of preferences (Blackwell, 1953).

Prior beliefs come into play when the entrepreneur considers gathering information and whether the information is sufficient to revise their decision.[4] There are two cases of interest where an entrepreneur is "predisposed to exploit" and "not predisposed to exploit." In the predisposed case, $R(\mu_E) > 0$, implying that, without information, the entrepreneur will launch the venture. In this case, the entrepreneur only finds it worthwhile to acquire information if:

$$(\lambda_1 \mu_E + (1 - \lambda_0)(1 - \mu_E))\,(R(\tilde{\mu}_E(1)) - C) - c \geq R(\mu_E)$$

$$\Leftrightarrow \lambda_1 \mu_E V - (\lambda_1 \mu_E + (1 - \lambda_0)(1 - \mu_E))C - c \geq \mu_E V - C$$

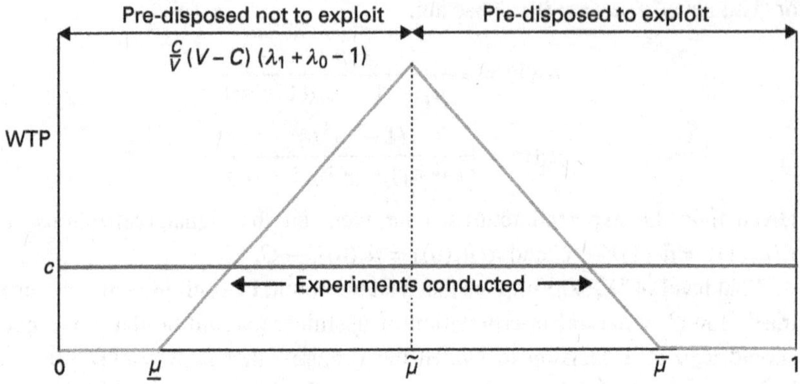


**Figure 1.1**
Willingness to pay (WTP) for experiments.

$$\Leftrightarrow \bar{\mu}_E(c) \equiv \mu_E \leq \frac{C\lambda_0 - c}{C\lambda_0 + (1-\lambda_1)(V-C)}.$$

Intuitively, if an entrepreneur will launch even if they receive bad news, it is not worth spending to obtain a signal, as it will not be decisive. For the not predisposed case, we have:

$$(\lambda_1\mu_E + (1-\lambda_0)(1-\mu_E))\,(R(\tilde{\mu}_E(1)) - C) - c \geq 0$$

$$\Leftrightarrow \lambda_1\mu_E V - (\lambda_1\mu_E + (1-\lambda_0)(1-\mu_E))C - c \geq 0$$

$$\Leftrightarrow \underline{\mu}_E(c) \equiv \mu_E \geq \frac{c + (1-\lambda_0)C}{(1-\lambda_0)C + \lambda_1(V-C)}.$$

Intuitively, an entrepreneur who is predisposed not to launch can only be persuaded by the signal to launch if their prior is sufficiently strong.

Given this, figure 1.1 shows the willingness to pay for an experiment with parameters $\{\lambda_1, \lambda_0\}$ as a function of the prior, $\mu$.[5] Notice that the willingness to pay peaks at $\mu_E = \tilde{\mu}_E$. At this point, $\tilde{\mu}_E V = C$, and so an uninformed entrepreneur is indifferent between pursuing the venture and not. At this point, the entrepreneur is willing to pay $C/V(V-C)(\lambda_1 + \lambda_0 - 1)$, which is the most for information that will determine whether the venture should go ahead and break the indifference. The experiment cost, $c$, determines the range of entrepreneurs in terms of their priors who will actually conduct an experiment, $\{\lambda_1, \lambda_0\}$. Note that as the informativeness of the experiment, $\lambda_1 + \lambda_0$, increases, the blue lines shift upwards, and an experiment is more likely to be conducted.

Thus, we see here that although priors do not play a role in determining what action to take in the face of signals that are informative (i.e., where $\lambda_1 + \lambda_0 > 1$), they play a role in determining whether entrepreneurs will gather information before launching a venture. In particular, priors will determine entrepreneurial demand for experiments.

### 1.4.3 Experimental Choice for Self-Informativeness

One of the tenets of Bayesian entrepreneurship is that entrepreneurs hold strong prior beliefs. However, it is also generally the case that entrepreneurs are resource constrained. That means that they need to secure resources from other people in order to launch and sustain their venture. By definition, those other people do not necessarily share the entrepreneur's beliefs. Thus, in addition to designing experiments to provide information regarding whether further investment in a venture is worth undertaking, the entrepreneur must consider whether experiments are designed in such a way that would persuade others who hold more pessimistic priors to devote resources to the venture.

Before considering the challenge associated with persuading others, it is useful to consider the entrepreneur's choice of experiment when they are seeking to persuade themselves to pursue the venture or not. To see this, we continue the model from section 1.3.1 by endogenizing experimental choice. The experimental choice set is $E = \{\lambda_1, \lambda_0 | \lambda_1 + \lambda_0 \leq \Lambda\}$. Here, $\Lambda$ parameterizes the scope or size of the experimental choice set. It is assumed that $\Lambda > 1$ so all experiments are informative and $\Lambda < 2$ so that the experimental set does not include the perfect experiment. The experimental choice set is depicted in figure 1.2 and is bounded by the red line. For simplicity, it is assumed here that $c = 0$ and that the entrepreneur is constrained to select one experiment only from the set.

Note that allowing the entrepreneur to choose experiments from this set is a more general model than is commonly assumed in other Bayesian contexts. For instance, Kamenica and Gentzkow (2011) and Bergemann and Morris (2019) consider the case of a sender that communicates information (i.e., the results of an experiment) to a receiver and assume that the set of experiments a receiver may choose from is Bayes plausible. Under this restriction, the expected value of the posteriors equals the prior.[6] In our context, note that the expected value of the posteriors is:

$$\begin{aligned}\mathbb{E}[\tilde{\mu}_i] &= \mu_i \Pr[s=1|V] + (1-\mu_i)\Pr[s=1|0] \\ &= \mu_i\lambda_1 + (1-\mu_i)(1-\lambda_0).\end{aligned}$$

This only equals the prior, $\mu_i$, if

$$\frac{\mu_i}{1-\mu_i} = \frac{1-\lambda_0}{1-\lambda_1}$$

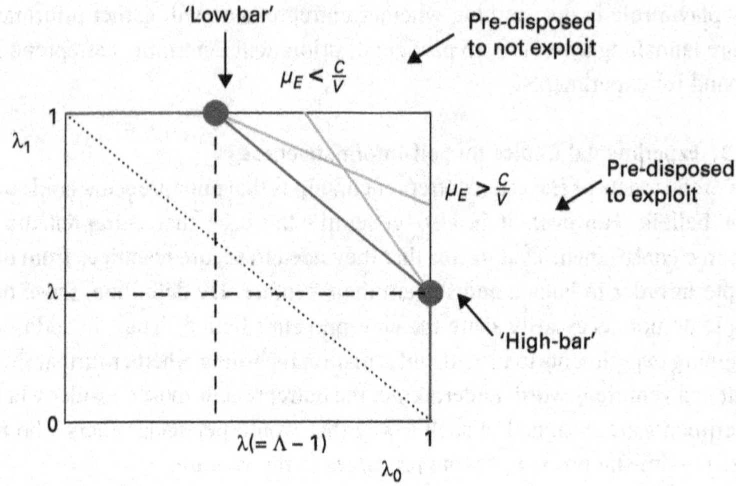


**Figure 1.2**
Choice of experiment.

that is, the ratio of false positives to false negatives equals the prior odds. Imposing this means that the set of experiments is constrained by the prior itself. Hence, it is not possible to consider how priors change observed experimental choices because that is part of the experimental choice set constraint. However, if this condition does not hold, the signal is biased. As we will now demonstrate, it is possible that entrepreneurial decision-makers might prefer signals that are biased, depending on the level of their priors. Thus, priors will determine the type of experiments employed.

To see this, consider figure 1.2. The blue lines represent two indifference lines for the entrepreneur's expected payoff. That expected payoff is:

$$\Pi \equiv \lambda_1 \mu_E V - (\lambda_1 \mu_E + (1-\lambda_0)(1-\mu_E))\, C.$$

Rearranging for experimental space, we have:

$$\lambda_1 = (1-\lambda_0)\frac{(1-\mu_E)C}{\mu_E(V-C)}.$$

Note that the marginal rate of substitution between $\lambda_1$ and $\lambda_0$ is:

$$MRS_{\lambda_1,\lambda_0} = -\frac{(1-\mu_E)C}{\mu_E(V-C)}.$$

Thus, if $\mu_E V > (<) C$, $MRS_{\lambda_1,\lambda_0} > (<) -1$.

**Table 1.3**
Extremal experiments

| $\{\lambda_1, \lambda_0\}$ | **Low bar** $\{1, \Lambda - 1\}$ | **High bar** $\{\Lambda - 1, 1\}$ |
|---|---|---|
| **Name** | "Best Foot Forward" | "Raising the Bar" |
| **Description** | Easy to pass | Hard to pass |
| **Bias** | No false negatives | No false positives |
| **Start-up formation rate** | $(\Lambda - 1)\mu$ | $\mu + (2 - \Lambda)(1 - \mu)$ |
| **Start-up failure rate** | $(2 - \Lambda)(1 - \mu)$ | 0 |
| **Example** | Target lead users (Moore, 2014) | Minimum viable product (Ries, 2011) |

Note that, depending upon whether the entrepreneur is predisposed to exploit or not in the absence of an experiment, the highest indifference curve will be tangent with the experimental set at opposite and extreme points. When the entrepreneur is predisposed to exploit, they choose $\{\lambda_1, \lambda_0\} = \{1, \lambda\}$ (where $\lambda = \Lambda - 1$). Gans (2023) terms this a "low-bar" experiment. This is because, for this experiment, it is very unlikely to generate bad news, but if it does, that news sends a clear (indeed perfect) signal that it is not worthwhile to exploit the opportunity. Building a high-quality product to establish a beachhead is a strategy emphasized by Moore (2014). The idea is to target customer segments where you believe you can gain traction among lead users. This is a low-bar experiment because if you build a sufficiently strong product, if that product does not succeed in that segment, this is a strong signal that it will not succeed more broadly (see table 1.3 for a characterization).

By contrast, when the entrepreneur is predisposed not to exploit, they choose $\{\lambda_1, \lambda_0\} = \{\lambda, 1\}$. This represents a high bar because, for this experiment, it is very unlikely to generate good news, but if it does, that news sends a clear (indeed perfect) signal that it is worthwhile to exploit the opportunity. An example of a high-bar experiment is a minimum viable product because it is launched in a way that is stacked against success (Ries, 2011).

As noted already, entrepreneurs are presumed to have a higher prior, $\mu_E$. Thus, we have demonstrated that:

**Proposition 1.1.** An entrepreneur for whom $\mu_E > C/V$ has an optimal experiment of $\{\lambda_1, \lambda_0\} = \{1, \Lambda - 1\}$.

The proof is in Gans (2023). The result that decision-makers should "bias" their information signals toward their priors was most clearly demonstrated in Che and Mierendorff (2019). Their application was a dynamic process; however, it was considered biased in how media outlets presented information to

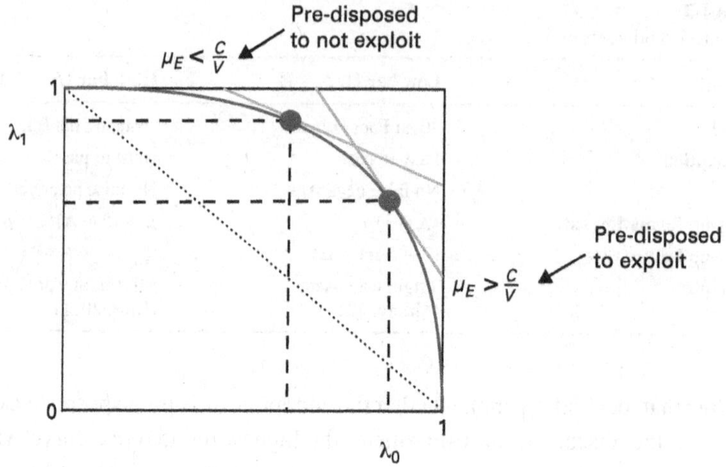


**Figure 1.3**
Strictly concave experimental frontier.

users. Antecedents also focusing on media and politics while not deriving clear own-bias results include Calvert (1985), Suen (2004), Burke (2008), and Damiano et al. (2020). In all of those cases, however, the experimental choice set was binary and confined to extremal choices. Here, it is demonstrated that these are the optimal choices across all of the suitably constructed experimental choices.

Finally, the construction of the experimental choice set in the example analyzed here is somewhat arbitrary. In particular, $\Lambda(\lambda_1, \lambda_0)$ could be a concave boundary, as depicted in figure 1.3. Notice that the optimal experiment is selected where the slope of the indifference curve, which depends on $\mu_E$, is tangent to the $MRS_{\lambda_1,\lambda_0}$. In this situation, extremal experiments are not chosen, but it can be seen that the same trade-off in terms of the bias of the optimal experiment and its relationship to $\mu_E$ remains.

### 1.4.4 Experimental Choice to Inform Others

We now turn to consider the experimental choice when an entrepreneur has to convince another agent to support the project. The typical situation would involve a venture capitalist supplying $C$ to the venture.

The issue of contrasting self-informative versus other-informative experimental choice is explored by Gans (2022). To consider this, let $\mu_O$ be the prior of others who are resource holders or, as we will refer to them, investors. As per our founding assumption, it is assumed that $\mu_E \geq \mu_O$. Given this, we can now define two experimental benchmarks:

- Entrepreneurial-optimal experiment: $\arg\max_{\{\lambda_1,\lambda_0\}\in E} \lambda_1\mu_E V - (\lambda_1\mu_E + (1-\lambda_0)(1-\mu_E))C$
- Investor-optimal experiment: $\arg\max_{\{\lambda_1,\lambda_0\}\in E} \lambda_1\mu_O V - (\lambda_1\mu_O + (1-\lambda_0)(1-\mu_O))C$

These optimization problems are linear in $\lambda_1$ after substituting in the constraint $\lambda_0 = \Lambda - \lambda_1$. Therefore, the entrepreneurial-optimal experiment involves $\lambda_1 = 1$ if $\mu_E \geq C/V$ whereas the investor-optimal experiment involves $\lambda_1 = 1$ if $\mu_O \geq C/V$. Thus, these choices will coincide if, absent information, both the entrepreneur and others will choose to invest or not choose to invest. By contrast, if $\mu_E > C/V > \mu_O$, the entrepreneur will favor an experiment with $\lambda_1 = 1$, whereas others will favor an experiment with $\lambda_0 = 1$.[7]

This conflict can be resolved in several ways. One way would be to run more than one experiment. This, however, would involve additional experimental costs, $c$, that may be prohibitive, given the marginal value of a second experiment is less than the value of the first one alone. The other way this is solved is by considering the equity position of investors in the venture. Suppose that the entrepreneur gives the others an equity share of any resulting value of $1-\alpha$ (where $\alpha \in [0, 1]$) in return for their contribution of $C$. For simplicity, it will be assumed that the entrepreneur can finance the experiment cost, $c$, out of pocket and will determine equity levels after the experiment.[8] Finally, suppose that there is a competitive investment market, which implies that as long as investors expect to earn at least 0, they will accept the deal.

The following proposition demonstrates that the entrepreneur is better off offering the investor-optimal experiment.

**Proposition 1.2.** Suppose that $\mu_E > C/V > \mu_O$. The entrepreneur chooses an experiment $\{\lambda_1, \lambda_0\} = \{\Lambda - 1, 1\}$.

The proof involves a simple comparison of conditions.[9] Intuitively, although the entrepreneur prefers the entrepreneur-optimal experiment, the amount by which they can reduce the equity ceded to the others is greater if they adopt the investor-optimal experiment. The second factor outweighs the first.

It is instructive to illustrate how this impacts the rate of exploitation as well as the rate of observed entrepreneurial success. To examine this, we need to make an assumption regarding what the true prior, $\hat{\mu}$, of success is. There are two interesting cases: (i) $\hat{\mu} = \mu_E$ or (ii) $\hat{\mu} = \mu_O$.

1. If the entrepreneur is correct ($\hat{\mu} = \mu_E$), then following Proposition 1.2, the venture is launched with probability $\mu_E + (2-\Lambda)(1-\mu_E)$, and it never fails.

2. If the venture capitalist is correct ($\hat{\mu} = \mu_O$), then following Proposition 1.2, the venture is launched with probability $\mu_O + (2 - \Lambda)(1 - \mu_O)$, and it never fails.

By contrast, if the entrepreneur's optimal experiment is selected, then:

1. If the entrepreneur is correct ($\hat{\mu} = \mu_E$), then following Proposition 1.1, the venture is launched with probability $(\Lambda - 1)\mu_E$, and it fails with probability $(2 - \Lambda)(1 - \mu_E)$.
2. If the venture capitalist is correct ($\hat{\mu} = \mu_O$), then following Proposition 1.1, the venture is launched with probability $(\Lambda - 1)\mu_O$, and it fails with probability $(2 - \Lambda)(1 - \mu_O)$.

This shows that when the entrepreneur is free to launch the ventures without having to convince others, ventures are more likely to be launched and more likely to fail compared with the case where other-informativeness is required. This is a testable implication of this framework.

This illustrates just one way in which experiment design might be used in a communication context. Gans et al. (2019) and Agrawal et al. (2021) explore others in which experiments might be altered by changing tested strategies alongside a fixed core idea when both are subject to uncertainty. In this case, there is a rationale for continued experimentation following some signal realizations. Specifically, Agrawal et al. demonstrate that an experiment that tests strategies with a lower expected prior first in a sequence can provide a better signal of the underlying idea as opposed to choosing the strategy with the highest prior.

### 1.4.5 A Comment on Bayesian Persuasion

How do you get someone to do something that is in your interests but not necessarily in the immediate interest of that person? One way is to provide incentives. Another is to make investments that reduce the cost that a person faces in doing what you want them to do. A final way is to persuade them. Persuasion involves providing information that causes people to change their decisions. Such persuasion is Bayesian when the process by which those people use that information is to update their priors according to Bayes' rule.

Entrepreneurs face challenges in providing incentives and changing costs. That is why persuasion can potentially be a means of getting others to take action to improve the venture's prospects. Above, we demonstrate how a clearly defined experiment can be designed to be persuasive. In economics, a literature nominally termed "Bayesian persuasion" has arisen that deals with the use of information to change decisions when the provider of that information may have incentives to selectively choose how information is generated and transmitted. In other words, this literature deals with situations when the receiver of such information does not know the underlying data-generating process of the signals they are receiving. Unlike older literatures that deal with cheap talk, verification opportunities, or plain-old signaling, this newer literature allows the sender far more commitment power.

The new literature was founded by Kamenica and Gentzkow (2011); (see Kamenica, 2019, for a review).[10] The basic model assumes that there is a receiver (e.g., a venture capital investor) who has a payoff function $u(a, x)$, which is a function of the action they take, $a$ (say, invest or not invest), and a state of the world, $x$ (say, the success of the venture). The sender (or entrepreneur) has their own payoff, $\pi(a, x)$. The sender chooses an experiment from a set, $E$. The receiver knows the structure of the experiment. Then, the experiment is run, generating signals that, in turn, influence the action the receiver takes. It is by knowing the structure of the experiment that the receiver is able to use Bayes' rule to update their prior.

The focus then is on what type of experiment the sender chooses.[11] In the entire Bayesian persuasion literature, it is always assumed that the experimental set is Bayes plausible, meaning, where

$$\frac{\mu}{1-\mu} = \frac{1-\lambda_0}{1-\lambda_1}.$$

As already noted, this does not leave much room for a consideration of priors. Expanding this to the unconstrained set, $E$, we already saw a version of that above when the entrepreneur chose the investor-optimal experiment in order to influence their own payoff, including the amount of equity the entrepreneur had to cede. However, the Bayesian persuasion literature is about more complicated experiments that, for instance, are designed to gather some types of evidence and not others. But a key feature is that it is the receiver's posterior beliefs (which, as we know, depend on the receiver's prior) that are the focus of the sender's attention. The contribution of the literature is to provide ways of computing the optimal persuasive experiment in various circumstances.

Another research line that follows from this is: What happens when some agents do not know the precise data-generating structure of an experiment?[12] How does this impact other-informativeness and also on incentives to distort persuasive signals? The work of Andrews and Shapiro (2021) makes a start on this interesting question.[13]

## 1.5 Priors and Experiments Under Deep Bayesian Learning

Under simple Bayesian learning, priors are given without any consideration of how they are formed. Moreover, the choice of experiments focuses solely on the stochastic properties of experiments rather than on their precise interpretation. This represents a discord between formal theory and the underlying descriptive structure of that theory. That is, a particular assumption on a data-generating process might motivate the levels of $\lambda_1$ and $\lambda_0$, but that assumption is embedded solely in those parameters. Moreover, the experiments were aimed at resolving some uncertainty about the likelihood of particular states using a given distribution where choices between experiments might alter the precision of signals in certain directions. In reality, experiments might inform about particular processes and also about the data-generating process—that is, entrepreneurial experiments may cause entrepreneurs to revisit and alter their theories about the world. This section examines what happens when we model entrepreneur prior formation in a deep rather than simple fashion.

### 1.5.1 Theory-Based Prior Formation

To set the stage, consider an entrepreneurial venture where $V$ will be generated in profit if two events occur based on two attributes that are modeled as the realization of two random variables, $x_t \in \{S, W\}$ and $x_m \in \{Y, N\}$. Specifically, profit is $V$ only if $x_m = Y$. $x_t$ is a random variable that captures the strength of a technological opportunity as strong ($S$) or weak ($W$), whereas $x_m$ is a random variable that captures the existence of a particular new market opportunity as existing ($Y$) or not ($N$). The total state space is defined by the Cartesian product of these individual attributes. Here, $X = x_t \times x_m$ or:

$$X = \{(S, Y), (S, N), (W, Y), (W, N)\}.$$

$x_t$ represents the logical antecedent that drives the prior of interest. It can be a predictor or truly a "cause," i.e., a hypothetical intervention of the entrepreneur (Pearl and Mackenzie, 2018). for which the entrepreneur sets the parameter. Differently from the other attributes, in this case conditional probabilities unambiguously reflect causality and not correlations. Given this, the probability that the venture yields a profit is:

$$\Pr[Y] = \Pr[S, Y] + \Pr[W, Y]. \tag{1.1}$$

Because the state space is *subjective*, the probabilities of the four states are also subjective and, potentially, rely on sparse data. In the appendix, we operationalize our approach using Dirichlet probability distributions following Camuffo,

Gambardella and Pignataro (2024). Entrepreneurs complement their sparse information about the state space with "pseudo-observations," which they build in their head "bootstrapping" form abductive reasoning or using analogies. Hence, they fit preliminary probability distributions on X stemming from synthetic empirical frequencies that occur in worlds that decision makers envision in their mind.

A simple approach to Bayesian learning would focus on the prior of interest—namely, $\mu = \Pr[Y]$. However, here we have posited another attribute, $x_t$, that potentially drives the level of that prior being its logical antecedent. A deep approach to Bayesian learning would take that into account and ensure that any updating that occurs as the result of experiments incorporates the structure of the relationship between $x_m$ and $x_t$.

It is at this point that there is a role for theorizing on the part of the entrepreneur. This is a process that respects the potentially sparse nature of information to assist in informing a prior as well as the role that theories and the confidence with which they are held play a role in the strength of those priors. When entrepreneurs build priors using causal reasoning and thinking, they strengthen their beliefs using logical arguments (Carroll and Sorensen, 2021; Ehrig and Schmidt, 2022; Felin and Zenger, 2017). They hold strong beliefs about the value of their venture in the sense of being "logically persuaded" that their understanding of how value will be created is plausible.

The challenge facing entrepreneurs is not simply to learn about not simply probabilities but also states. In entrepreneurial environments, both are potentially unknown. For example, an entrepreneur might think that the development of a *new* technology will generate a *new* market. In order to believe in this idea, they need to build the future state space and form a prior belief (subjective probability distribution) about it. We follow Camuffo, Gambardella and Pignataro (2024) to illustrate how prior beliefs can be formed as theories of value and then tested and updated through experiments.

To define the future state space, entrepreneurs start by identifying *attributes* of the problem, which are elements of the future state space with uncertain realizations. Among the many attributes that they can choose, entrepreneurs focus on the attributes that they believe affect the likelihood of occurrence of a state of interest, such as the existence of a new market. The theory-based approach requires that entrepreneurs explicitly consider the identification of attributes and causal relationships among them as a forward-looking process. Entrepreneurs envision future states and attach subjective probabilities (beliefs) to them. In our example, the entrepreneur wonders whether a new market will exist and thinks that if they can develop an appropriate technology, the rise of this market will be more likely.

Entrepreneurs' uncertainty derives not only from the randomness of their subjective probabilities but also from the uncertainty about the parameters of their probability distributions. A specific set of parameters, $\theta$, identifies one distribution and therefore, in our example, one expected probability of new market rise, $\nu(\theta)$. But entrepreneurs do not know the probability distribution of the parameters either. Thus, they also form, cognitively, a subjective probability distribution, $\mu(\theta)$, of the parameters. This defines a family of probability distributions (see appendix A for formal derivation) identified by the different $\theta$ in the space of all the possible realizations of the parameters in the example.

Causal reasoning or theories (Karni, 2022) allow us to select a set of probability distributions on which entrepreneurs can concentrate their beliefs. In our example, the entrepreneur believes that the availability of the new technology increases the probability of the emergence of a new market. This logical causal chain can be represented as a Bayesian network (Pearl, 2000), which, in our simple example, is:

$$X_t \longrightarrow X_m$$

and $\Pr[Y|S] > \Pr[Y|W]$. This is the entrepreneur's *theory*, which we define as a restriction of the parameters of the probability distribution or, equivalently, as a restriction of the set of probability distributions that the entrepreneur considers for their future state of interest, $\Pr[Y]$.[14]

**Priors and expected values of theories** Using Bayes' rule, we can decompose equation 1.1 to find the expected prior of interest, $\mathbb{E}\Pr[Y]$ as:

$$\mathbb{E}[\Pr[Y]] = \mathbb{E}[\Pr[Y|S]\Pr[S]] + \mathbb{E}[\Pr[Y|W](1-\Pr[S])] \tag{1.2}$$

$$= \mathbb{E}\Pr[Y|W] + \mathbb{E}[\Pr[S](\Pr[Y|S] - \Pr[Y|W])]. \tag{1.3}$$

We can use this to posit various theories that capture the underlying parameters of the distribution. For instance, consistent with causal reasoning expressed above, let $\Theta$ denote the theory that $\Pr[Y|S] > \Pr[Y|W]$—that is, the market opportunity is more likely to exist if the technological opportunity is strong. A natural alternative theory or null hypothesis is that there is no relationship between the technological opportunity and the existence of the market. Let $\Theta'$ denote that alternative, whereby $\Pr[Y|S] = \Pr[Y|W]$. Each theory has an expected value, which is the expected value of the future states of interest under the theory:[15]

$$\begin{aligned} V_\Theta &\equiv \big(\mathbb{E}_\Theta[\Pr[Y|W]] + \mathbb{E}_\Theta[\Pr[S](\Pr[Y|S] - \Pr[Y|W])]\big)V > V_{\Theta'} \\ &\equiv \mathbb{E}_{\Theta'}[\Pr[Y|W]]V. \end{aligned}$$

Under the theory, the parameters outside $\Theta$ have probability 0. Noteworthy, the more concentrated the parameter set on parameters that imply high expected probability, the higher the subjective probability of occurrence of the future states of interest (in our example, the rise of a new market). This, given $\mathbb{E}_{\Theta}[\Pr[Y]]$, makes the entrepreneur's prior stronger because, thanks to their theory, they are persuaded that the states of interest are more likely to occur. This depends on their reasoning—that is, on a choice of more powerful attributes or causal links (a "better Bayesian network") that increases, to a greater extent, the likelihood of occurrence of the states of interest.[16] In this way, entrepreneurs build, through logical thinking and causal reasoning, prior beliefs about an idea or opportunity.

In our example, the potential development of the new technology raises the probability of the emergence of a new market in the entrepreneur's mind. This theory is subjective and idiosyncratic. It determines the entrepreneur's confidence or optimism, leading them to believe what others might not. This process contributes to explaining why, given individual characteristics and contextual factors, priors across entrepreneurs might be heterogeneous, even about the same opportunity and with the same information available.

Noteworthy, the presence of a null hypothesis is important, as it allows the causal nature of theories to be fully captured.

In our example, special cases of the theory $\Pr[Y|S] > \Pr[Y|W]$ are: (a) if $x_t = S$ happens, then $x_m = Y$ happens—that is, $x_t = S$ is a *sufficient condition* for $x_m = Y$; and (b) if $x_t = S$ does *not* happen, then $x_m = Y$ does *not* happen—that is, $x_t = S$ is a *necessary condition* for $x_m = Y$. In the former case, entrepreneurs theorize that a strong technology implies the emergence of the market—that is, $\Pr[Y|S] = 1$—and that other attributes may also generate high demand—that is, $\Pr[Y|W] > 0$. In the latter case, they recognize that $x_m = Y$ cannot occur without $x_t = S$—that is, $\Pr[Y|W] = 0$—but may not occur, even if $x_t = S$—that is, $\Pr[Y|S] < 1$—which means that other attributes can increase the probability of $x_m = Y$. An even more special case is the one in which $x_t = S$ is a necessary and sufficient condition for $x_m = Y$—that is, $x_m = Y$ happens if and only if $x_t = S$ happens. In this case, entrepreneurs theorize that a strong technology implies the emergence of the market and that the new market will emerge only in presence of a strong technology—that is, $\Pr[Y|S] = 1$—and that no other attributes may also make the new market emerge—that is, $\Pr[Y|W] = 0$.

Null hypotheses and alternative theories represent the way in which entrepreneurs incorporate counterfactual thinking (Pearl and Mackenzie, 2018) in the prior formation process.

**Priors on priors (or theories)** Entrepreneurs might be aware that the future state space they build their prior beliefs on might not be true. This equates to acknowledging that there are low-probability states they do not consider or unforeseen contingencies their prior beliefs do not account for.

In this case, entrepreneurs are aware of being unaware (Karni and Vierø, 2017) and might have "priors on priors," which are priors on whether their theories are true. We represent this prior on prior as a probability, $\omega \in (0, 1)$. Priors on priors imply that entrepreneurs define a null hypothesis, as we have done with $\Theta'$ above. This is a different configuration of the parameters against which entrepreneurs compare their theories. In our example, this corresponds to the probability that the new market will rise absent the entrepreneur's attribute and causal link or, equivalently, to the probability the new market will rise under all other imaginable alternative theories.

The expected value of the theory is then:

$$V = \omega V_{\Theta} + (1 - \omega) V_{\tilde{\Theta}}, \tag{1.4}$$

which takes into account that the entrepreneur is not sure whether their theory—that is, the Bayesian network of their theory—is the right one (i.e., the entrepreneur has methodic doubt). This is the prior belief formed by an entrepreneur who is also aware that their theory might be false.[17]

### 1.5.2 Experimental Choice and Design

Under simple Bayesian learning, experiments are easy to conceive: they are signals with a known structure that inform the entrepreneur—and others—about the true state. Deep Bayesian learning is different. It does more than update prior distributions; it can prompt a reformulation or even a rejection of the very theories that produced those priors. Crucially, experiments vary in their capacity to do this. As we will show, experiments can be designed to distinguish among alternative parameterizations of a given distribution, but the testing of theories themselves typically occurs only within a theory—usually with the goal of refuting that theory—rather than by enabling clean comparisons across competing theories. Nevertheless, we argue that experimentation conducted within a theory can still help entrepreneurs choose which theory to rely on when making key decisions.

**Testing within a theory** As already noted, a theory involves not only distributional assumptions on a state space but also assumptions regarding the states under consideration and the particular causal structure between states. In addition, an entrepreneur does not simply hold a theory but also has methodic doubt and so will consider there is some probability, $1 - \omega$, that the particular set of parametric restrictions do not hold and an alternative hypothesis (i.e. other sets of attributes and causal links) is true. Thus, when the entrepreneur conducts experiments within the theory, it can inform on the distribution, state space, and also parametric restrictions.

According to Ortoleva (2012), when decision-makers face *drastically* contradicting evidence that disrupts dynamic coherence, which is a set of coherent beliefs that they hold, they will update their prior on prior—that is, the prior on the future state space or decision problem they are working on. These coherent beliefs are analogous to the above-defined entrepreneurial theories in that Ortoleva has in mind logical links contradicted (or possibly supported) by evidence. In our framework, this implies that when the distance between the two distributions is higher than some threshold, entrepreneurs also change their prior on prior, $\omega$, to, say, $\omega'$. Intuitively, this says that if an experiment does not change expectations about parameters radically, entrepreneurs only change their distributions of the parameters. However, beyond the threshold, not only do $V_\Theta$, $V_{\Theta'}$, and $V$ change but also entrepreneurs update $\omega$. An entrepreneur who gains extremely negative evidence (beyond a threshold they have in mind) will not only adjust their prior but also question the state space at hand. They will update negatively their prior on prior (i.e., prior $\omega$ on $\Theta$).

When this occurs, entrepreneurs do not simply update, based on experimental evidence, their beliefs on the idea but also update their beliefs on the underlying conceptual structure (i.e. their theory. This means that the experiment has either raised the probability of states that were previously neglected, leading the entrepreneur to consider them in their theory or provided evidence about the existence of new attributes, i.e. of states originally unforeseen by the entrepreneur.

Such "upper level" updating can occur in two ways. The first, defined as "reverse Bayesianism" (Karni and Vierø, 2013, 2017), posits that as entrepreneurs obtain awareness that unforeseen contingencies are possible, they will shift probabilities of foreseen contingencies proportionally to them. Specifically, entrepreneurs change their prior beliefs over (i) new states that may emerge when they obtain awareness about previously unforeseen actions or consequences and (ii) previously null states that had been previously considered as not possible. Reverse Bayesianism enables entrepreneurs to adjust the state space they envision (their theory/opportunity) for unknown events. This process is consistent with the methodic doubt of scientists, who acknowledge there might be other explanations/theories to what they study, which they do not know (Camuffo et al., 2023). We can think of reverse Bayesianism as forward-looking Bayesianism.

The second, defined as the "hypothesis-testing model" (Ortoleva, 2012), posits that entrepreneurs use standard Bayesian updating but might have non-Bayesian reactions to low-probability events or unforeseen contingencies. Under the hypothesis-testing model, entrepreneurs follow Bayes' rule if they receive signals to which they assign a probability above a threshold. Otherwise, they look at a prior over priors and update these second-level priors using Bayes' rule. Such updating of second-level priors corresponds to forming, testing, and updating beliefs over theories (Karni, 2022) or future state spaces and corresponds to the process underlying the changes in an entrepreneur's confidence that their theory/future state space is true. As a result, Bayesian entrepreneurs form, test, and update their beliefs about their ideas and the generating mechanisms underlying such ideas and their choices.

After the experiment, the expected value under the theory (the prior belief on the opportunity or idea) changes to values $V'_\Theta$ and $V'_\Theta$ based on updated distributional assumptions. The conditional expectation $V'_\Theta$ can be larger or smaller than $V_\Theta$, depending on whether the information from the experiment upholds or contradicts prior beliefs. Of course, there will be also an updated expected value under the null hypothesis, $V'_{\Theta'}$, developed analogously to $V_{\Theta'}$. Following the experiment, the expected value of the theory becomes:

$$V' = \omega' V'_\Theta + (1 - \omega') V'_{\Theta'}, \tag{1.5}$$

where the update $\omega'$ occurs only if the evidence is sufficiently contradictory to question the current future state space.

**Testing when there are competing theories** There is a subtlety that arises when testing in the context of holding a theory—that is, when entrepreneurs hold $\omega = 1$. If entrepreneurs use all the information they have to form their prior beliefs, then before running an experiment on any subset of parameters, $\theta$, of their theory, their expected posterior (updated probability distribution) on a theory is equal to the current prior belief—specifically, $\mathbb{E}[\omega'] = \omega$. Of course, after the experiment, when they observe its outcome, they make a positive or negative update. But before the experiment, they weigh positive or negative updates in such a way that the expected updated probability distribution and the expected updated prior on prior remain the same as their current beliefs.

If this is not the case, entrepreneurs must have, before the experiment, information that puts greater weight on favorable or unfavorable information about the theory. However, in this case, their current expected probability distribution and prior must be different from their current beliefs. In other words, any information available to the entrepreneur before the experiment will be incorporated in the current probability distributions of the theory and in the prior belief on it. In turn, this implies that, using equation 1.4, the expected update of $V$ is $V$. Therefore, if the experiment is even minimally costly, it is not worth running it because it does not provide, in expectation, additional information.[18]

This suggests that testing within a theory is, at best, useful for understanding the limitations of that theory itself. It does not create information that would necessarily allow for updating of competing or distinct theories. In part, this is a result of the way we have set up the problem; for instance, two theories may be based on common assumptions about the value of a technological opportunity, and so information collected regarding one may inform others (see the discussion in Gans et al., 2019). But it is also a conclusion that changes when the entrepreneur has engaged in the cognitive processes to construct more than one competing theory. As Gans et al. (2024) argue (i.e., "test two, choose one"), there is value to entrepreneurs considering two distinct choices or plans—here, two alternative theories—precisely so that they compare them. The second theory, in a sense, comes with an alternative set of actions that can be taken, the expected value of which is an outside option when evaluating the initial theory.

Thus, experiments, even if informing only of the value of a specific theory, can be valuable when alternative theories have been articulated.[19] An alternative theory comprises a different state space characterized by different attributes or causal links. Using the relevant parameters and probabilities, the entrepreneur calculates $V_2$, the unconditional expected value of the alternative theory, in the same way as $V$, which, for clarity, we call $V_1$. If entrepreneurs have an alternative theory with expected value $V_2$, then important changes in $V_1$—such as those induced by updating priors or priors on priors—are more likely to imply $V_1' < V_2$ and thus a switch to the alternative theory $V_2$. Experiments that generate unexpected (surprising) outcomes—for instance, observations that represent unforeseen contingencies (Karni and Vierø, 2013, 2017), or regarding low-probabilities states (Ortoleva, 2012)—may nudge the entrepreneur to consider novel attributes and causal links and hence be conducive of novel theories. Therefore, entrepreneurs use experiments not only to test their prior beliefs about a future state space, aiming at higher $V_1$, but also as a potential source of anomalous observations that can give rise to alternative theories (Mullainathan and Rambachan, 2023).

To the entrepreneur, the value of experimentation increases with the number of theories. A higher $V_1$ (higher expected value of theory $\Theta$) makes it more valuable either to experiment with it or to commit to it. If $V_1$ is relatively small compared to $V_2$, the same applies to the alternative theory. As figure 1.4 shows (see the appendix for the formal proposition), there are two exploration zones: one for $\Theta$ and the other one for the alternative theory.

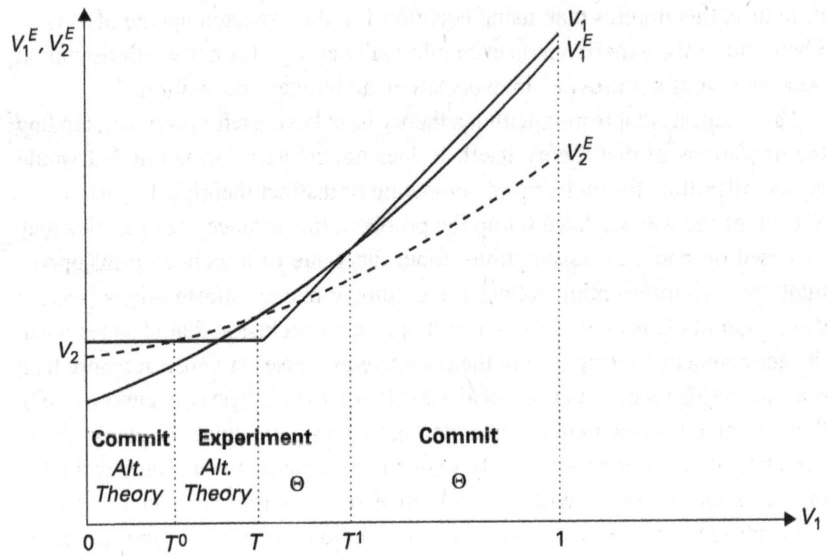


**Figure 1.4**
Value of experimentation and exploration zones.

Camuffo, Gambardella and Pignataro (2024) show under which conditions it is optimal for entrepreneurs to choose a given theory or to experiment. They also show that more uncertain theories (i.e., theories with larger spreads in parameters or with higher entropy priors) increase their relative space of exploration, the overall space of exploration, and the value of experimenting with it. These experiments are more informative because they might yield a higher information upside, while the alternative theory protects against the potential negative outcomes of the experiment. This greater value of more uncertain theories supports Felin and Zenger's (2017) conjecture about the value of testing contrarian beliefs. In addition, this framework is compatible with testing weaker priors—that is, theories such that, initially, $V_1 < V_2$. Thus, it can be optimal to test not only more uncertain theories but also weaker priors.

Furthermore, uncertain theories are complementary—that is, experimenting with a more uncertain theory increases the benefits of experimenting with other more uncertain theories. The intuition is that a more uncertain alternative theory raises the value of the outside option of an experiment on theory Θ, which raises the value of this experiment. This widens the exploration zone and the expected value of theories.

## 1.6 Specific Research Topics

In this final section, we outline two areas where we believe a Bayesian approach can improve specific research areas in entrepreneurship.

### 1.6.1 A Bayesian Approach to Entrepreneurial Strategy

Entrepreneurial strategy is the sequence of choices a founding team makes to test specific value creation and capture hypotheses when entrepreneurial experimentation requires partial commitment (Gans, Scott and Stern, 2018; Gans et al., 2019; Gans, Scott, and Stern, 2024). Specifically, as entrepreneurs explore a given idea, they face many alternatives that cannot be pursued at once, and so must adopt (implicitly or explicitly) a process for testing their priors about, and ultimately choosing among, entrepreneurial strategies for moving forward their chosen entrepreneurial opportunity. The presence of deep uncertainty and limitations on commitment-free learning has the implication that entrepreneurial experimentation provides only partial insight into the relative value of alternative strategies. As a result, most early-stage entrepreneurial experiments are not singularly decisive, with founders facing trade-offs between three competing elements of experimental design (see figure 1.5):

- Criticality—the relative importance of the hypothesis being tested in the overall entrepreneurial strategy
- Fidelity—the degree to which a test of a given level of criticality provides meaningful and informative feedback to the founders about the hypothesis being tested
- Opportunity cost—the overall cost, including resources, time, and strategic commitments, required to conduct a test of a given fidelity with a given level of criticality

When choosing to conduct an entrepreneurial experiment, the single most important element to consider is whether a successful test would meaningfully reduce the most significant drivers of uncertainty associated with the entrepreneurial opportunity. Put simply, entrepreneurial experimentation is only valuable if the experiment is testing something that actually matters. Concretely, for any given idea and strategy, there is a set of hypotheses and sub-hypotheses that must turn out to be true in order for the idea and strategy to succeed. These include, among other factors, the presence of a paying customer, feasible technology and organization, and a commercialization path that allows for value creation and capture on a sustainable basis. The first step in the process of testing an entrepreneurial strategy is to delineate the most critical elements of the

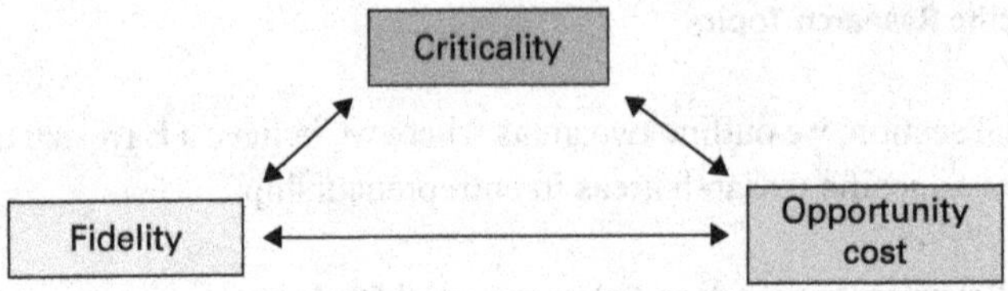


**Figure 1.5**
A strategic approach to entrepreneurial experimentation.

strategy to be tested. The most concrete consequence of stating these hypotheses clearly and crisply is to help identify both the underlying opportunity and most critical risks facing a venture. Being clear and sharp about the most critical hypotheses facing the venture allows the founding team to effectively test their entrepreneurial strategy and focus their time and attention on ensuring that those critical hypotheses turn out to be true.

Of course, simply knowing that a hypothesis is critical is not enough. Instead, one must be able to test that hypothesis in a way that is informative. For a test to have fidelity, it has to have the potential to meaningfully shape the future choices of the founders. To do so, the test, therefore, needs to meaningfully reduce uncertainty, providing validation (or not) to the overall idea or to a particular strategy. Entrepreneurial experiments must drive the entrepreneurs' choice of whether to pursue a given idea and choice of the entrepreneurial strategy to realize that opportunity. But, as with criticality, simply exhorting entrepreneurs to engage in high-fidelity, informative experiments is much simpler than actually designing and conducting those experiments. Two key obstacles—noise and bias—stand in the way.

First, even if one expends significant time and resources, the outcome of an entrepreneurial experiment might simply be noisy. Without a sufficiently high signal-to-noise ratio, it will be difficult to draw out meaningful insights that inform entrepreneurial choice. Particularly at the earliest stages of a venture, it is possible to engage in a wide range of "experiments" that nonetheless leave founders with data that yields neither holistic themes nor a consistent body of evidence. At the conclusion of such experiments, despite often considerable effort, founders have gained little to no clarity on the validity of their hypotheses or the path ahead. The challenge is often not the lack of centrality of the hypotheses but rather the lack of fidelity of the experiment. When exploring a new idea, any data—qualitative or quantitative—is likely to have significant variation, and being cognizant of how noisy data can be allow founders to avoid leaping too quickly to a conclusion on the basis of very preliminary evidence.

The second potential obstacle to experimental learning is experimental bias. Bias occurs when the data or feedback from an experiment is distorted by the ways in which data are gathered or analyzed. For example, if founders first test their product or service idea on their friends and family (a sensible, low-cost, low-commitment approach), they are likely to simultaneously receive too much positive feedback from acquaintances who are simply trying to be polite and fail to gain accurate feedback from potential beachhead customers if there is limited overlap between that beachhead and their social network or other accessible settings for entrepreneurial experimentation (Cao et al., 2023). As a consequence, bias has the potential to encourage founders to believe in an assumption that is, in fact, false or dismiss a critical assumption that, in fact, holds.

The impact of noise and bias does not negate the value of systematic experimentation and learning. Instead, the very fact that many experiments are noisy and biased simply reinforces how little entrepreneurs actually know about their most critical hypotheses when they are choosing a given entrepreneurial opportunity and the entrepreneurial strategy to pursue that idea. The key is to take a proactive and disciplined approach to generate accurate and actionable information about the most critical hypotheses. The field continues to contribute a range of tools and frameworks that improve the fidelity of entrepreneurial experimentation, given the natural constraints entrepreneurs face (see, among others, Murray and Tripsas, 2004; Nanda and Rhodes-Kropf, 2016; Cohen et al., 2019; Koning et al., 2022).

The final critical factor impacting the value of entrepreneurial experimentation and learning is the opportunity cost of a given experiment. All else equal, the cheaper and faster one can make an entrepreneurial experiment, the more valuable it will be in terms of being able to inform the core choices of a new venture. Although this point might seem obvious, it is also easy to underestimate the value that arises from less costly or more timely testing. Most notably, when the uncertainty around an opportunity is significant, the simple process of clarifying the most critical assumptions or how to design a high-fidelity test is both challenging and time-consuming. However, identifying low-cost and speedy ways to explore the underlying idea, potential beachhead markets, or potential commercialization paths can sharply reduce uncertainty.

The relationship between cost, speed, and commitment is subtle. On the one hand, when the main cost of experimentation, is the time cost of the founders, lower cost, rapid experimentation, and low commitment all go together; interviewing more potential customers or running more experiments per day simply accelerates the process of learning (and does not necessarily commit the venture to one path over the other). This natural inverse relationship between cost and speed is at the heart of the lean start-up philosophy: "The only way to win is to learn faster than anyone else" (Ries, 2011). When entrepreneurs can focus their efforts and scarce resources on maximizing learning, rapid learning can serve as a potential source of competitive advantage. On the other hand, although rapid learning and low cost go together when the main resource being used is founder time, the relationship between speed and cost can be reversed if one is willing to, for example, enhance speed by incurring a higher upfront cost. For example, doubling the rate of customer interviews might involve hiring (perhaps temporary) employees who would need to be paid in cost or equity; more rapid product design iterations might involve investing in costly machinery (e.g., 3D printers). As such, it is important for entrepreneurs to clarify the relationship between cost and speed for the experiments they undertake and, in particular, how they can best make use of their own time (i.e., enhance the virtuous cycle of low-cost and rapid learning) while also being judicious in expending resources in order to allow learning to go faster (i.e., by expending money for more rapid feedback cycles).

However, in many cases, the most important "cost" of an experiment is neither time nor resources but instead a strategic commitment cost. Two particular types of strategic commitments are particularly salient in shaping the opportunity cost of an entrepreneurial experiment. First, many potentially informative entrepreneurial strategy tests involve committing to particular stakeholders, including potential customers, collaboration partners, or early employees. For example, launching a crowdfunding campaign, such as on Kickstarter or Indiegogo, is an excellent way to gauge the preliminary level of demand for an early-stage product idea, but such a test comes with the commitment to actually build and deliver the product for those who contribute to the campaign. However, as documented by Ethan Mollick, the vast majority of successful crowd funding projects are unable to meet their promised timelines, resulting in significant negative feedback from early users. As a result, the use of crowdfunding as a tool for entrepreneurial experimentation brings with it a high likelihood of ending up with a negative reputation with precisely those customers who initially expressed the greatest interest. Said otherwise, although a minimal viable product allows the entrepreneur to learn a great deal about customers, it also allows customers to learn about the product and venture, which should be pursued with awareness as to the commitments the venture is making to key stakeholders. Second, many entrepreneurial experiments involve the disclosure of information—either knowledge about the underlying idea or statements of future strategic intention—that reduce the scope for future strategic choice. Perhaps most notably, if one reveals information publicly about a novel technology, it is possible that the potential scope for formal intellectual property rights such as patents might be reduced. The strategic opportunity costs involved in an entrepreneurial experiment may limit the ability of an entrepreneur to effectively pursue that tested path, even if that path would have otherwise been viable based on the learning that occurred.

In other words, the costs of an experiment are not simply time but also the new resources and potential strategic commitments that come along with a given experiment. Rather than simply weighing the value of learning versus the value of time, entrepreneurs will have to consider more carefully whether that learning outweighs the overall opportunity cost of the proposed experiment. Calculating the potential value of alternative experiments is central to the effective design and sequencing of experiments in the process of entrepreneurial learning. Understanding the interplay between criticality, fidelity, and opportunity cost is central to this process. In an ideal case, it would be possible to quickly identify and implement experiments and approaches to entrepreneurial learning that simultaneously tested the most critical hypotheses with a high level of fidelity and at minimal opportunity cost. Most tests of entrepreneurial strategy involve significant trade-offs between the three dimensions—testing the most critical assumptions may require the highest cost or the best available data to inform hypotheses that are of only modest importance. Figuring out how to navigate the natural and inherent trade-offs between criticality, fidelity, and opportunity cost is among the most challenging yet rewarding elements of taking a strategic approach to entrepreneurship. A Bayesian approach to entrepreneurship will allow the field to further guide entrepreneurs in designing their entrepreneurial experiments and information on their choices as to whether to pursue a given idea and the entrepreneurial strategy to realize that opportunity.

### 1.6.2 The Scientific Approach for Deep Bayesian Learning

Another example of an entrepreneurial method that embodies Bayesian learning is the scientific approach (Camuffo et al., 2020). This approach encourages entrepreneurs to form prior beliefs on future state spaces using theories, test such beliefs through experiments, and use the associated evidence to update beliefs (Zellweger and Zenger, 2023). Prior formation through theorizing is the first step of the Bayesian learning cycle and involves engaging in deliberate cognitive efforts to formulate entrepreneurial problems, ideas, or opportunities as causal conceptual structures (Camuffo, Gambardella, Messinese et al., 2024; Ehrig and Schmidt, 2022; Felin and Zenger, 2017). Prior testing occurs through experiments that, as previously highlighted, are shaped by priors. Experiments elicit signals that are incorporated, through Bayes' rule, into updated beliefs. This allows entrepreneurs to make more informed decisions about whether to continue with the current project, pivot to a modified or different project, or terminate the venture.

As discussed above, experiments can either test prior beliefs within a given future state space (simple Bayesian learning) or test "priors on priors"—that is, question which future state space entrepreneurs should attend to when forming priors (deep Bayesian learning on theories of value). Likewise, the information produced by experiments can prompt straightforward Bayesian updating, or—if the evidence is sufficiently surprising—lead entrepreneurs to challenge their theories of value and revise their priors on priors (deep Bayesian learning). Notably, viewing the scientific process as deep Bayesian learning provides a theoretical foundation for entrepreneurial pivoting and helps explain the various typologies proposed in the literature (Kirtley and O'Mahony, 2023).

Below, we explain how the experimental evidence on the adoption of the scientific approach to entrepreneurial decision-making points to two mechanisms at work in generating more quitting, more focused pivoting, and higher performance: search efficiency, which corresponds to simple Bayesian learning, and methodic doubt, which corresponds to deep Bayesian learning.

The evidence coming from fourteen randomized control trials (RCTs) involving approximately three thousand start-ups around the world shows that the Bayesian learning embodied in a scientific approach to entrepreneurial decision-making increases the probability of terminating projects and of efficient pivoting and, conditional on survival, leads to higher performance (revenues) (Camuffo et al., 2020; Camuffo, Gambardella, Messinese et al., 2024). The evidence also shows that theories and experiments have separate, identifiable effects (Agarwal et al., 2025), and that scientific entrepreneurs better navigate uncertainty accounting for unknown events (Camuffo, Gambardella and Messinese, 2023). As previously highlighted, this occurs in ways compatible with reverse Bayesianism (Karni and Vierø, 2013, 2017) and theory-based decision-making (Ortoleva, 2012; Karni, 2022).

The scientific approach affects entrepreneurial decision-making in two ways (Camuffo, Gambardella, Messinese et al., 2024). On the one hand, entrepreneurs who adopt it use more clearly defined theory and validation procedures when they iteratively explore their ideas. This makes their search process more (*efficient search*). This raises the probability of success in earlier stages because it combines a higher probability of success stemming from theories and experiments and the ability to identify successful ideas earlier. This is because theories enable entrepreneurs to envision opportunities, form priors, and rank them better at the outset of the process. Entrepreneurs will then work first on priors that they expect to have a higher probability of success. When one of the ideas they are working on proves unsuccessful, they pivot to other ideas, but these will have a lower expected probability of success. Nonscientific entrepreneurs, absent theories, are instead more likely to learn about the value of their ideas as they work on them rather than ex ante in a trial-and-error fashion. Thus, they start with an idea/prior, and if it proves to be unsuccessful, the new idea they pivot to will not have an expected probability of success as low as in the case of scientific entrepreneurs because ideas are considered sequentially as they emerge in subsequent stages of exploration and not ranked beforehand based on theories.

As a result, efficient search allows entrepreneurs to work on more solid priors earlier in their exploration process. However, if efficient search was the only mechanism associated with the scientific approach, entrepreneurs who adopted it would simply be more likely to commit to their ideas rightaway, less likely to terminate, and more likely to pivot.

However, as described above, the scientific approach also instills in entrepreneurs a methodic doubt. The very fact that entrepreneurs who adopt this approach develop a theory for why their ideas may be successful makes them aware that this might be imperfect, false, or less plausible than others. As a result, they are more cautious (less optimistic about their ideas). Again, if methodic doubt was the only mechanism associated with the scientific approach, entrepreneurs who adopted it would be more likly to terminate and less likely to pivot or commit to their ideas.

It can be easily seen that efficient search and methodic doubt go in opposite directions, and it is an empirical matter to establish which of the two mechanisms dominates the other. The birth and early developments of ØSense—one of the start-ups accelerated in the above-mentioned RCTs—help to illustrate how the scientific approach embodies a process of Bayesian learning. The founder's initial idea was to develop existing technology (matching algorithms) to build a peer-to-peer rental service for household items, inspired by the success of Airbnb in the accommodation sector. The company was named YouRent, and its theory was that the concept underlying the sharing economy could be extended, by analogy, to everything, including everyday products and small items, allowing people—through technology—to rent items instead of buying them, and that this would create a new market with the value proposition of saving money while promoting sustainability and reducing waste. The theory summary was "Imagine Airbnb for everyday items." The founder was optimistic about the project, but the scientific approach learned during the RCT's intervention led him to doubt and wonder whether his beliefs were strong for good, logically grounded reasons. Hence, he set out to run an experiment in the form of field interviews with potential customers to test his priors. He ran several interviews at vintage fairs—a higher likelihood of individuals inclined toward reuse/rental of used goods—and set ex ante a threshold of 60 percent to accept their hypotheses. This was how he operationalized running an efficient experiment and getting the likelihood function necessary to apply Bayes' rule. The interviews did not support the necessity or willingness to engage in a rental platform for low-value items as hypothesized for YouRent. The founder negatively updated his prior and decided to terminate the initial project.

However, from those very interviews, a recurring theme of sustainability emerged, spurring a new potential future state space (a new potential theory and prior). The founder conceptually developed the insights from the interviews into WERI, a platform aimed at enhancing corporate car fleet sharing with a gamified experience that tracks and rewards sustainability efforts. Here, the founder believed that incentivizing sustainable behavior would also encourage the sharing of underutilized corporate fleet vehicles, leading to cost savings and reduced environmental impact.

Another Bayesian cycle started. Field interviews were conducted with fleet managers and sustainability officers to validate WERI. Also in this case, the theory underlying WERI did not get corroborated. However, the interviews pointed to a strong and generalized concern for sustainability and the need for reliable data to measure the carbon footprint of corporate activities, mainly driven by the pressure coming from top management teams (pushed by EU policy constraints). The founder refined this conceptual causal structure and crafted a new theory in which he envisioned his technology as an artificial intelligence platform to collect and reduce carbon emissions across the entire value chain. This led to Øsense (https://osense.ai/), a startup that helps firms "activate a real carbon reduction strategy by leveraging Artificial Intelligence to measure, simulate, reduce, and report your environmental footprint at scale." The pivot from WERI to Øsense was a total change of the theory. In the founder's words, "I reevaluated my entire theory. This new piece of information seemed initially disconnected and 'far' from my original framework. Through the process of questioning and engaging in cognitive efforts to explore logical connections, I managed to recognize the value of this opportunity and incorporate it into my understanding."

The case illustrates how Bayesian learning occurs for entrepreneurs adopting the scientific approach at two levels. First, scientific entrepreneurs form, test, and update their beliefs about the success or value of their ideas. Second, they form, test, and update their beliefs about the generating mechanisms underlying such ideas.

### 1.6.3 A Bayesian Approach to the Market for Judgment

A Bayesian approach to the market for judgment builds on the foundational understanding that both entrepreneurs and investors operate under potentially different priors about the viability and potential of new ventures. Judgment, a critical component of decision-making, is the application of experiential knowledge to prioritize the priors that most urgently require updating and how to do so most effectively (Agrawal et al., 2021). Experienced entrepreneurs possess well-honed judgment derived from past successes and failures, enabling them to prioritize the actions that maximize the likelihood of venture success.

In a Bayesian framework, better judgment lowers the cost of experimentation by better sequencing and designing experiments. Better judgment enables better assessing the return on investment associated with updating which priors for which actors. In addition, better judgment increases the return on investment of the experiment itself by enhancing its design to generate a more effective signal at a lower cost. Experienced entrepreneurs, with their superior judgment, can devise experiments that provide effective signals that update the targeted actor(s). For instance, they might recommend conducting pilot studies or market tests that yield high-quality data on customer acceptance, technological feasibility, or competitive positioning. By leveraging their judgment, they increase the likelihood that each experiment is designed to minimize ambiguity and maximize the information gained to address the priors that are in most urgent need of updating.

However, the market for judgment faces significant frictions due to the difficulty of transferring experiential knowledge. Unlike tangible assets, judgment cannot be easily bought or sold. Programs such as Creative Destruction Lab (CDL) aim to mitigate these frictions by facilitating interactions between experienced and novice entrepreneurs (Lakhani et al., 2019; Luo, 2022). By creating structured environments where judgment can be shared, programs such as the CDL help new entrepreneurs access the insights and guidance of those who have successfully navigated the commercialization process before (Gans, 2018; Sariri, 2024). This exchange of judgment allows new entrepreneurs to update their priors and the priors of potential investors, customers, and partners more effectively. The Bayesian approach to the market for judgment may be implemented through such institutions, enabling the transfer of experiential knowledge in order to design experiments that more efficiently generate signals to address the highest-priority priors.

When introducing a new product based on a recent scientific invention, different actors—such as the entrepreneur, investors, and potential customers—are likely to have varying priors about the product's viability. Entrepreneurs, driven by their vision and understanding of the underlying science, might have a more optimistic prior about the product's potential. Investors, who assess the product's marketability and financial prospects, may hold more skeptical priors, balancing enthusiasm with caution. Potential customers, who are crucial for market validation, often base their priors on perceived utility and credibility of the product. These divergent perspectives necessitate targeted strategies to align beliefs and facilitate the product's successful commercialization.

Consider the case of a new medical device designed to detect early-stage cancer using a novel biomarker discovered in academic research. The entrepreneur, typically a scientist involved in the discovery, believes strongly in the device's efficacy and potential to revolutionize cancer diagnostics. However, investors might be wary due to the high failure rate of medical start-ups and the lengthy, costly regulatory approval process. Potential customers, such as hospitals and clinics, may doubt the device's accuracy and reliability compared with established methods. These differing priors highlight the challenge of securing investment and market adoption.

The urgency to update priors depends on the stage of product development and the most urgent barriers to overcome. In the early stages, it may be more critical to update the entrepreneur's prior, especially if their optimism leads to overestimation of the product's readiness or market fit. Accurate self-assessment is crucial to avoid costly missteps and to design feasible development plans. Conversely, once a prototype is available, updating investors' priors becomes more urgent, as their funding is essential for further development and scaling. Investors' confidence can be bolstered by credible demonstrations of the product's potential, reducing their perceived risk.

In our medical device example, in the initial phase, the entrepreneur might rigorously test the device within controlled environments to ensure that their confidence in its efficacy is justified. Suppose initial laboratory tests show promising results, indicating that the device accurately detects cancer biomarkers. At this point, updating the investors' priors becomes the priority. A different signal is now required. In other words, there may be an optimal sequence for generating different signals for updating the priors of different actors.

Once the sequence for updating priors is established, the entrepreneur must design experiments to provide the necessary signals as effectively and efficiently as possible. The experiment must be tailored to address the specific doubts and questions of the actor whose prior is most in need of adjustment. For entrepreneurs, this might involve extensive prototype testing and iterative refinement. For investors, it may involve pilot studies, market analyses, and validation from respected third-party entities. The goal is to produce robust, credible data that either confirm or challenge the prevailing beliefs, facilitating informed decision-making.

In our example, after initial laboratory validation, the next step might involve a small-scale clinical trial to update investors' priors. This trial should be designed to provide clear, reliable data on the device's performance in a real-world setting. For instance, the trial could involve multiple hospitals and a diverse patient population to demonstrate the device's accuracy and usability across different conditions. Positive results from such a trial would significantly reduce investor skepticism, providing strong evidence that the device is ready for further development and scaling. This approach ensures that the most critical priors are addressed efficiently, paving the way for successful commercialization.

Thus, a Bayesian approach to the market for judgment provides a structured framework for aligning the diverse priors of different actors involved in the commercialization of scientific innovations. By leveraging the judgment of experienced entrepreneurs, this approach enhances the efficiency of experiments aimed at updating critical priors. Programs such as the CDL facilitate the transfer of experiential knowledge, helping novice entrepreneurs design and execute experiments that yield clear, actionable signals. Through this iterative process, entrepreneurs can systematically address the doubts of investors, potential customers, and other stakeholders, thereby increasing the likelihood of securing the necessary resources and market acceptance. Ultimately, the Bayesian framework for the market for judgment increases the likelihood that the commercialization process is guided by well-informed decisions, maximizing the potential for success in bringing scientific advancements to market.

## 1.7 Conclusions and Next Steps

This paper has outlined the key tenets of a Bayesian approach to entrepreneurship, emphasizing the central role of heterogeneous priors, the relatively strong priors held by entrepreneurs, and the process of belief updating through Bayesian learning. We argue that embracing this Bayesian perspective provides a unifying framework to connect and enhance diverse streams of research on entrepreneurial decision-making, including the formation of entrepreneurial beliefs, purposeful entrepreneurial experimentation, and the persuasion of key stakeholders.

By focusing attention on the interplay between an entrepreneur's prior beliefs and the potential for learning through experimentation to shape those beliefs, a Bayesian approach provides insight into the choices that entrepreneurs make regarding which opportunities to pursue and how to design experiments to reduce uncertainty and ultimately improve entrepreneurial performance. Critically, the presence of heterogeneous priors provides a natural explanation for why entrepreneurs agree to disagree about the value of a given opportunity, and the systematic stronger priors of entrepreneurs choosing to pursue a given idea provides a micro-foundation for observed entrepreneurial action.

Moreover, by distinguishing between simple versus deep Bayesian learning approaches, we highlight how entrepreneurial theories about value creation and capture can themselves evolve as experiments yield outcomes that challenge fundamental assumptions. This provides a more nuanced understanding of entrepreneurial pivoting behavior, grounded in how entrepreneurs interpret potentially surprising signals.

Building upon this foundation, we see several promising avenues for future research. First, empirical studies can leverage the Bayesian framework to develop sharper predictions and measures related to entrepreneurial belief formation, experimental choices, and pivoting behavior. This could involve both qualitative work examining how entrepreneurs articulate and update their beliefs, as well as quantitative analyses linking measures of prior beliefs to specific strategic choices and venture outcomes.

Second, we believe there is significant potential to develop richer formal models of entrepreneurial decision-making within a Bayesian framework. This could include game-theoretic models of entrepreneurial competition and cooperation under heterogeneous priors, optimal stopping models of entrepreneurial experimentation and abandonment, and persuasion models of entrepreneurial communication with key stakeholders. Such models could yield new insights and predictions to guide empirical work.

Third, the Bayesian approach suggests opportunities to reexamine and potentially reframe findings from behavioral approaches to entrepreneurship. Although behavioral models often emphasize cognitive biases and heuristics as deviations from rationality, a Bayesian perspective suggests some of these patterns may, in fact, reflect entrepreneurs' priors and how those priors shape their interpretation of new information. Empirical studies could seek to disentangle these mechanisms.

Finally, from a practical perspective, the Bayesian framework offers a structured approach to guide entrepreneurial decision-making. Pedagogical research could explore how to most effectively teach a Bayesian approach to entrepreneurship students and practitioners, potentially building on research related to the scientific approach. Tools and frameworks to help entrepreneurs articulate key assumptions, design informative experiments, and pivot based on surprising information could have significant impact.

In conclusion, we believe that the Bayesian approach to entrepreneurship provides a promising foundation for integrating key strands of entrepreneurship research and opening up productive new avenues for scholarly inquiry and practical impact. By placing the dynamic interplay of prior beliefs and experiential learning at the center of the entrepreneurial process, this perspective offers a compelling lens to understand the entrepreneurial judgments and choices that drive the creation of new value in the economy.

## 1.8 Appendix: Details of Deep Bayesian Learning

### 1.8.1 Models, Causal Links, and Theories

When entrepreneurs build theories, they envision future state spaces for which they (and, to the extent to which the theory is novel, others) have limited information. Hence, about this state space, they have sparse empirical observations and pseudo-observations they subjectively develop, for example using analogies or "bootstrapping" from abductive reasoning. Therefore, they ultimately fit preliminary probability distributions on the state space stemming from "soft" rather than "hard" empirical frequencies that occur in worlds that decision-makers envision in their mind.[20] To operationalize our approach, let's assume that entrepreneurs' probabilities are distributed as Dirichlet with parameters $n_{IJ} > 0, I = Y, N; J = Y, N$:

$$\propto Pr(Y,Y)^{n_{YY}-1} \cdot Pr(Y,N)^{n_{YN}-1} \cdot Pr(N,Y)^{n_{NY}-1} \cdot Pr(N,N)^{n_{NN}-1}. \tag{1.6}$$

In a Dirichlet distribution:

$$\mathbb{E}Pr(I,J) = \frac{n_{IJ}}{n},$$

with $n \equiv n_{YY} + n_{YN} + n_{NY} + n_{NN}$. These parameters represent subjective empirical distributions of the four states envisioned by the entrepreneurs in the example.[21]

It helps to redefine these parameters as:

$$\theta_{YY} \equiv \frac{n_{YY}}{n_{YY}+n_{NY}}, \quad \theta_{YN} \equiv \frac{n_{YN}}{n_{YN}+\eta_{NN}}, \quad \text{and} \quad \theta_{Y} \equiv \frac{n_{YY}+n_{NY}}{n},$$

and we define $\theta \equiv \{\theta_{YY}, \theta_{YN}, \theta_{Y}\}$ to be the set of parameters of our distribution. As discussed by Marinacci (2015, p. 1037), we can then write the statistical model:

$$\mathbb{E}Pr(Y) \equiv v(\theta) = \theta_{YY}\theta_{Y} + \theta_{YN}(1-\theta_{Y}) = \theta_{YN} + \theta_{Y}(\theta_{YY} - \theta_{YN}). \tag{1.7}$$

Entrepreneurs acknowledge that their entrepreneurial decisions (pursue or not the opportunity) are affected not only by the randomness of their subjective probabilities $Pr(I,J)$ but also by the uncertainty about the parameters of their probability distributions (equation 1.6). A specific set of parameters, $\theta$, identifies one Dirichlet distribution and, therefore, in our example, one expected probability of new market rise, $v(\theta)$. But entrepreneurs do not know the probability distribution of the parameters either. They realize that their empirical distributions are also random variables because they know that $n_{IJ}$ is not an objective frequency of the state $(I, J)$. Thus, they also form, in their head, a subjective probability distribution $\mu(\theta)$ of the parameters. This defines a family of probability distributions (equation 1.6) identified by the different $\theta$ in the space of all the possible realizations of the three parameters in the example.

### 1.8.2 Extent of Experimentation

We model the extent of entrepreneurial experimentation using models of optimal information acquisition (Moscarini and Smith, 2001; Camuffo, Gambardella and Pignataro, 2024). Consider an entrepreneur who has two theories, $\Theta$ and $\Gamma$, with expected values, respectively, $V_1$ and $V_2$. Without loss of generality, we focus on an experiment on theory $\Theta$. Let $c > 0$ be a fixed cost of running an experiment, $\rho \in (0, 1)$ a discount factor accounting for the fact that the experiment delays the final decision, and $H(V_1')$ the cumulative probability distribution of the update $V_1'$ of $V$ from the experiment on $\Theta$. Specifically, $V_1'$ is the expected value of committing to $\Theta$ after the experiment or of running a new experiment on $\Theta$.

Entrepreneurs can take a stream of sequential experiments, and at the beginning of each stage of experimentation, the condition for running the experiment is:

$$V_1^E \equiv \rho\left[\int_{V_2^*}^{1} V_1' dH(V_1') + V_2^* H(V_2^*)\right] - c > \Pi_{V1} \tag{1.8}$$

where $V_2^* \equiv max(V_2, V_2^E)$, $\Pi_{V1)} \equiv max(V_1, V_2^*)$, and $V_2^E$ is the equivalent value of the experiment on the alternative theory. This condition says that: (1) if after the experiment entrepreneurs observe $V_1' > V_2^*$, they will either commit to $\Theta$ or run a new experiment on $\Theta$ obtaining $V_1'$; (2) otherwise, they either commit to or run an experiment on the alternative theory $\Gamma$ and obtain $V_2^*$.[22] The key insight is that entrepreneurs keep experimenting until the present value of the experiment is higher than all the other options, $V_1$ , $V_2$, and $V_2^E$.

**Proposition 1.3.** There are three thresholds such that: (a) If $V_1 \leq T^0$, entrepreneurs commit to the alternative theory $\Gamma$. If $V_1 \in (T^0, T)$, they run an experiment on the alternative theory $\Gamma$. If $V_1 \in (T, T^1)$, they run an experiment on theory $\Theta$. If $V_1 \geq T^1$, they commit to theory $\Theta$ and obtain $V$.

*Proof.* Using equation 1.8, and the fact that $\mathbb{E}V_1' = V_1$, rewrite $V_1^E$ as:

$$V_1^E \equiv \rho\left[V_1 + V_2^* H(V_2^*) - \int_0^{V_2^*} V_1' dH(V_1')\right] - c = \rho\left[V_1 + \int_0^{V_2^*} H(V_1')dV_1'\right] - c. \tag{1.9}$$

where the second equality stems from integration by parts. This expressions establishes that $V_1^E$ increases with $V_1$.

We first establish that:

$$\frac{\partial V_1^E}{\partial V_1} > \frac{\partial V_2^*}{\partial V_1} \geq 0.$$

Let $D = 1 - \rho^2 H(V_2^E)K(V_1^E)$, where $K$ is the cumulative distribution of the update $V_2'$ of $V_2$ if decision-makers run an experiment on the alternative theory. Take the differentials $dV_1^E$, $dV_2^*$, and $dV_1$ in (equation 1.9) and, if $V_2^* = V_2^E$, in the equivalent expression for $V_2^E$. Solving the system, or setting $dV_2^* = 0$ if $V_2^* = V_2$, obtain:

$$\frac{\partial V_1^E}{\partial V_1} = D^{-1}\rho > \frac{\partial V_2^*}{\partial V_1},$$

which is equal to 0 or $D^{-1}\rho^2 K(V_1^E)$, depending on whether $V_2^* = V_2$ or $Q^* = Q^E$.

When $V_1 \leq V_2^*$, $\Pi = V_2^*$, and the fact that $V_1^E$ increases with $V_1$ faster than $V_2^*$ is a necessary condition to state that there is a threshold $T$ such that $V_1^E$ switches from smaller to higher than $V_2^*$, this condition is sufficient for appropriate values of $\rho$ or $c$. Since $V_2^E$ grows with $V_1$, unlike $V_2$, $V_2^*$ could switch from $V_2$ to $V_2^E$ at a threshold $T^0 < T$.

When $V_1 > V_2^*$, $\Pi = V_1$, and $V_1^E$ increases at a lower rate than $V_1$ if $D^{-1}\rho < 1$. Since $V_1, V_1^E < 1$, the $V_1^E$ curve will cut the $V_1$ curve from above at a threshold $T^1$ such that if $V_1 > T^1$, entrepreneurs commit to theory $\Theta$. The slower growth of $V_2^E$ than $V_1^E$ with respect to $V_1$ is a necessary condition for $V_2^E < V_1^E$ at $V_1 = T^1$. □

## Notes

1. For example, Melnikoff and Strohminger (2024).
2. This requires each $\mu_i$ is continuous with respect to $\tilde{\mu}$ so that there exists a set of subjective beliefs that the information is correct such that $\tilde{\mu}_i = \mu_i$ for all $i \in N \cup 0$.
3. See, for instance, Gans (1995).
4. In the information design literature, this condition, where a signal determines the action, is known as "obedience"; see Bergemann and Morris (2019).
5. Arora and Fosfuri (2005) derive this willingness to pay function and explore different types of pricing structures for that information.

6. It has also been implicitly assumed in work on entrepreneurial experimentation in finance; e.g., Nanda and Rhodes-Kropf (2016).

7. Gans (2025, ch. 10) shows that the value of an experiment for a decision-maker with a given prior is always maximized at an extremal choice, implying that an extremal choice always maximizes the aggregate willingness to pay of a population of decision-makers where the priors are distributed uniformly on [0, 1].

8. The notion of experimentation prior to larger finance commitments has been explored extensively by Nanda and Rhodes-Kropf (2017), Kerr et al. (2014), and Ewens et al. (2018).

9. The other investors' participation constraint for a given experiment is:

$$(1-\alpha)\frac{\lambda_1\mu_O}{\lambda_1\mu_O+(1-\lambda_O)(1-\mu_O)}V \geq C,$$

which implies that

$$\alpha \leq \frac{\lambda_1\mu_O V-(\lambda_1\mu_O+(1-\lambda_O)(1-\mu_O))C}{\lambda_1\mu_O V}.$$

The entrepreneur chooses an experiment to maximize:

$$\alpha\frac{\lambda_1\mu_E}{\lambda_1\mu_E+(1-\lambda_O)(1-\mu_E)}V$$

subject to the investor's participation constraint. Substituting this constraint and the experiment boundary gives:

$$\frac{\mu_E}{\mu_O}\frac{\lambda_1\mu_O V-(\lambda_1\mu_O+(1-\Lambda+\lambda_1)(1-\mu_O))C}{\lambda_1\mu_E+(1-\Lambda+\lambda_1)(1-\mu_E)}.$$

Note that the second derivative of this with respect to $\lambda_1$ is positive, so there is a corner solution. For $\lambda_1=1$, we have:

$$\frac{\mu_E}{\mu_O}\frac{\mu_O V-(\mu_O+(2-\Lambda)(1-\mu_O))C}{\mu_E+(2-\Lambda)(1-\mu_E)}.$$

For $\lambda_1=\Lambda-1$, we have $V-C$.

Note that the expected entrepreneurial profit, with $\lambda_1=\Lambda-1$ is greater than the profit with $\lambda_1=1$ if and only if:

$$\mu_O(V-C) \geq \frac{\mu_E}{\mu_E+(2-\Lambda)(1-\mu_E)}(\mu_O(V-C)-(2-\Lambda)(1-\mu_O)C),$$

which can be easily shown to hold.

10. This new literature also has deep connections linking persuasion to the behavioral concept of multiple selves over time; see Jakobsen (2021).

11. There is an important difference between the approach here and the Bayesian persuasion literature. That literature supposes that senders can commit to arbitrary signals. Here, it is assumed that the sender can only commit to an experiment, and the receiver can see the outcome. As Ball and Espín-Sánchez (2022) argue, this is closer to applications and also leads to different outcomes than the full Bayesian persuasion approach.

12. One paper that has explored these differing understandings of signals is Chavda et al. (2024).

13. There is a voluminous literature on Bayesian persuasion now. Some papers that are of relevance to the Bayesian entrepreneurship agenda include work that takes into account heterogeneous priors (Alonso and Câmara, 2016), work on persuasion with anecdotes (i.e., small samples; Haghtalab et al., 2021), non-Bayesian persuasion (de Clippel and Zhang, 2022), and price theoretic approaches (Dworczak and Martini, 2018).

14. More generally, an entrepreneur's goal is to concentrate the subjective probability distribution on $x_m \in \{Y, N\}$—that is, to become more persuaded whether the state they are interested in will occur. Alternatively, they could also reason counterfactually (Pearl and Mackenzie, 2018) and build a counter-theory that increases $\Pr[N]$ using a falsificationist approach.

15. There are often several $\mu_\Theta(\theta)$ compatible with the entrepreneur's partial knowledge and information. How entrepreneurs elicit a prior probability distribution from empirical distributions—their partial information and knowledge—is another driver of priors' heterogeneity. Among the various

methods suggested in Bayesian statistics (see Kass and Wasserman, 1996, for a review), the principle of maximum entropy—that is, a prior probability assignment that incorporates the fewest assumptions on the data (Jaynes, 1968)—is particularly appealing to Bayesian entrepreneurs, as it approximates objectivity, ensures the highest learning potential, and reduces potential bias.

16. A more plausible theory is correlated but does not strictly imply that the theory is more valuable. This depends on the decision-maker's goals and the actions they choose, *given the theory that they choose.*

17. Chavda et al. (2024) consider a related notion of "theory-based" experimentation based on comparing experimental choices when one type of entrepreneur has a deeper theoretical understanding of the environment than other entrepreneurs, whereas Agrawal et al. (2021) examine the role of conflated signals and how a theory regarding the relationship between random variables determines experimental sequencing.

18. As highlighted above, the condition under which the expected value of the posteriors equals the prior corresponds to Bayes plausible experiments (Kamenica and Gentzkow, 2011). However, contingent on the level of their priors, entrepreneurs might prefer signals that are biased.

19. We assume that entrepreneurs can compare and rank the expected value of the theories.

20. Different approaches have been used to model the subjective process of building distributions for novel state spaces out of sparse information. For example, Gilboa et al. (2020) propose a model that couples case-based reasoning (i.e., sparse past data) and theories to form prior beliefs. Our approach is similar, although we do not limit the role of theories to analogical reasoning.

21. $n$ = number of observations included in the empirical distribution of the decision-maker comprising both actual observations (facts and data) and pseudo-observations.

22. We model a general exploration stage of a dynamic problem. At every exploration stage, entrepreneurs redefine $V_1, V_1^E, V_2, V_2^E$ . Since we also assumed decreasing information gains over time from experimentation, $V_1^E$ and $V_2^E$ will get smaller over time so that entrepreneurs will eventually commit to one of the two theories.

## References

Acs, Z. J., Stam, E., Audretsch, D. B., and O'Connor, A. (2017). The lineages of the entrepreneurial ecosystem approach. *Small Business Economics*, 49, pp. 1–10.

Agarwal, R., Bacco, F., Camuffo, A., Coali, A., Gambardella, A., Msangi, H., Sonka, S., Temu, A., Waized, B., and Wormald, A. (2025). Does a theory-of-value add value? Evidence from a randomized control trial with Tanzanian entrepreneurs. *Organization Science*, 36(2), pp. 601–625.

Agrawal, A., Gans, J. S., and Stern, S. (2021). Enabling entrepreneurial choice. *Management Science*, 67(9), pp. 5510–5524.

Alonso, R., and Câmara, O. (2016). Bayesian persuasion with heterogeneous priors. *Journal of Economic Theory*, 165, pp. 672–706.

Amabile, T. M. (1997). Entrepreneurial creativity through motivational synergy. *The Journal of Creative Behavior*, 31(1), pp. 18–26.

Andrews, I., and Shapiro, J. M. (2021). A model of scientific communication. *Econometrica*, 89(5), pp. 2117–2142.

Arora, A., and Fosfuri, A. (2005). Pricing diagnostic information. *Management Science*, 51(7), pp. 1092–1100.

Åstebro, T., Chen, J., and Thompson, P. (2011). Stars and misfits: Self-employment and labor market frictions. *Management Science*, 57(11), pp. 1999–2017.

Åstebro, T., Herz, H., Nanda, R., and Weber, R. A. (2014). Seeking the roots of entrepreneurship: Insights from behavioral economics. *Journal of Economic Perspectives*, 28(3), pp. 49–70.

Aumann, R. (1976). Agreeing to disagree. *Annals of Statistics*, 4, pp. 1236–1239.

Ball, I., and Espín-Sánchez, J. (2022). *Experimental persuasion.* Cowles Foundation Discussion Paper No. 2298. https://ssrn.com/abstract=3906869

Baron, R. (1998). Cognitive mechanisms in entrepreneurship: Why and when entrepreneurs think differently than other people. *Journal of Business Venturing*, 13(4), pp. 275–294.

Baum, J., Frese, M., and Baron, R. (Eds). (2014). *The psychology of entrepreneurship*. Psychology Press.

Baum, J., and Locke, E. (2004). The relationship of entrepreneurial traits, skill, and motivation to subsequent venture growth. *Journal of Applied Psychology*, 89(4), p. 587.

Bergemann, D., and Morris, S. (2019). Information design: A unified perspective. *Journal of Economic Literature*, 57(1), pp. 44–95.

Bhide, A. (1994). How entrepreneurs craft strategies that work. *Harvard Business Review*, 72(1994), p. 3.

Blackwell, D. (1953). Equivalent comparisons of experiments. *The Annals of Mathematical Statistics*, 24(3), pp. 265–272.

Blume, B., and Covin, J. (2011). Attributions to intuition in the venture founding process: Do entrepreneurs actually use intuition or just say that they do? *Journal of Business Venturing*, 26(1), pp. 137–151.

Burke, J. (2008). Primetime spin: Media bias and belief confirming information. *Journal of Economics and Management Strategy*, 17(3), pp. 633–665.

Calvert, R. L. (1985). The value of biased information: A rational choice model of political advice. *The Journal of Politics*, 47(2), pp. 530–555.

Camuffo, A., Cordova, A., Gambardella, A., and Spina, C. (2020). A scientific approach to entrepreneurial decision-making: Evidence from a randomized control trial. *Management Science*, 66(2), pp. 564–586.

Camuffo, A., Gambardella, A., and Messinese, D. (2023). Bayesianism and unforeseen events: Empirical evidence from two field experiments. *Academy of Management Proceedings*, Vol. 2023, 1.

Camuffo, A., Gambardella, A., Messinese, D., Novelli, E., Paolucci, E., and Spina, C. (2024). A scientific approach to entrepreneurial decision-making: Large-scale replication and extension. *Strategic Management Journal*, 45(6), pp. 1209–1237.

Camuffo, A., Gambardella, A., and Pignataro, A. (2024). Theory-driven strategic management decisions. *Strategy Science*, 9(4), pp. 382–396.

Carroll, G. R., and Sorensen, J. (2021). *Making great strategy: Arguing for organizational advantage*. Columbia University Press.

Chavda, A., Gans, J. S., and Stern, S. (2024). Theory-based entrepreneurial search. *Strategy Science*, 9(4), pp. 397–415.

Che, Y., and Mierendorff, K. (2019). Optimal dynamic allocation of attention. *American Economic Review*, 109(8), pp. 2993–3029.

Clippel, G. de, and Zhang, X. (2022). Non-Bayesian persuasion. *Journal of Political Economy*, 130(10), pp. 2594–2642.

Cohen, S., Bingham, C., and Hallen, B. (2019). The role of accelerator designs in mitigating bounded rationality in new ventures. *Administrative Science Quarterly*, 64(4), pp. 810–854.

Damiano, E., Li, H., and Suen, W. (2020). Learning while experimenting. *The Economic Journal*, 130(625), pp. 65–92.

Delgado, M., Porter, M., and Stern, S. (2010). Clusters and entrepreneurship. *Journal of Economic Geography*, 10(4), pp. 495–518.

Dushnitsky, G. (2010). Entrepreneurial optimism in the market for technological inventions. *Organization Science*, 21(1), pp. 150–167.

Dworczak, P., and Martini, G. (2018). The simple economics of optimal persuasion. *Journal of Political Economy*, 127(5), pp. 1993–2048.

Eesley, C., and Roberts, E. (2012). Are you experienced or are you talented?: When does innate talent versus experience explain entrepreneurial performance? *Strategic Entrepreneurship Journal*, 6(3), pp. 207–219.

Ehrig, T., and Schmidt, J. (2022). Theory-based learning and experimentation: How strategists can systematically generate knowledge at the edge between the known and the unknown. *Strategic Management Journal*, 43(7), pp. 1287–1318.

Eisenmann, T., Ries, E., and Dillard, S. (2012). *Hypothesis-driven entrepreneurship: The lean startup*. Harvard Business School Entrepreneurial Management Case (812-095).

Evans, D. S., and Jovanovic, B. (1989). An estimated model of entrepreneurial choice under liquidity constraints. *Journal of Political Economy*, 97(4), pp. 808–827.

Ewens, M., Nanda, R., and Rhodes-Kropf, M. (2018). Cost of experimentation and the evolution of venture capital. *Journal of Financial Economics*, 128(3), pp. 422–442.

Felin, T., and Zenger, T. (2009). Entrepreneurs as theorists: On the origins of collective beliefs and novel strategies. *Strategic Entrepreneurship Journal*, 3(2), pp. 127–146.

Felin, T., and Zenger, T. (2017). The theory-based view: Economic actors as theorists. *Strategy Science*, 2(4), pp. 258–271.

Fraser, S., and Greene, F. (2006). The effects of experience on entrepreneurial optimism and uncertainty. *Economica*, 73(290), pp. 169–192.

Frese, M., and Gielnik, M. (2014). The psychology of entrepreneurship. *Annual Review of Organizational Psychology and Organizational Behavior*, 1(1), pp. 413–438.

Gans, J. (1995). Evolutionary selection of beliefs. *Economics Letters*, 49(1), pp. 13–17.

Gans, J. (2018). *A better way to bring science to market*. MIT Sloan Management Review.

Gans, J. (2022). *A theory of visionary disruption*. NBER Working Paper (No. 30091).

Gans, J. (2023). Experimental choice and disruptive technologies. *Management Science*, 69(11), pp. 7044–7058.

Gans, J. (2025). *The microeconomics of artificial intelligence*. MIT Press.

Gans, J., Kearney, M., Scott, E., and Stern, S. (2021). Choosing technology: An entrepreneurial strategy approach. *Strategy Science*, 6(1), pp. 39–53.

Gans, J., Scott, E., and Stern, S. (2018). Strategy for start-ups, *Harvard Business Review*, 96(3), 44–51.

Gans, J., Scott, E., and Stern, S. (2024). *Entrepreneurship: Choice and strategy*. W. W. Norton & Company.

Gans, J., Stern, S., and Wu, J. (2019). Foundations of entrepreneurial strategy. *Strategic Management Journal*, 40(5), pp. 736–756.

Geanakoplos, J., and Polemarchakis, M. (1982). We can't disagree forever. *Journal of Economic Theory*, 28(1), pp. 192–200.

Gilboa, I., Minardi, S., and Samuelson, L. (2020). Theories and cases in decisions under uncertainty. *Games and Economic Behavior*, 123, pp. 22–40.

Gimeno, J., Folta, T., Cooper, A., and Woo, C. (1997). Survival of the fittest? Entrepreneurial human capital and the persistence of underperforming firms. *Administrative Science Quarterly*, 40(4), pp. 750–783.

Guzman, J., Oh, J., and Sen, A. (2020). What motivates innovative entrepreneurs? Evidence from a global field experiment. *Management Science*, 66(10), pp. 4808–4819.

Haghtalab, N., Immorlica, N., Lucier, B., Mobius, M., and Mohan, D. (2021). *Persuading with anecdotes*. NBER Working Paper (No. 28661).

Hamilton, B. (2000). Does entrepreneurship pay? An empirical analysis of the returns to self-employment. *Journal of Political Economy*, 108(3), pp. 604–631.

Harsanyi, J. (1967). Games with incomplete information played by "Bayesian" players, parts I, II and III. *Management Science*, 14(5), pp. 159–182.

Hmieleski, K., and Baron, R. (2009). Entrepreneurs' optimism and new venture performance: A social cognitive perspective. *Academy of Management Journal*, 52(3), pp. 473–488.

Hsieh, T. (2010). Zappos's CEO on going to extremes for customers. *Harvard Business Review*, 88(7–8), pp. 41–45.

Hurst, E., and Lusardi, A. (2004). Liquidity constraints, household wealth, and entrepreneurship. *Journal of Political Economy*, 112(2), pp. 319–347.

Jakobsen, A. (2021). An axiomatic model of persuasion. *Econometrica*, 89(5), pp. 2081–2116.

Jakobsen, A. (2023). Coarse Bayesian updating. Working Paper, Northwestern University, Evanston, IL.

Jaynes, E. (1968). Prior probabilities. *IEEE Transactions on Systems Science and Cybernetics*, 4(3), pp. 227–241.

Kamenica, E. (2019). Bayesian persuasion and information design. *Annual Review of Economics*, 11, pp. 249–272.

Kamenica, E., and Gentzkow, M. (2011). Bayesian persuasion. *American Economic Review*, 101(6), pp. 2590–2615.

Karni, E. (2022). A theory-based decision model. *Journal of Economic Theory*, 201, p. 105444.

Karni, E., and Vierø, M. (2013). Reverse Bayesianism: A choice-based theory of growing awareness. *American Economic Review*, 103(7), pp. 2790–2810.

Karni, E., and Vierø, M. (2017). Awareness of unawareness: A theory of decision making in the face of ignorance. *Journal of Economic Theory*, 168, pp. 301–328.

Kartik, N., Xu Lee, F., and Suen, W. (2021). Information validates the prior: A theorem on bayesian updating and applications. *American Economic Review: Insights*, 3(2), pp. 165–182.

Kass, R. E., and Wasserman, L. (1996). The selection of prior distributions by formal rules. *Journal of the American Statistical Association*, 91(435), pp. 1343–1370.

Kerr, W., Nanda, R., and Rhodes-Kropf, M. (2014). Entrepreneurship as experimentation. *The Journal of Economic Perspectives*, 28(3), pp. 25–48.

Kirtley, J., and O'Mahony, S. (2023). What is a pivot? explaining when and how entrepreneurial firms decide to make strategic change and pivot. *Strategic Management Journal*, 44(1), pp. 197–230.

Knight, F. (1921). *Risk, uncertainty and profit*. Houghton Mifflin.

Koning, R., Hasan, S., and Chatterji, A. (2022). Experimentation and startup performance: Evidence from A/B testing. *Management Science*, 68(9), pp. 6434–6453.

Lakhani, K., Luo, H., and Katsnelson, L. (2019). *Market for judgement: The Creative Destruction Lab*. Harvard Business School Case (No. 719-479).

Landier, A., and Thesmar, D. (2008). Financial contracting with optimistic entrepreneurs. *The Review of Financial Studies*, 22(1), pp. 117–150.

Lazear, E. (2004). Balanced skills and entrepreneurship. *American Economic Review*, 94(2), pp. 208–211.

Liedtka, J. (2018). Why design thinking works. *Harvard Business Review*, 96(5), pp. 72–79.

Lindholm-Dahlstrand, Å., Andersson, M., and Carlsson, B. (2019). Entrepreneurial experimentation: A key function in systems of innovation. *Small Business Economics*, 53, pp. 591–610.

Luo, H. (2022). *Market for judgement: The Creative Destruction Lab*. Harvard Business School Teaching Note (No. 722-419).

Manso, G. (2016). Experimentation and the returns to entrepreneurship. *The Review of Financial Studies*, 29(9), pp. 2319–2340.

Marinacci, M. (2015). Model uncertainty. *Journal of the European Economic Association*, 13(6), pp. 1022–1100.

Martin, B., McNally, J., and Kay, M. (2013). Examining the formation of human capital in entrepreneurship: A meta-analysis of entrepreneurship education outcomes. *Journal of Business Venturing*, 28(2), pp. 211–224.

Masters, B., and Thiel, P. (2014). *Zero to one: Notes on start ups, or how to build the future*. Random House.

Maurya, A. (2022). *Running lean*. O'Reilly Media.

Melnikoff, D. E., and Strohminger, N. (2024). Bayesianism and wishful thinking are compatible. *Nature Human Behaviour*, 8, pp. 692–701.

Minniti, M., and Bygrave, W. (2001). A dynamic model of entrepreneurial learning. *Entrepreneurship Theory and Practice*, 25(3), pp. 5–16.

Moore, G. (2014). *Crossing the chasm*. Harper Business.

Moscarini, G., and Smith, L. (2001). The optimal level of experimentation. *Econometrica*, 69(6), pp. 1629–1644.

Mullainathan, S., and Rambachan, A. (2023). *From predictive algorithms to automatic generation of anomalies*. NBER Working Paper (No. 32422).

Murray, F., and Tripsas, M. (2004). The exploratory processes of entrepreneurial firms: The role of purposeful experimentation. *Business strategy over the industry lifecycle*. Emerald, pp. 45–75.

Nanda, R., and Rhodes-Kropf, M. (2016). Financing entrepreneurial experimentation. *Innovation Policy and the Economy*, 16(1), pp. 1–23.

Nanda, R., and Rhodes-Kropf, M. (2017). Financing risk and innovation. *Management Science*, 63(4), pp. 901–918.

Ortoleva, P. (2012). Modeling the change of paradigm: Non-Bayesian reactions to unexpected news. *American Economic Review*, 102(6), pp. 2410–2436.

Osterwalder, A., and Pigneur, Y. (2010). *Business model generation: A handbook for visionaries, game changers, and challengers*, volume 1. John Wiley & Sons.

Patvardhan, S., and Ramachandran, J. (2020). Shaping the future: Strategy making as artificial evolution. *Organization Science*, 31(3), pp. 671–697.

Pearl, J. (2000). *Causality: Models, reasoning, and inference*. Cambridge University Press.

Pearl, J., and Mackenzie, D. (2018). *The book of why: The new science of cause and effect*. Hackhette UK.

Ries, E. (2011). *The lean startup: How today's entrepreneurs use continuous innovation to create radically successful businesses*. Currency.

Roach, M., and Sauermann, H. (2015). Founder or joiner? The role of preferences and context in shaping different entrepreneurial interests. *Management Science*, 61(9), pp. 2160–2184.

Robson, A. J., and Samuelson, L. (2011). The evolutionary foundations of preferences. *Handbook of social economics 1*, North-Holland, pp. 221–310.

Rosenberg, N. (1992). Economic experiments. *Industrial and Corporate Change*, 1(1), pp. 181–203.

Samuelson, L. (2004). Modeling knowledge in economic analysis. *Journal of Economic Literature*, 42(2), pp. 367–403.

Sarasvathy, S. D. (2001). Causation and effectuation: Toward a theoretical shift from economic inevitability to entrepreneurial contingency. *Academy of Management Review*, 26(2), pp. 243–263.

Sariri, A. (2024). *Are angel investors more likely than venture capitalists to drive entrepreneurial experimentation?* Purdue University Working Paper.

Shane, S. (2000). Prior knowledge and the discovery of entrepreneurial opportunities. *Organization Science*, 11(4), pp. 448–469.

Shane, S. (2003). *A general theory of entrepreneurship: The individual–opportunity nexus*. Edward Elgar.

Stevenson, H. H. (1983). *A perspective on entrepreneurship*, Vol. 13. Harvard Business School Working Paper #9-384-131.

Stuart, T., and Sorenson, O. (2005). Social networks and entrepreneurship. *Handbook of entrepreneurship research: Interdisciplinary perspectives*, Springer, pp. 233–252.

Suen, W. (2004). The self-perpetuation of biased beliefs. *The Economic Journal*, 114(495), pp. 377–396.

Sull, D. (2004). Disciplined entrepreneurship. *MIT Sloan Management Review*, 46(1), p. 71.

Tartari, V., and Stern, S. (2021). *More than an ivory tower: The impact of research institutions on the quantity and quality of entrepreneurship*. NBER Working Paper (No. w28846).

Unger, J. M., Rauch, A., Frese, M., and Rosenbusch, N. (2011). Human capital and entrepreneurial success: A meta-analytical review. *Journal of Business Venturing*, 26(3), pp. 341–358.

Van de Ven, A., Sapienza, H., and Villanueva, J. (2007). Entrepreneurial pursuits of self- and collective interests. *Strategic Entrepreneurship Journal*, 1(3–4), pp. 353–370.

Van den Steen, E. (2001). Essays on the Managerial Implications of Differing Priors. PhD thesis, Stanford University.

Van den Steen, E. (2005). Organizational beliefs and managerial vision. *Journal of Law, Economics, and Organization*, 21(1), pp. 256–283.

Van den Steen, E. (2011). Overconfidence by Bayesian-rational agents. *Management Science*, 57(5), pp. 884–896.

Yildiz, M. (2000). Essays on Sequential Bargaining. PhD thesis, Stanford University.

Zellweger, T., and Zenger, T. (2023). Entrepreneurs as scientists: A pragmatist approach to producing value out of uncertainty. *Academy of Management Review*, 48(3), pp. 379–408.

# 2

## Homo Entrepreneuricus

Alfonso Gambardella, Luca Gius, and Scott Stern

### 2.1 Introduction

Entrepreneurs routinely perceive opportunities that others either fail to appreciate or deem too risky to pursue. Since others do not see the same opportunity as the entrepreneur, the behaviors that result from pursuing opportunity can appear puzzling from a standard economic perspective (Åstebro et al., 2014). Indeed, a central assumption in mainstream economic theory—the common prior assumption—is that if rational agents share the same information, they should converge on a common probability assessment of any uncertain outcome (Aumann, 1976; Morris, 1995). Persistent disagreement on the value of opportunities is even more puzzling when one considers that the "irrational" pursuit of opportunity will, sooner or later, drive the entrepreneur out of business (Surowiecki, 2014). Yet, the reality of entrepreneurship is littered with examples of successful founders who held idiosyncratic beliefs that diverged from conventional wisdom and persisted in their ventures, despite widespread skepticism (Masters and Thiel, 2014; Agrawal et al., 2025). Moreover, many of the entrepreneurs who pursue opportunities that others consider unpromising do not simply charge ahead in the absence of evidence; instead, many seek to actively learn about and experiment with how to pursue the opportunity in a systematic way. To be clear, some founders seem to simply rely on learning by doing (learning as they go without a systematic plan for experimentation), but others adopt a more systematic, hypothesis-driven approach to uncovering the underlying mechanics of an opportunity (Murray and Tripsas, 2004).

This essay brings together these dual observations—the presence (and persistence) of subjective priors and the presence (and persistence) of proactive experimentation—to explore an archetype for entrepreneurial decision-making, *Homo entrepreneuricus*. *Homo entrepreneuricus* holds subjective priors that allow an individual to "agree to disagree" with other market participants (Aumann, 1976; Van den Steen, 2001) while at the same time being willing to undertake deliberate, theory-based, "scientific" learning to test critical assumptions (Felin and Zenger, 2017; Camuffo et al., 2020). The presence of subjective priors explains why entrepreneurs can perceive—and persist in pursuing—prospects that others with the same factual information would dismiss (Shane and Venkataraman, 2000). But subjective priors alone, in the absence of a purposeful learning architecture, can yield stasis or misguided conviction. *Homo entrepreneuricus*, by contrast, treats uncertainty as a domain for structured exploration, shaping and revising beliefs through carefully planned experimentation. Their subjective framework renders them unusually alert to certain signals, whereas scientific experimentation enables them to gather and interpret new evidence systematically, even if that evidence ultimately challenges their prior optimism.

Our approach is broader than "strict" Bayesian updating. The focus on subjective beliefs and other elements of our framework enable us to relegate the distinction between Bayesian and non-Bayesian approaches to learning under the same general umbrella: Our approach emphasizes the management of uncertainty, information processing, and choice. Our model acknowledges the limitations of perfect rationality—some disagreement, after all, can arise from biases and heuristics—while recognizing the structured yet flexible approach entrepreneurs take to navigate unknowns. Our purpose is to explore in an introductory way the distinctive attributes of *Homo entrepreneuricus*, blending subjective priors with theory-based purposeful experimentation.

We begin, in section 2.2, by situating *Homo entrepreneuricus* within a two-by-two framework that contrasts how entrepreneurs form priors (shared vs. subjective) and how they learn (emergent vs. theory-based). This framework reveals four archetypes—neoclassical, effectuation, real options, and bayesian—clarifying what sets *Homo entrepreneuricus* apart. In section 2.3, we focus on the foundations of subjective priors, showing how divergent worldviews—or what biologists call an *umwelt*—allow entrepreneurs to interpret shared information through distinct mental models. In section 2.4, we turn to the rationale for a broadly Bayesian approach to entrepreneurial experimentation, highlighting why *Homo entrepreneuricus* goes beyond trial and error to adopt systematic, hypothesis-driven tests. When uncertainty is high and the cost of uninformed commitment is steep, a structured program of "scientific" experimentation enables more effective updating, better resource allocation, and a deeper capacity to pivot in response to new evidence. We then integrate these insights in section 2.5 by drawing out the centrality of "choice" for *Homo entrepreneuricus* (Gans et al., 2019). Contrary to the notion that entrepreneurs—under similar conditions—should mechanically converge on a single "best" path, *Homo entrepreneuricus* surfaces multiple viable strategic paths. With no universally accepted ground truth, each strategic decision reflects not only commonly available information, but also the entrepreneur's unique subjective beliefs. Finally, in section 2.6, we explore several implications of this approach. We consider how entrepreneurs tailor experiments to specific audiences or beliefs, how contrarian ventures can attract resources without sparking immediate competitive imitation, and how subjective priors can mitigate classic concerns about disclosing ideas. *Homo entrepreneuricus* offers the foundations for a rich yet testable model of how entrepreneurial action emerges when conviction meets curiosity—illuminating both the promise and the perils of building new ventures in worlds where agreement is far from guaranteed.

## 2.2 Who Is Homo Entrepreneuricus?

An enduring question in entrepreneurship research is how certain individuals recognize and act upon opportunities that others fail to see or dismiss as untenable. Standard economic theory has long assumed that different economic actors, provided the same information, should converge on a common understanding of an opportunity's underlying potential. This assumption is not without cause: Many of the most celebrated results in game theory and information economics depend on the idea that, if information is freely shared among agents, the agents will share a common perspective on the likelihood of different events (i.e., such an assumption allows economic theorists to "zero in on purely informational issues"—Aumann, 1987; see also Aumann, 1976; Morris, 1995).

However, a growing strand of literature challenges this assumption and explores the impact of allowing heterogeneous priors on economic decision-making (Morris, 1995; Van den Steen, 2001). Not simply a theoretical matter, empirical observation suggests that entrepreneurs often "agree to disagree" with others, holding unique interpretations of opportunities, even when faced with seemingly common facts. But that does not mean that entrepreneurs are immune to new learning. Entrepreneurs engage in learning about an opportunity, sometimes in unstructured trial-and-error ways and sometimes through highly systematic, even scientific, processes. To capture these dual dimensions—whether entrepreneurs share or diverge in priors about the viability of a venture, and whether they learn by emergent adaptation or deliberate testing—we propose a two-by-two framework. The framework highlights four archetypes of entrepreneurial decision-makers and clarifies what sets apart the "Bayesian entrepreneur," or what we call *Homo entrepreneuricus*, from other entrepreneurial types (table 2.1).

The vertical axis of this framework distinguishes between shared priors and subjective priors. The notion of shared priors is rooted in the premise that if

**Table 2.1**
Entrepreneurial decision-making archetypes

| | **Learning Approach** | |
|---|---|---|
| | **Emergent** | **Theory-Based** |
| **Shared Priors** | Neoclassical | Real Options |
| **Subjective Priors** | Effectuation | Homo Entrepreneuricus |

two individuals are given the same information, they should arrive at the same probabilistic assessment of an opportunity's viability. Indeed, standard models of entry and exit suggest that if two individuals observe equivalent signals—say, about the size of a new market, the cost structure of production, or the strength of consumer demand—they will converge on a common estimate of the likelihood that a new venture will succeed (Jovanovic, 1982).

Yet, real-world entrepreneurship highlights a more intriguing phenomenon, often labeled "agreeing to disagree." Entrepreneurs can hold subjective priors that depart significantly from those of peers, potential investors, and even industry experts. Such priors might be grounded in personal experience, tacit knowledge, social networks, or heuristics that shape how signals are interpreted. Even when two parties observe the same factual information, they may update their beliefs differently. One might find the signals encouraging, whereas another sees them as noisy and unconvincing. In effect, the subjective priors lens recognizes that differences in how entrepreneurs form and update beliefs can be consequential—even more so than differences in the information they observe (Van den Steen, 2001).

The horizontal axis of the framework separates emergent or "learning by doing" approaches from theory-based or "learning by testing" processes. Learning by doing refers to a largely experiential approach: An entrepreneur enters a market or launches a product with only a rudimentary sense of the opportunity's potential, then updates as real-world feedback unfolds. This viewpoint dominates classic industry evolution models. Jovanovic's (1982) influential work, for example, portrays firms as learning about their cost efficiency through operating in the marketplace. Over time, high-cost producers either exit or adapt, whereas more efficient firms survive and grow. The learning process is reactive: Entrepreneurs (or firms) respond to the signals they receive from day-to-day activities, adjusting beliefs based on practical, hands-on experience.

In contrast, theory-based learning involves the *deliberate* design of experiments or tests to validate specific hypotheses about an opportunity's viability. Under this approach, entrepreneurs do not simply stumble upon useful information while running their venture; rather, they carefully seek it out. They formulate explicit theories about how value might be created and captured, then systematically collect data to confirm or disprove those theories. "Scientific" entrepreneurship, in practice, might involve techniques such as randomized controlled trials (RCTs) of a new product feature, customer interviews structured around a clear hypothesis, or pilot programs designed to validate pricing strategies (Camuffo et al., 2020). The point is to reduce uncertainty in a purposeful, efficient manner rather than relying solely on the slow grind of operational feedback.

Placing these two dimensions together produces four archetypes of entrepreneurial decision-makers. Neoclassical entrepreneurs embody the *shared priors + emergent learning* quadrant. They accept that if a given piece of information is publicly available, others will interpret it the same way. Having entered the market, they learn primarily from day-to-day experience, discovering their own costs, demand conditions, and competitive advantages over time. These entrepreneurs mirror the logic of Jovanovic's (1982) model: Under the pressures of market competition, the better informed (or more efficient) persist, whereas the less efficient exit. Any idiosyncrasy in their decision-making stems more from private information or cost structures than from divergent belief updating.

Moving to the *subjective priors + emergent learning* quadrant, we encounter effectuation entrepreneurs. Effectuation, as advanced by Sarasvathy (2001) and reflected in Amar Bhidé's writings (e.g., Bhidé, 1986), asserts that entrepreneurs often hold unique, deeply personal beliefs about new opportunities—beliefs that may not be universally shared or validated at the outset. Yet, these entrepreneurs do not typically design methodical tests to reduce uncertainty. Rather, they adapt and pivot as they experiment in real time with products, markets, and business models. An effectual entrepreneur's approach is improvisational: They "start with who they are, what they know, and whom they know," and let these resources guide early action. Feedback from that action then triggers adjustments to their evolving strategy. Here, subjective priors help explain why these individuals see an opportunity where others do not, and their reliance on emergent, hands-on learning explains how they adapt (and either grow or abandon their venture) over time.

In the *shared priors + theory-based learning* quadrant, we find a "real options" entrepreneur, a label that captures much of the logic behind staged financing and systematic risk reduction in venture capital settings. In this scenario, an entrepreneur and a financier (e.g., a venture capitalist) broadly agree on the initial probability of success—but they also appreciate that uncertainty remains. Thus, they work together to design milestone-based experiments, each of which will reveal crucial information about feasibility or market traction.

Thomas Hellmann, Josh Lerner, and others have described how venture capital contracts and governance arrangements facilitate these experiments (Hellmann, 1998; Gompers and Lerner, 2001); Kerr et al. (2014) explicitly frame entrepreneurship as experimentation. The entrepreneur, along with the venture capitalist, invests a certain amount of capital to test a specific hypothesis about the business (e.g., whether there is sufficient demand at a particular price point). If the test yields encouraging results, the venture continues; if not, the project is terminated. Because the priors are shared, both parties interpret the experimental outcomes similarly, and decisions about continued funding become a matter of updating a commonly held probability estimate. The "financial guillotine" falls when the evidence points convincingly toward failure.

Finally, in the *subjective priors + theory-based learning* quadrant, we arrive at *Homo entrepreneuricus*—the individual who both agrees to disagree with the crowd and takes a purposeful, systematic approach to reducing uncertainty. This is the core of Bayesian entrepreneurship, in which entrepreneurs rely on idiosyncratic signals or interpretations of shared signals to form an initial belief about an opportunity. Precisely because that belief diverges from what others see, *Homo entrepreneuricus* is often motivated to pursue a project that, in the eyes of most observers, looks too risky or not lucrative enough. Yet, unlike the effectual entrepreneur who operates more improvisationally, *Homo entrepreneuricus* recognizes the uncertainty inherent in their own beliefs and sets out to test them. Rather than waiting passively to see how real-world conditions unfold, they devise experiments—whether they be market pilots, small-scale prototypes, or even computational simulations—that systematically probe the validity of the underlying theory of the venture.

One crucial insight from this perspective is that the demand for experimentation depends on the entrepreneur's subjective prior. If an entrepreneur's optimistic belief about the opportunity is extremely high, they might only be dissuaded by a very powerful negative signal. The type of experiment they deem useful is dramatically different from that required if they are moderately skeptical, in which case a small bit of favorable evidence could propel them forward. Differences in how entrepreneurs run and interpret experiments—even around the same idea—are the result not of sloppy reasoning but rather of rigorous Bayesian updating, which integrates both prior belief and newly observed information. This quadrant, therefore, represents a unique blend of entrepreneurial boldness and scientific rigor. *Homo entrepreneuricus*, guided by a deeply held yet idiosyncratic interpretation of an opportunity, invests in carefully designed experiments to probe that opportunity's true potential. If the results are favorable, they double down; if not, they are willing—at least in principle—to adjust or abandon the idea.

By juxtaposing these four archetypes—neoclassical, effectual, real options, and Bayesian entrepreneurs—we can highlight the pivotal role of both priors and learning styles in shaping entrepreneurial behavior. Some of the most enduring puzzles in entrepreneurship, such as why certain founders persist in the face of widely perceived improbability and how others reduce uncertainty so rapidly, become more tractable when we account for how entrepreneurs form and update their beliefs. *Homo entrepreneuricus* unites bold, sometimes contrarian beliefs with a systematic, "scientific" approach to testing those beliefs. Guided by subjective priors yet disciplined in gathering evidence, these entrepreneurs refine their visions through deliberate experiments, deciding whether to double down or pivot as results emerge. This synthesis of conviction and curiosity explains why certain founders persist despite widespread doubts, as well as how they can rapidly reduce uncertainty without resorting to blind leaps of faith or purely mechanistic methods. By integrating personal interpretations with rigorous validation, *Homo entrepreneuricus* offers a powerful lens on how entrepreneurs navigate uncharted opportunities in a world of limited and often ambiguous information.

## 2.3 The Foundation of Subjective Priors

Why do some people see the proverbial glass as half-full while others insist it is half-empty—even after they have all examined the same glass, under the same light, from the same angle? In many areas of life, and especially in entrepreneurship, people often agree to disagree about the very same information (Aumann, 1976). This phenomenon poses a fundamental question: If information is shared, why don't people simply converge on a common understanding of the future? The notion of subjective priors offers one avenue for resolving this puzzle. Entrepreneurs who hold subjective priors do not necessarily reject rational analysis. Quite the contrary, they can be rational *within* the interpretive framework—or model—they use to understand the world. The critical point is that entrepreneurs may construct and apply *different* models of the world, leading them to form and update beliefs in ways that diverge from those of others (Van den Steen, 2004).

### 2.3.1 The Umwelt of Homo Entrepreneuricus

Since entrepreneurs, almost by definition, must assess probabilities that are inherently novel (otherwise, the opportunity would not exist), the choices of which attributes to include and how they interrelate are shaped by each entrepreneur's worldview. In biology, the term *umwelt* is used to describe how different organisms inhabit different perceptual worlds based on their sensory apparatus and evolutionary history (Von Uexkull, 1909; Yong, 2022). By analogy, each entrepreneur's umwelt is the lens through which they identify and interpret what is possible and important. One entrepreneur might see government regulation as crucial, another might ignore regulation altogether and focus instead on consumer trends, while a third is fixated on potential technological breakthroughs. These differing perspectives lead to distinct state spaces—sets of attributes and their potential realizations—that serve as the foundation for probabilistic yet individualistic reasoning.

At the heart of this perspective is the central concept of *model uncertainty* (Marinacci, 2015; Camuffo, Gambardella, and Pignataro, 2024). In many domains, and especially those at the frontiers of technology or new market creation, entrepreneurs cannot simply look up probabilities in a well-defined manual, nor can they rely on stable historical frequencies to gauge the likelihood of specific future events. Instead, entrepreneurs may craft an internal representation—or model—of the attributes they believe matter most. These attributes might include future demand for a new product, the likelihood that a novel technology will work as intended, or the intensity of competition in the target market.

This perspective is especially valuable for considering how actors make decisions about "low-frequecy, high-impact" choices (Camuffo, Gambardella, and Pignataro, 2024). In these circumstances, the prior of the entrepreneur depends not simply on the observable information that might be publicly shared with others but also on the *logical links* that the entrepreneur prioritizes to assess which future scenarios are plausible or implausible. In such a scenario, the entrepreneur builds a framework of specific logical links that restrict the set of subjective probabilities that are implied by the logical model they have constructed. Specifically, entrepreneurs focus on *attributes* they believe are central elements of the problem. For example, they may think that developing a workable technology is key to launching a new business. Then, they establish *causal links* such as if "technology works" then "demand is likely high." This imposes a logical restriction on the probability distribution they consider valid: In their eyes, any distribution that assigns a high probability to "technology works and demand is low" obtains with a much lower probability.

Importantly, because they are dealing with uncertain situations, entrepreneurs are aware that the attributes of the problem and the causal links formed at the initial stage may not in fact be correct. A hallmark of *Homo entrepreneuricus*—the Bayesian entrepreneur—is a willingness to question and refine not just the probabilities assigned to a fixed state space but also the *composition* of that state space itself. Unlike a purely neoclassical agent, who might treat the attributes of a problem as given, *Homo entrepreneuricus* entertains the possibility that entirely new attributes could become relevant. A wave of user complaints, for example, might reveal that user experience design is actually more critical than the technology's performance. Or an unexpected regulatory shift could render market demand analysis moot if certain energy sources are suddenly banned or subsidized. In each of these cases, *Homo entrepreneuricus* does not merely adjust their probability estimates—they *reshape* their model. Experimentation itself refines and shapes the umwelt of the entrepreneur.

An important distinction within such a model is the presence of fundamental uncertainty, whereby the entrepreneur (at least as an initial condition) either places zero probability or cannot assign a probability to certain events. To deal with these zero-probability events, *Homo entrepreneuricus* could have not only subjective priors on the likelihood of the logical links but also "priors on priors" (Ortoleva, 2012, 2024)—that is, priors on which theories are most plausible. They espouse the theory they find most plausible, possibly after running experiments—that is, after acquiring additional information on the plausibility of one theory over the other. The experiments, and the ensuing information they gain, can then inform them about the plausibility of logical links as well as the relative plausibility of one theory over another (Camuffo, Gambardella, and Pignataro, 2024). All this shapes the subjective probabilities that entrepreneurs attribute to different states and theories, effectively defining the boundary between "what could happen" and "what can't happen" in their worldviews. This is one way—as emphasized by Gans (2025), building on Bewley (2002)—to incorporate Knightian uncertainty, subjective probabilities, and proactive Bayesian learning and still retain a well-defined model.

To be clear, the presence of subjective priors comes with what seems to be anomalous behavior from the perspective of an outside observer: The entrepreneur will insist that a product or strategy has a high chance of success, despite all evidence to the contrary. Yet, this apparent intransigence can be understood once we appreciate that entrepreneurs may perceive themselves as operating in a different decision space altogether. Two individuals might watch the same user tests or read the same market research reports but form divergent conclusions because they interpret the data via different mental models.

### 2.3.2 Subjective Priors Can Be (But Are Not Necessarily) Rational

Allowing for subjective priors does not resolve whether entrepreneurs are indeed "rational." On the one hand, there is nothing "irrational" about the fact that, when facing a novel situation, the probabilities assigned to the likelihood of certain events (the model) is grounded in the individualized umwelt of that focal entrepreneur. Indeed, given that it is a novel circumstance, how could it be otherwise. Because each entrepreneur's umwelt highlights some attributes while downplaying others, entrepreneurs construct different possible futures in their minds.

At the same time, the presence of an individualistic model (with associated logical links) does not ensure against the presence of biases or heuristics. Indeed, differences in worldviews often stem from behavioral biases such as overconfidence (Hayward et al., 2006), and previous work has recognized that heterogeneity in the entrepreneurs' cognitive processes plays an important role in value creation and capture (Alvarez and Busenitz, 2001). Some entrepreneurs begin with an incomplete or inconsistent model of the world—one that diverges from existing evidence without their awareness. Even when they integrate new information systematically, their biased starting point can delay understanding and foster disagreement with peers who are either unbiased or biased in different ways. Alternatively, entrepreneurs might misinterpret the information they gather, overvaluing the signals they observe or the experiments they conduct—even when their initial intuition is correct. This overreliance on strong preconceptions or experimental outcomes can lead them to forego cause-and-effect thinking altogether. For example, in effectuation, entrepreneurs can be seen *as if* operating within an idiosyncratic, narrow worldview—one in which each contingency plan is limited to a finite set of options (much like a cook restricted by the ingredients on hand—Sarasvathy, 2001).

What we emphasize, however, is that subjective beliefs are consistent with rationality. By definition, under uncertainty, there exists a variety of mental models, each consistent with the available data, that reasonable people could adopt. If different entrepreneurs rely on different mental models, they will filter and interpret data through their respective theories. Even if everyone gathers data from the same source and updates their beliefs rationally, the differences in their models mean that their subjective priors do not necessarily converge.

Imagine two equally knowledgeable Alzheimer's researchers who both have deep expertise in developing novel therapies. She, an entrepreneur, firmly believes in the amyloid theory, seeing it as the central driver of disease progression. This conviction leads her to view a new experimental treatment targeting amyloid plaques as having huge potential for rapid breakthroughs. He, equally seasoned in the field but focused on tau pathology, is more skeptical of amyloid's primacy. Given his theoretical framework, he interprets the same early-stage results with caution, foreseeing significant hurdles in reaching effective outcomes. Both apply Bayesian reasoning—updating their expectations as new data emerge and calculating the probability of different scenarios—yet arrive at divergent conclusions because they start from contrasting scientific models, each shaping a different prior belief.

To be clear, allowing for subjective priors does not imply that *anything goes*. We do not expect entrepreneurs to have just *any* prior potentially conceivable: Physically impossible theories of value, for example, will sooner or later be rejected and their proponents driven out of the marketplace. And even highly idiosyncratic entrepreneurs rely on various forms of reasoning, evidence, or experience to rule out certain possibilities. They could, for instance, perform a rudimentary technical feasibility assessment to decide that a particular combination of attributes is nearly impossible (e.g., "Zero chance the product can work with no R&D investment") or rely on an intuitive gut feeling that the market will only respond to technology once it is fully integrated into a user-friendly platform. In each case, the entrepreneur is applying a personal lens to the signals at hand, translating them into structured beliefs about how the world works. Similarly, we should be careful not to attribute *every* difference in entrepreneurial behavior to ad hoc differences in beliefs (as opposed to differences in incentives, ability to process information, cognitive biases, knowledge, or information), focusing only on belief differences that are plausible and empirically testable.

### 2.3.3 The Role of Theories in Entrepreneurial Strategy

Much of the power of subjective priors lies in how they inform entrepreneurial strategy. If you believe, for instance, that making the technology work will almost guarantee strong demand, you are likely to invest heavily in R&D or product development while placing less emphasis on early marketing. Conversely, if you think the technology is only weakly connected to demand and that branding or distribution channels are the key to success, you might start by hiring marketers or forming strategic partnerships—even if you have not yet perfected the underlying product. These choices stem from distinct theories of how attributes link together and which attributes matter most (Felin and Zenger, 2017).

Far from an abstract exercise, these theories guide concrete decisions about resource allocation, business model design, and growth strategies. Differences in theories can lead two founders in the same sector to tackle the same consumer need with radically different approaches. One might focus on building a user base first (betting on network effects), while another channels resources into technical superiority. Although they share knowledge of the problem domain, they do not necessarily share the same logic about *why* certain strategies will succeed.

Another way to understand how subjective priors channel entrepreneurial learning is to draw an analogy from machine learning, where a "regularizer" serves as a probabilistic prior that prevents overfitting by penalizing overly complex or unconstrained models (Torralba and Isola, 2024). Translated to an entrepreneurial setting, subjective priors guide learning: They constrain the space of solutions that an entrepreneur finds plausible, thereby restricting which signals are taken seriously and which are dismissed. This filtering can help entrepreneurs avoid wasting time on marginal paths, but it also risks causing them to overlook genuinely valuable information that falls outside their worldview.

Taken together, these insights into subjective priors reaffirm that disagreement among entrepreneurs is not merely a by-product of differences in (cap)abilities, knowledge, or contradictory incentives. Instead, it can reflect genuine differences in how individuals model the world, define the attributes they believe relevant, and interpret signals through their distinctive theories. This divergence can be productive: If one entrepreneur's worldview is particularly well suited to an emerging opportunity, their venture may succeed precisely because they perceived the situation differently from everyone else.

## 2.4 The Rationale for a (Broadly) Bayesian Approach to Entrepreneurial Experimentation

Confronted with ambiguity over both an idea's intrinsic value and the strategies that might bring that idea to fruition, entrepreneurs face the difficult task of reducing uncertainty before fully committing scarce resources. In this section, we compare experiential and experimental approaches to reducing uncertainty and lay out a conceptual framework for understanding when and how a purposeful (Bayesian) experimentation strategy can enhance entrepreneurial decision-making.

A traditional "learning by doing" approach to reducing uncertainty might suggest plunging in—building prototypes, hiring a team, and launching the product—then seeing what happens (Ries, 2011). Over time, real-world outcomes would offer clues about whether to pivot, persist, or abandon the venture. This approach partly draws on classic economic literature that treats learning as a "black box," relegating the intricate strategic challenges of experimentation to an unmodeled optimization process that offers limited, actionable lessons to entrepreneurs. In part, it draws on a strand of literature that dismisses the need for an explicit experimentation strategy altogether, arguing that entrepreneurs should generate information simply by "doing" rather than by planning. But for entrepreneurs operating in fast-moving or resource-constrained contexts, *doing* is often expensive, time-consuming, and risky, and crucial insights might trickle in too slowly or become muddled by overlapping changes to the product and market. A better strategy might be to "open" the experimentation black box—explicitly addressing and managing the learning process—instead of abandoning it entirely.

*Homo entrepreneuricus* learns *deliberately*, treating experimentation as a strategic priority. This perspective marks the essence of a (broadly) Bayesian approach to entrepreneurial experimentation (Agrawal et al., 2025). Entrepreneurs who embrace it begin with subjective priors—those initial hunches or theories about how markets, technologies, and consumers behave—but understand that these priors, however promising, remain incomplete. Although they might be positively inclined toward their venture, they also recognize substantial unknowns. What if the technology's performance falters? What if their target customer segment fails to respond? Rather than immediately pushing forward with a full-scale launch, such entrepreneurs pause to ask: Which hypothesis is most critical to test right now, and how can we learn about it effectively most efficiently?

This desire to learn systematically reflects what we might call a "Bayesian spirit": Even if an entrepreneur's initial beliefs deviate from perfect rationality or incorporate personal biases, they nonetheless commit to updating those beliefs when exposed to meaningful new evidence. The term "Bayesian" here should be interpreted broadly. No one expects entrepreneurs to run textbook-perfect experiments in which each new data point triggers a precise mathematical update of a carefully calibrated probability distribution. Yet, they do follow the *logic* of Bayesian reasoning in that they design and execute experiments to glean the kinds of evidence most likely to inform their next move. In short, they treat learning as a proactive, hypothesis-driven process rather than as a purely emergent by-product of daily operations. In so doing, they effectively adopt a *scientific approach* to entrepreneurial experimentation, formulating clear hypotheses, designing tests that can confirm or invalidate those hypotheses, and adjusting their strategies based on the results.

Even though not *all* entrepreneurs necessarily behave as perfectly rational Bayesian updaters, a growing body of empirical evidence underscores the normative appeal of the "Bayesian spirit"—that is, the tangible value of embracing a scientific approach. In an RCT by Camuffo et al. (2020), participants were trained to apply a hypothesis-driven, experimental mindset in launching or developing their ventures. The results showed that those who adopted rigorous, theory-based experimentation methods outperformed their counterparts along various dimensions. They were more likely to terminate, possibly ruling out false positives, and performed better conditional upon survival. Such findings suggest that a structured, deliberate approach to entrepreneurship—one that tests assumptions systematically, evaluates data objectively, and updates beliefs accordingly—offers tangible performance benefits. Camuffo,

Gambardella, Messinese, et al. (2024) then showed with a far larger RCT, covering 759 entrepreneurs, that key mechanisms for the observed performance of entrepreneurs adopting a scientific approach were (a) a *methodical doubt* who—pushed to question their assumptions and test alternative theories and (b) a more *efficient search* in the sense of greater ability to identify new causal links that led to better solutions to their entrepreneurial problem.

The notion that *Homo entrepreneuricus* is open to learning is *not* at odds with strong prior beliefs. By contrast, experiments are useful when they *can* spur action. Only entrepreneurs who—while uncertain—at least entertain the possibility that the venture will be successful have reasons to substantiate (or refute) those convictions. Formal, theory-based testing gives them a chance to identify their blind spots and reduce the chance of chasing a mirage: *No prior is completely definitive*. If the experiment reveals a major flaw—say, that the product is technologically unworkable or that prospective customers find the offering underwhelming—then the entrepreneur can pivot early or redirect resources before it is too late. Conversely, if the test shows strong evidence of feasibility or unexpected enthusiasm among users, the entrepreneur's subjective prior is reinforced, emboldening them to commit more fully or scale up. In each case, the point of a hypothesis-driven experiment is to produce timely, decision-relevant feedback—information that profoundly shapes the entrepreneur's subsequent path.

Nor is a "Bayesian spirit" at odds with the notion that entrepreneurs deal with great unknowns. The humility underlying a Bayesian approach rewards entrepreneurs with *structure*. In contrast to purely effectual or emergent approaches—where entrepreneurs might rely heavily on serendipity, iterative product tweaks, or trial-and-error processes that can be slow to yield actionable clarity—those who treat their beliefs as hypotheses to be tested can systematically target the greatest unknowns. An entrepreneur might ask, for instance, whether customers truly desire a specialized feature set. If so, an experiment that focuses on validating this feature becomes a high-priority step—often executed via a low-cost mock-up or a limited-release pilot. Such targeted probes minimize the entrepreneur's opportunity cost and reduce the risk of conflating multiple factors (technology, pricing, distribution channels, etc.) at once. Moreover, a broadly Bayesian mindset encourages iteration in the entrepreneur's understanding of *which* questions matter most. If an initial test fails to confirm a key assumption, the entrepreneur may pivot to explore an adjacent market or a simpler technological variant. If a pilot indicates demand in an unexpected user segment, that newfound insight might become the basis for a second, more focused experiment. This ongoing loop—pose hypothesis, test, update, refine or pivot—exemplifies how Bayesian thinking thrives on dynamic environments. It treats each setback or surprise not merely as a reason to persist or quit but also as a potential catalyst for deeper insight.

And, to reiterate, the "Bayesian" label does not insist on perfect rationality. Entrepreneurs might well incorporate heuristics, derive their strong priors from overconfidence, or discount certain types of data that conflict with their preconceptions. Indeed, as outlined earlier, subjective priors can diverge starkly across different individuals precisely because they rely on personal experiences and unique mental models. Entrepreneurs could also engage in alternative forms of learning that differ from "pure" experimentation, such as parallel play (McDonald and Eisenhardt, 2019). Yet, if an entrepreneur is open to learning—meaning they do not disregard new evidence merely because it is inconvenient—then their updating process captures the spirit of Bayesianism. They will, in effect, shift probabilities around in their head, however informally, as they accumulate fresh insights. In this sense, "broadly Bayesian" captures the reality that entrepreneurs work under uncertainty with limited time, resources, and cognitive bandwidth but nonetheless aim to integrate relevant signals in a disciplined way.

Finally, entrepreneurs who articulate clear hypotheses and corresponding success metrics can better communicate their rationale to potential investors, cofounders, or employees. It can be easier to secure buy-in for a staged investment or pilot study when decision-makers share a transparent logic for what the experiment aims to discover. Furthermore, the results of well-designed tests are often clearer and more credible to outsiders than the patchwork learning that emerges from simply "giving it a go." In other words, a hypothesis-driven experiment can not only sharpen the entrepreneur's own updates but also persuade stakeholders.

That said, employing a more theory-based approach does involve certain costs. Formulating experiments demands time and resources, as well as a measure of methodological rigor. Entrepreneurs must decide which hypotheses are worth testing first, design protocols to glean useful data, and analyze results before moving on. In periods of intense market competition, these steps can feel cumbersome. Nonetheless, the potential payoff is avoiding large, unproductive commitments down the road, reducing the chance that a venture sinks huge sums of capital or months of labor into a direction that could have been invalidated (or validated) more efficiently.

In sum, the rationale for a (broadly) Bayesian approach to entrepreneurial experimentation flows from the recognition that ventures are rife with uncertainty and high stakes. Founders do not want to remain in the dark about the viability of their product, the reality of market demand, or the fundamental soundness of their chosen strategy. Even if they believe passionately in their ideas, stepping back to design well-targeted experiments can clarify whether that passion aligns with underlying market and technological realities. If it does not, they can update their priors to accommodate this new evidence, set forth a revised hypothesis about how best to serve that niche, and run fresh tests accordingly. Thus, purposeful learning begets further purposeful learning, setting in motion a virtuous cycle of discovery, adaptation, and recalibrated commitment.

## 2.5 The Centrality of Choice

In an entrepreneurial setting marked by uncertainty and resource constraints, learning about an idea's ultimate viability often comes with significant costs—and cannot be fully completed before some commitment of time, money, or reputation. If learning were cheap and yielded immediate, definitive answers, priors would play little role: One could simply experiment at low cost and either proceed with the venture or abandon it without loss. Conversely, if entrepreneurship were purely about freewheeling experimentation, then only those with minimal costs to explore would be drawn in, making priors irrelevant. Yet, most real-world ventures require founders to commit scarce resources in the face of substantial unknowns. In this environment of *noisy learning*, entrepreneurs' subjective priors become a critical factor: Strong beliefs help them "stomach" the commitment before full information is available.

As a result, choice lies at the heart of entrepreneurial activity. Entrepreneurs do not merely discover opportunities (Alvarez and Barney, 2007); they must also decide *how* to pursue them. Even once they have uncovered a compelling idea—whether inspired by deeply held beliefs, refined through proactive learning, or discovered serendipitously—they confront multiple potential paths, each with its own strategic implications, resource requirements, and risks. In many cases, these paths are not straightforward substitutions, where one is clearly superior to the other. Instead, each path has its own advantages and drawbacks, and the entrepreneurs' ultimate decision often reflects not just objective profitability (as shaped by their capabilities and incentives), but also their subjective worldview.

Recent advances in entrepreneurial strategy underscore this point. In a series of studies (e.g., Gans et al., 2019; Agrawal et al., 2021; Scott et al., 2025), scholars argue that purposeful learning about an opportunity can reveal *multiple* viable ways to bring that opportunity to fruition. Imagine, for instance, a start-up developing a cutting-edge medical device. One strategic path might entail collaborating with established hospitals and device manufacturers, leveraging their distribution channels but conforming to established regulatory pathways. Another path might favor a more autonomous route, wherein the entrepreneur raises significant venture capital and builds in-house expertise to compete directly in the value chain. If the underlying technology is indeed promising, experimentation may confirm that *both* strategies are feasible, albeit with different trade-offs. Given the inherent resource constraints most founders face, they cannot fully pursue both paths at once. Thus, *choice* becomes central: Which path best aligns with the founder's capabilities and subjective beliefs about the future?

This framing contrasts with two other well-known approaches: *effectuation* and *real options*. An effectual entrepreneur often relies on what is "close at hand," forging a single path through improvisational processes and incremental learning. This path may prove fruitful, but effectuation theory generally prioritizes near-term resources and partnerships over systematically unearthing all major paths to market. In the real-options view, an entrepreneur (often alongside investors) tests a single, plausibly optimal course of action through milestone-based experimentation. The assumption is that once the "best" strategy is identified—given shared priors about what ultimately works—further testing and subsequent funding stages simply refine or validate that path.

By contrast, a world where simple experiments cannot deliver a simple, objective "ground truth"—one of noisy learning—is the natural *habitat* of *Homo entrepreneuricus*. It is a world where entrepreneurs engage in proactive, hypothesis-based exploration *before* locking into a single plan; a world where subjective priors allow entrepreneurs to see aspects of the opportunity that others might miss and, at the same time, to acknowledge that uncertainty is high and others might *not* share their worldview; and a world where their interpretation of new evidence causes entrepreneurs to update their priors and refine their understanding of the idea's potential—but does not necessarily resolve all ambiguities in favor of one clearly superior strategy.

*Homo entrepreneuricus's* choice is thus more than a mechanical calculation of the objective, expected profit arising from an opportunity. For the same reason that an experiment might leave a founder indifferent between two potential paths—uncertainty cannot be fully resolved without commitment—two founders can navigate the same uncertainty in drastically different ways, following their subjective beliefs. One founder might gravitate to the cooperative path, identifying personal strengths or networks well-suited to alliance building and integration. Another might choose the competitive route, motivated by a strong desire for autonomy or a conviction that incumbents are vulnerable.

Moreover, exploring multiple paths is itself a sign of *robustness*. First, *Homo entrepreneuricus* implicitly agrees to disagree with others: The inability to imagine at least one alternative route—one that other entrepreneurs with different but plausible beliefs might pursue—can signal a short circuit in judgment. Second, if rigorous testing reveals more than one workable strategy, the entrepreneur gains a measure of strategic flexibility: If the chosen path encounters unexpected obstacles—regulatory delays, partner conflicts, or an unanticipated shift in consumer preferences—the entrepreneur can pivot to the alternative. In a world governed by subjective priors, where experimental feedback is filtered through the entrepreneur's mental models, having multiple validated routes provides an extra layer of security. Rather than proving a single strategy "correct," experimentation may offer the reassurance that the core idea is substantial enough to succeed in more than one way.

This perspective highlights the centrality of choice in entrepreneurial decision-making. Entrepreneurs, armed with private theories and shaped by a proactive, scientific approach to learning, craft their own sense of what is possible. Although data-driven tools (such as decision trees or real-options analyses) can inform the probabilities and payoffs of each pathway, it is ultimately the entrepreneur's subjective priors and personal preferences that drive the final commitment. The journey from idea to impact is not only about validating whether an opportunity can work but also about selecting the strategic direction that resonates most closely with the founder's worldview and capabilities.

Thus, for *Homo entrepreneuricus*, decision-making goes beyond determining a single best route. Instead, it becomes a process of systematically surfacing multiple viable options, testing them thoroughly, then choosing which path to follow based on both objective insights and subjective convictions. Choice is not a one-time event; it is an ongoing commitment that may need revisiting as fresh evidence emerges. By embracing the possibility that more than one strategy can succeed, *Homo entrepreneuricus* maintains the nimbleness to pivot or adapt. In doing so, they embody the hallmark of the Bayesian entrepreneur: a willingness to learn broadly, update flexibly, and forge ahead confidently once they have identified (and chosen) their path in an uncertain world.

## 2.6 Exploratory Implications

The preceding discussion has shown how subjective priors and proactive experimentation can shape not only how entrepreneurs discover and refine opportunities but also how they interact with other market participants. In this section, we illustrate three broader insights that emerge when *Homo entrepreneuricus* enters a world where beliefs are not universally shared and where purposeful testing becomes a critical mechanism for reducing (although never eliminating) uncertainty. These insights suggest new ways of thinking about experimental design, competitive dynamics, and even foundational economic problems such as the disclosure dilemma.

### 1. Tailoring Experiments to One's Audience (and Prior)

A key feature of *Homo entrepreneuricus* is the recognition that experiments are not one-size-fits-all (Agrawal et al., 2021). Because each individual—founder, venture capitalist, potential employee—possesses different priors, the same piece of evidence can be interpreted in radically different ways, and different individuals will prefer different experiments, advisers, and news sources (Suen, 2004; Che and Mierendorff, 2019). An entrepreneur who already feels bullish about a new technology may design experiments that put the "best foot forward." By ensuring conditions are conducive to success (for instance, perfecting a prototype's user interface before testing), the entrepreneur gains greater confidence if the experiment succeeds—yet runs a substantial risk of updating negatively if it fails. Conversely, to persuade more skeptical parties, entrepreneurs might choose leaner tests, such as a minimum viable product trial, which only succeeds if the underlying idea is truly solid. A strong result from this stripped-down test can be more persuasive to outsiders who assume a high failure rate for contrarian ventures.

These design choices highlight how subjective priors and audience perspectives jointly influence experimental strategy. Entrepreneurs grapple with the question: Should I conduct experiments that reassure me, or that convince others? *Homo entrepreneuricus* recognizes that their own worldview may not be shared. Thus, they tailor the testing environment—whether by emphasizing features that matter to them or by showcasing the product under stark conditions that might allay external doubts. This underscores a critical point: Experimentation is not merely about generating a "true" signal of viability, but also about crafting signals that align with the beliefs and biases of different stakeholders.

### 2. Polarized Beliefs as a Source of Competitive Advantage

Second, individuals with subjective priors pursue opportunities they are disproportionately optimistic about relative to others (Agrawal et al., 2025). An implication of this simple observation is that opportunities will attract individuals with similarly optimistic beliefs. In other words, the easiest people to persuade to partner with an entrepreneur, join a start-up, or invest in the venture are precisely those who are as optimistic as the entrepreneur about the underlying opportunity. Entrepreneurs pursuing polarizing, even contrarian opportunities—that is, only appreciated by a *small* minority—might have a harder time finding like-minded partners, employees, and investors. The good news, however, is that some of their staunchest skeptics will be potential competitors who—although working in the same broader industry—disbelieve the underlying logic or remain wedded to a mainstream approach. Such a dynamic can make it hard for contrarian entrepreneurs to initially gather resources, but once the initial roadblock has been cleared, it can eventually grant them a valuable "head start."

Gius (2025), for instance, documents that in artificial intelligence contests, contestants advancing unconventional algorithms can attract significant attention and resources if they place highly, yet often fail to face a swarm of copycat rivals. Entrenched players, who regard the contrarian solution as improbable or unsustainable, may continue plowing resources into their preferred method. This asymmetry of belief provides a window in which a contrarian entrepreneur can consolidate progress, recruit key personnel, and refine the business model before mainstream competitors decide to respond—if they ever do. What emerges is an environment where skepticism can actually help the pioneer, protecting them from immediate competitive pressure even as they scale.

Even abstracting from competition, differences in beliefs can have a nuanced effect on the valuation enjoyed by contrarian ideas. In private markets (and even public markets, absent short selling), even ideas rejected by most can be potentially valued exceptionally highly, reflecting (or even exceeding) the valuation of the most optimistic investors (Harrison and Kreps, 1978). This, in turn, can rationalize why disagreement among venture competition judges predicts start-up success (Gius, 2024), why venture capital funds favor "champion-based mechanisms" when selecting early-stage ideas (Malenko et al., 2024), and more generally a (mostly practioner-based) literature on the value associated with pursuing polarizing, even "contrarian" opportunities (Masters and Thiel, 2014).

### 3. Revisiting Arrow's Disclosure Paradox

Finally, the "disclosure paradox," first articulated by Arrow (1962), highlights a perennial tension: Entrepreneurs need to share their ideas to gain resources, yet by disclosing the idea, they risk enabling imitation at minimal cost to others. Although this logic remains powerful in settings where information is interpreted uniformly, subjective priors mitigate the effect. If potential imitators do not share the entrepreneur's interpretation of the idea's true value, they may discount its commercial potential—even when the entrepreneur openly discusses it.

For instance, a founder might reveal preliminary results about a novel medical device to potential collaborators. Because these collaborators view the world through a different lens—perhaps holding entrenched doubts about the device's ultimate market or viability—they do not rush to duplicate the project. Indeed, the contrarian nature of the innovation may reinforce outsiders' skepticism, particularly if their beliefs diverge sharply from the entrepreneur's. Thus, the entrepreneur retains a relative advantage in exploiting the opportunity, despite having divulged core elements of its design. Over time, if the venture does gain traction, outsiders may update their beliefs—but by then, the entrepreneur can be well established, limiting the scope for imitative entry. In this way, subjective priors offer a subtle channel through which disclosure (or other informational spillovers) need not equate to forfeiting competitive advantage.

Moreover, a large strategy literature has long argued that agents with superior resources can "hack" factor markets with superior information, perhaps exploiting less sophisticated and boundedly rational sellers (Barney, 1986). Although hacking is hard to explain with simple differences in information, it is easily rationalizable in a setting where buyers and sellers simply agree to disagree on the value of a resource—allowing entrepreneurs to hoard the resources they believe in.

## 2.7 Concluding Thoughts

These insights highlight the broader significance of *Homo entrepreneuricus*. Once we abandon the assumption of common priors, experimentation becomes more than just a mechanical learning process for uncovering "true" probabilities. Instead, it becomes a carefully orchestrated process—one that entrepreneurs can strategically shape, depending on whose beliefs they seek to influence and how those beliefs align with their own. Likewise, competitive dynamics and knowledge dissemination take on new contours when people can persistently "agree to disagree." Rather than immediately converging on a universally recognized best practice, markets may see more (and benefit from) sustained diversity of approaches. By recognizing the central role of subjective beliefs, we open the door to a more nuanced understanding of entrepreneurial action and the multifaceted nature of competitive advantage in uncertain markets.

Thus, a key challenge for researchers and practitioners will be to better understand the origin and evolution of subjective priors and the impact of proactive learning and experimentation. Which experiences, forms of expertise, or social networks foster the creation of particularly valuable theories? How does an entrepreneur remain open enough to external feedback to adapt their worldview when necessary, yet sufficiently confident to act on convictions that diverge from the consensus? And how do we assess whether an emerging divergence in worldviews is a sign of visionary insight or a harbinger of costly miscalculation?

Examining these implications will require a range of mixed methods that capture not only outcomes but also the deeper beliefs and thought processes that shape entrepreneurial action. For instance, survey-based approaches can help elicit individual beliefs and reveal their evolution in response to feedback

(Puri and Robinson, 2007). Randomized controlled trials can further identify which individuals benefit most from interventions designed to cultivate a more "Bayesian" style of entrepreneurship and who might remain resistant to belief revision. Finally, observational studies that construct proxies of "revealed beliefs" from entrepreneurs' actions could help explore the implications more broadly (Malmendier and Tate, 2005).

In answering these questions, we have the opportunity to advance our understanding not just of how entrepreneurs think but also of how innovation itself unfolds in the face of deep uncertainty. *Homo entrepreneuricus* reminds us that the pursuit of novel ventures is seldom a purely mechanical process of plugging data into a formula. It is instead a creative, model-building enterprise in which different people imagine and evaluate the future in ways that can—and often do—persistently deviate from one another. By acknowledging the legitimacy of these diverse worldviews, we gain a richer appreciation for the role of subjective priors in fostering entrepreneurial action and, ultimately, driving progress in markets and society.

## References


Agrawal, A., Camuffo, A., Gambardella, A., Gans, J. S., Scott, E. L., and Stern, S. (2025). The foundations of Bayesian entrepreneurship. *Bayesian entrepreneurship*. MIT Press.

Agrawal, A., Gans, J. S., and Stern, S. (2021). Enabling entrepreneurial choice. *Management Science*, 67(9), pp. 5510–5524.

Alvarez, S. A., and Barney, J. B. (2007). Discovery and creation: Alternative theories of entrepreneurial action. *Strategic Entrepreneurship Journal*, 1(1–2), pp. 11–26.

Alvarez, S. A., and Busenitz, L. W. (2001). The entrepreneurship of resource-based theory. *Journal of Management*, 27(6), pp. 755–775.

Arrow, K. (1962). Economic welfare and the allocation of resources for invention. *The rate and direction of inventive activity: Economic and social factors*. Princeton University Press, pp. 609–626.

Åstebro, T., Herz, H., Nanda, R., and Weber, R. A. (2014). Seeking the roots of entrepreneurship: insights from behavioral economics. *Journal of Economic Perspectives*, 28(3), pp. 49–70.

Aumann, R. (1976). Agreeing to disagree. *Annals of Statistics*, 4, pp. 1236–1239.

Aumann, R. J. (1987). Correlated equilibrium as an expression of Bayesian rationality. *Econometrica*, 55(1), p. 1.

Barney, J. B. (1986). Strategic factor markets: Expectations, luck, and business strategy. *Management Science*, 32(10), pp. 1231–1241.

Bewley, T. F. (2002). Knightian decision theory, part I. *Decisions in Economics and Finance*, 25(2), pp. 79–110.

Bhidé, A. (1986). Hustle as strategy. *Harvard Business Review*, 64(5), pp. 59–65.

Camuffo, A., Cordova, A., Gambardella, A., and Spina, C. (2020). A scientific approach to entrepreneurial decision-making: Evidence from a randomized control trial. *Management Science*, 66(2), pp. 564–586.

Camuffo, A., Gambardella, A., Messinese, D., Novelli, E., Paolucci, E., and Spina, C. (2024). A scientific approach to entrepreneurial decision-making: Large-scale replication and extension. *Strategic Management Journal*, 45(6), pp. 1209–1237.

Camuffo, A., Gambardella, A., and Pignataro, A. (2024). Theory-driven strategic management decisions. *Strategy Science*, 9(4), pp. 382–396.

Che, Y.-K., and Mierendorff, K. (2019). Optimal dynamic allocation of attention. *American Economic Review*, 109(8), pp. 2993–3029.

Felin, T., and Zenger, T. (2017). The theory-based view: Economic actors as theorists. *Strategy Science*, 2(4), pp. 258–271.

Gans, J. (2025). Knightian uncertainty and Bayesian entrepreneurship. *Bayesian entrepreneurship*. MIT Press.

Gans, J., Stern, S., and Wu, J. (2019). Foundations of entrepreneurial strategy. *Strategic Management Journal*, 40(5), pp. 736–756.

Gius, L. (2024). Disagreement predicts startup success: Evidence from venture competitions. *Strategy Science*, 10(2), pp. 93–108.

Gius, L. (2025). Commercializing contrarian ideas: Evidence from AI contests. Working Paper.

Gompers, P., and Lerner, J. (2001). The venture capital revolution. *Journal of Economic Perspectives*, 15(2), pp. 145–168.

Harrison, J. M., and Kreps, D. M. (1978). Speculative investor behavior in a stock market with heterogeneous expectations. *The Quarterly Journal of Economics*, 92(2), pp. 323–336.

Hayward, M. L. A., Shepherd, D. A., and Griffin, D. (2006). A hubris theory of entrepreneurship. *Management Science*, 52(2), pp. 160–172.

Hellmann, T. (1998). The allocation of control rights in venture capital contracts. *The RAND Journal of Economics*, 29(1), p. 57.

Jovanovic, B. (1982). Selection and the evolution of industry. *Econometrica*, 50(3), pp. 649–670.

Kerr, W., Nanda, R., and Rhodes-Kropf, M. (2014). Entrepreneurship as experimentation. *The Journal of Economic Perspectives*, 28(3), pp. 25–48.

Malenko, A., Nanda, R., Rhodes-Kropf, M., and Sundaresan, S. (2024). Catching outliers: Committee voting and the limits of consensus when financing innovation. *Journal of Finance, forthcoming*.

Malmendier, U., and Tate, G. (2005). CEO overconfidence and corporate investment. *The Journal of Finance*, 60(6), pp. 2661–2700.

Marinacci, M. (2015). Model uncertainty. *Journal of the European Economic Association*, 13(6), pp. 1022–1100.

Masters, B., and Thiel, P. (2014). *Zero to one: Notes on startups, or how to build the future*. Random House.

McDonald, R. M., and Eisenhardt, K. M. (2019). Parallel play: Startups, nascent markets, and effective business-model design. *Administrative Science Quarterly*, 65(2), pp. 483–523.

Morris, S. (1995). The common prior assumption in economic theory. *Economics and Philosophy*, 11(2), pp. 227–253.

Murray, F., and Tripsas, M. (2004). The exploratory processes of entrepreneurial firms: The role of purposeful experimentation. *Business strategy over the industry lifecycle*, Emerald, pp. 45–75.

Ortoleva, P. (2012). Modeling the change of paradigm: Non-Bayesian reactions to unexpected news. *American Economic Review*, 102(6), pp. 2410–2436.

Ortoleva, P. (2024). Alternatives to Bayesian updating. *Annual Review of Economics*, 16(1), pp. 545–570.

Puri, M., and Robinson, D. (2007). Optimism and economic choice. *Journal of Financial Economics*, 86(1), pp. 71–99.

Ries, E. (2011). *The lean startup: How today's entrepreneurs use continuous innovation to create radically successful businesses*. Currency.

Sarasvathy, S. D. (2001). Causation and effectuation: Toward a theoretical shift from economic inevitability to entrepreneurial contingency. *Academy of Management Review*, 26(2), pp. 243–263.

Scott, E., Gans, J., and Stern, S. (2025). *Entrepreneurship: Choice and strategy*. W.W. Norton & Company.

Shane, S., and Venkataraman, S. (2000). The promise of enterpreneurship as a field of research. *The Academy of Management Review*, 25(1), p. 217.

Suen, W. (2004). The self-perpetuation of biased beliefs. *The Economic Journal*, 114(495), pp. 377–396.

Surowiecki, J. (2014). Epic fails of the startup world. *New Yorker*.

Torralba, A., and Isola, P. (2024). *Foundations of computer vision*. MIT Press.

Van den Steen, E. (2001). Essays on the managerial implications of differing priors. PhD thesis. Stanford University.

Van den Steen, E. (2004). Rational overoptimism (and other biases). *American Economic Review*, 94(4), pp. 1141–1151.

Von Uexkull, J. (1909). *Umwelt und innenwelt der tiere*. 1st ed. Springer.

Yong, E. (2022). *An immense world*. Random House.

# 3

## Knightian Uncertainty and Bayesian Entrepreneurship

Joshua Gans

### 3.1 Introduction

New ideas are a domain where Knightian uncertainty naturally applies. Ideas are often contained within the mind of a single person and have yet to be explored or tested in any way. The newness of ideas encompasses a new event or state that has not been conceived by others, let alone a probability assessment of their reasonableness or future prospects. One recent example was the discovery of generative artificial intelligence by Vaswani et al. (2017). This new methodology led directly to ChatGPT, one of the most successful products in terms of adoption speed in history. ChatGPT, when it was released in November 2022, was a surprise advance even to those familiar with the scientific advance of five years earlier. Thus, for those working in this area, it is reasonable to propose that they faced Knightian uncertainty regarding whether large language models (or LLMs) would be of practical use prior to November 2022, which was resolved substantially shortly afterward.

Knightian uncertainty is thus intimately tied to entrepreneurial research, with Knight (1921) himself proposing the famous distinction between risk and uncertainty as part of a theory to explain the existence of entrepreneurial profit. At the same time, Knightian uncertainty has presented a challenge to models of choice. Unlike the clear axiomatic foundations that underlie Bayesian decision-making, the presence of Knightian uncertainty undermines those axioms and, with it, the ability of such theories to generate empirical predictions, let alone robust predictions. In its wake, many entrepreneurial researchers forgo deductive formal reasoning and instead base analyses on inductive examinations of entrepreneurial choice processes.[1]

This paper challenges the notion that taking Knightian uncertainty seriously will undermine the ability to use formal models in entrepreneurial research. The context here is based on the agenda laid out by Agrawal et al. (2025) called "Bayesian entrepreneurship" that seeks to motivate and ground research questions in a Bayesian decision-making approach. Specifically, that research agenda rests on three premises, (i) that important agents in entrepreneurial environments have heterogeneous beliefs, (ii) that entrepreneurs hold stronger (some might say more optimistic) beliefs regarding venture success than other agents, and (iii) that all agents use Bayes' rule to update their priors upon receipt of new information. They argue that this can lead to a number of fruitful, positive examinations of entrepreneurial choice, as well as provide normative prescriptions that will improve entrepreneurial success.

What this paper shows is that the approaches and important empirical conclusions from the Bayesian entrepreneurship agenda are, and will continue to be, robust to a consideration of Knightian uncertainty. This is done by adopting the formal approach to Knightian uncertainty proposed by Bewley (2002) and studied by numerous economists to understand the impact of such uncertainty on financial markets (Rigotti and Shannon, 2005) and incentive contracts (Lopomo et al., 2011). That approach adopts a strong form of uncertainty aversion but has not hitherto been explored in the published literature on entrepreneurship.

This paper is structured as follows. Section 2 describes the Bewley (2002) approach and establishes its operation in a simple entrepreneurial context. Section 3 then examines three aspects of the Bayesian entrepreneurship agenda—the role of experiments, venture financing, and entrepreneurial strategy—to demonstrate that the empirical predictions of the Bayesian approach are robust to the inclusion of Knightian uncertainty. The reason for this is that although Knightian uncertainty generates a more challenging decision environment, it is still amenable to a consideration of the three pillars of the Bayesian approach listed above.

Section 4 then turns to consider the methodology in entrepreneurial research that, in recent years, has been most grounded on a foundation of Knightian uncertainty: effectuation theory, as developed by Sarasvathy (2001). Although not providing a comprehensive examination of that theory, the discussion demonstrates that it is difficult, although not impossible, to generate two key predictions of effectuation theory—that entrepreneurs will make choices based on resources and relationships at hand and that they will base decisions on affordable loss rather than on a full rate of return analysis—to be generated from the same set of underlying assumptions. That said, the analysis shows that using a formal model of Knightian uncertainty can act as a bridge to help clarify some of the distinctions between the methodologies of effectuation and those of Bayesian entrepreneurship.

A final section concludes.

## 3.2 Formal Representation of Knightian Uncertainty

It is with some trepidation that anyone purports to consider a formal treatment of Knightian uncertainty, especially in the field of entrepreneurship. The term itself is the subject of much controversy. My intention here is not to wade into that controversy but instead to adopt what I regard as a well-grounded, explored, and meaningful formal definition of uncertainty and explore its implications.

### 3.2.1 Origins of the Concept

Nonetheless, it is useful to understand the origins of the formal treatment I will adopt here. The term "Knightian uncertainty" has its origins with the distinction by Frank Knight (1921) between risk (where probabilities have an objective foundation, such as might arise in gambling games or from data-rich statistical estimations) and uncertainty (where that isn't the case):

> Uncertainty must be taken in a sense radically distinct from the familiar notion of risk, from which it has never been properly separated. . . . The essential fact is that "risk" means in some cases a quantity susceptible of measurement, while at other times it is something distinctly not of this character; and there are farreaching and crucial differences in the bearings of the phenomena depending on which of the two is really present and operating.... It will appear that a measurable uncertainty, or "risk" proper, as we shall use the term, is so far different from an unmeasurable one that it is not in effect an uncertainty at all. Knight (1921, p. 19)

Keynes (1921) made a similar distinction.[2] Knight's purpose, however, was directly relevant to the study of entrepreneurship. He related his distinction to the existence of markets that would allow entrepreneurs and businesses to purchase insurance. He contended that they could do so for risks but not for uncertainty.[3] As a consequence, Knight viewed entrepreneurs as being agents who bore uncertainty and that the supra-normal profits (if any) generated were a compensation for that bearing.

### 3.2.2 A Formal Definition

Formal decision theory emerged after the work of Knight, mostly notably with Von Neumann and Morgenstern (1944) and Savage (1954), who gave expected utility theory an axiomatic foundation. Anscombe and Aumann (1963) then extended this research program to encompass subjective utility, which, in effect, rendered some of Knight's concerns moot. That said, Aumann (1962) noted that if a decision-maker was unable to form even subjective utilities over states, their preference ordering would be incomplete and therefore incapable of being represented by a utility function or something similar.[4] This meant that when faced with Knightian uncertainty, there was an important gap in decision theory and its ability to generate behavioral hypotheses, including those that could be tested with data using revealed preference.

Meanwhile, work empirically testing various forms of decision theory revealed that some of these gaps may arise in common situations, notably Allais (1953) and Ellsberg (1961). This led to various new theories of choice that explored how to adjust the axioms of expected utility theory to explain or rationalize the empirical challenges. My purpose here is not to review these theories because, apart from the decision-making under an ambiguity approach (e.g., Gilboa and Schmeidler, 1989), few have dealt head-on with Knightian uncertainty or something related.

As noted in the introduction, my purpose here is to find a way of formally representing uncertainty, as opposed to risk, and to do so in a way that makes it capable of generating hypotheses regarding entrepreneurial choices. Given these criteria, than evaluate the broader merits of alternative decision-theoretic models, here I explore the so-called unanimity approach developed by Bewley (2002).[5] In contrast to other developments, this approach directly allowed for preferences to be incomplete in the Aumann (1962) sense. That is, when faced with uncertainty, a decision-maker was assumed not to be indifferent between uncertain outcomes but rather to have no preference ranking at all. In effect, the decision-maker was permitted to claim and/or believe that they "simply do not know" what to do.[6]

Bewley (2002) operationalized this in a natural way. To see this, suppose there is a decision-maker who can make a choice, $x$, from a set of actions, $X$. The decision-maker faces uncertainty with respect to a random variable, $\theta$, that is a state realization from a set, $\Theta$. Let $F(\theta)$ be a probability distribution function with respect to $\theta$ drawn from a set, $\mathcal{F}$. I assume the decision-maker's preference relation, $\succ$, for actions, is represented by a unique closed, convex set of probability distributions, $\mathcal{F}$, and a continuous, strictly increasing, concave function, $u : \mathbb{R}_+ \to \mathbb{R}$, unique up to positive affine transformations, such that:

$$x \succ y \text{ if and only if } \mathbb{E}_F[u(x,\theta)] > \mathbb{E}_F[u(y,\theta)] \text{ for all } F \in \mathcal{F}.$$

Here, $\mathbb{E}_F[u(x)]$ denotes the expected value with respect to the probability distribution $F$, and $u(x,\theta)$ is the decision-maker's payoff from choosing $x$ if the state is $\theta$. This definition of preferences says that the decision-maker will only strictly prefer $x$ over $y$ if their expected payoff is strictly higher in choosing $x$ versus $y$ for each possible probability distribution in $\mathcal{F}$. Thus, if there is one probability distribution for which the ranking using expected payoffs is reversed, preferences between the two options will be incomplete. This also means that if we observed such a decision-maker choosing $x$ over $y$, we could not say $x$ was revealed preferred to $y$, only that $y$ was not revealed preferred to $x$.

This is and should be interpreted as a strong condition regarding the decision-maker's aversion to uncertainty. Uncertainty is captured by the fact that there is a set of possible probability distributions, and the decision-maker cannot form preferences over actions unless it is completely clear that, regardless of the actual distribution, the ranking would be unchanged.[7] At the same time, this preference condition does not say, as can sometimes be assumed in the informal literature, that Knightian uncertainty (as seen as the absence of the ability to pin down probabilities for various states) implies that there is no room for choice. This definition allows for such choice when an option is undominated in the sense defined by the preference condition. It is this that allows us to consider the impact of uncertainty itself.

The size of $\mathcal{F}$ is itself a measure of the degree of uncertainty (and the decision-maker's uncertainty aversion). To see this, note that if $\mathcal{F}$ is a singleton, then the preference condition collapses to the standard expected utility theory preference relation where there is no uncertainty. Moreover, Rigotti and Shannon (2005) show that as the size of $\mathcal{F}$ becomes smaller, the preferences of the decision-maker converge to the expected utility theory outcome. Thus, the size of $\mathcal{F}$ is itself a measure of preference incompleteness—that is, if $\mathcal{P}_{\succ} = \{(x,y) \in \mathbb{R}_{+} \times \mathbb{R}_{+} | x \succ y\}$, then if $\succ'$ is the preference order under $\mathcal{F}'$, then if $\mathcal{F}' \subset \mathcal{F}$, $\mathcal{P}_{\succ'} \subset \mathcal{P}_{\succ}$ implying that $\succ'$ is more complete than $\succ$.[8]

There is an interesting issue as to whether there are any constraints on $\mathcal{F}$. Bewley (2002) allows for it to differ between agents. But the question is whether it can constitute uncertainty unless each agent places some nonzero probability of each state over which each element of $\mathcal{F}$ is defined. That is, could a distribution that assigns a probability to a state of 1 or, indeed, of 0 be permissible? To do so would allow for an agent to act as if they had no doubt as to what might or might not occur. This would invalidate the contention of Bhidé (2024), for example, that, operationally, Knightian uncertainty was a situation where agents faced some doubt as to what might occur. In what follows, I will be forced to adopt some practical bounds to $\mathcal{F}$. For instance, I could surely not permit entrepreneurial returns to be greater than all of the monies and wealth of Earth (although equal to it would be fine). And I would surely have to rule out those returns being 0, as an agent who acted on such a distribution or could only act as if that distributional outcome could possibly be true would be acting as if they had no doubt. The precise, reasonable bounds and implications of this are left for future research. Conveniently, as we will see, this allows me to use Bayes' rule with abandon in this environment.

It is instructive to consider all of this in a (very basic) entrepreneurial example. Suppose that a potential venture can be a success or a failure—that is, $\Theta = \{\text{success, failure}\}$. Suppose that success generates a payoff of $V$ and failure a payoff of 0 if a venture is launched (i.e., $x = 1$)—that is, $u(1, \text{success}) = V$ and $u(1, \text{failure}) = 0$. Finally, suppose that the elements of $\mathcal{F}$ can be expressed by a probability, $p = \Pr[\text{success}]$, where $p \in [\underline{p}, \bar{p}] \subset (0, 1)$; where $\underline{p} \equiv \arg\min_{p \in \mathcal{F}}$ and $\bar{p} \equiv \arg\max_{p \in \mathcal{F}}$.[9] Finally, suppose that it costs $C$ to launch a venture—that is, $u(0, \text{success}) = u(0, \text{failure}) = -C$. Then, the preference ordering between launching ($x = 1$) and not launching ($x = 0$) is:

$$\text{launch} \succ \text{not launch, if and only if } \underline{p}V > C$$

$$\text{not launch} \succ \text{launch, if and only if } \bar{p}V < C.$$

Note that if $\bar{p} > \underline{p}$, then it could easily be the case that neither of these inequalities hold—that is, $\underline{p}V \leq C \leq \bar{p}V$. In this case, the model cannot tell us what will be chosen.

This outcome is illustrated in figure 3.1. Observe that as $\bar{p} > \underline{p}$, there is a domain in $(C, V)$ space where the entrepreneur cannot rank the actions to launch and not launch. This preference incompleteness is a consequence of uncertainty. As $\bar{p} \to \underline{p}$, this set shrinks ultimately to a simple indifference condition defined by $p^* = C/V$. What this also highlights is that the analyst faces an issue: Since the

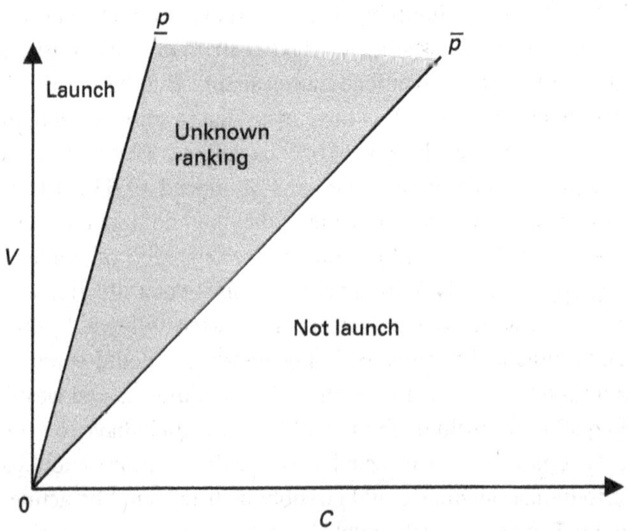


**Figure 3.1**
Entrepreneurial preferences.

entrepreneur will either launch or not launch the venture, given the uncertainty, what theoretical predictions can be made about which choice will be made? It is this issue that I will turn to next.

### 3.2.3 The Importance of Defaults

Allowing for preference incompleteness involves a cost; preferences alone cannot pin down likely decisions. To generate a theoretical prediction, we need an additional assumption. A natural way to achieve this is to posit that the uncertainty that emerges is the result of something "new." In the domain of entrepreneurship, this would be a new idea, with the status quo being whatever the decision-maker was doing prior to that emergence. This was the approach taken by Bewley (2002) when he proposed an *inertia* assumption.[10] He argued that the natural starting point was the decision-maker's choice "pre-newness." For instance, previously, the states might have had a probability of zero or the actions being considered were not available, making at least the payoff that the decision-maker was previously earning a natural benchmark from which to rank new alternatives.[11]

From this perspective, if the idea for the venture arose at a given date, then because the venture had, by definition, not been launched at that point, the appropriate default could be the payoff the entrepreneur receives if they did not launch a venture—that is, 0. Thus, in this case, the entrepreneur would only choose to launch a venture if:

$$\underline{p}V > C.$$

This creates a high bar in terms of $V$ relative to $C$ for a venture to be launched and is consistent with the sometimes expressed notion that uncertainty is a barrier to entrepreneurial activity. This can be termed the *no-launch* ($N$) default.

It can immediately be seen that this is not a very satisfying argument in the domain of entrepreneurship. Entrepreneurs are often imagined to be those who are more optimistic, less risk averse, and more committed to action than not (Gans et al., 2024). Indeed, as already noted, the entire premise of the Bayesian entrepreneurship agenda is that entrepreneurs have stronger priors in their assessment of the likelihood of venture success. This suggests that when a new entrepreneurial opportunity emerges, the appropriate default is a *launch* ($L$) default. In this case, the entrepreneur would choose to launch a venture if:

$$\bar{p}V \geq C.$$

An interpretation of this is that the entrepreneur is committed to pursuing the venture in the face of uncertainty and will only be deterred from this if $V$ is low relative to $C$.

This discussion implies that we need to be both explicit in our default assumptions and circumspect to the default assumption we make in analyzing entrepreneurial choices. Thus, in what follows, I will often examine outcomes from both $L$-default and $N$-default perspectives.[12]

It is worth mentioning at this point that there may be alternatives to simply assuming that one action or outcome is a default. From a modeling perspective, it might be possible to parameterize the decision-maker's preferences with regard to uncertainty. Following a suggestion by Arrow and Hurwicz (1972), suppose that, in the case of the entrepreneurial launch decision, weights were placed on the best and worst distributions such that it was assumed that:

$$L \succ N \text{ if and only if } \left(\omega \underline{p} + (1-\omega)\bar{p}\right) V > C$$

for some $\omega \in [0, 1]$, where $\omega = 1$ ($\omega = 0$) corresponds to the $N$-default ($L$-default) entrepreneurial cases. $\omega$ is a parameter that can be used to characterize uncertainty aversion on the part of the decision-maker. Although this parameterization is not used below, in some applications, it may prove fruitful for more precise empirical predictions.

### 3.2.4 Comparison with Theory-Based Decision-Making

One dimension of Bayesian entrepreneurship is where prior beliefs come from. Agrawal et al. (2025) consider developments in theory-based decision-making whereby entrepreneurs are presumed to use theoretical constructs and narratives to place more structure on probability distributions that impact entrepreneurial choice. This approach has been pushed most recently by Felin and Zenger (2017), who argue that such theories can be causal structures on stochastic processes. This examination has its formal roots in Ryall (2009) and has been developed further by Camuffo et al. (2024) and Ehrig et al. (2025). Montiel Olea et al. (2022) provide an alternative but potentially important approach making use of model specification.

Theory-based decision-making differs from Knightian uncertainty in that, ultimately, the probability distribution is knowable even if its parameters are pinned down by theoretical thought processes rather than empirical evidence per se. This is useful in modelling the decision-making sophistication of agents and how this might reveal itself through observational data. For instance, Chavda et al. (2024) consider an environment where there are two potential probability distributions for venture valuation, with some agents not considering that fact and acting as if signals are from a single distribution whereas others understand those signals can reveal the precise distribution itself. This leads to different search behavior and pivoting behavior.

The Knightian uncertainty approach stands apart here in that there is no avenue for theory to identify the correct distribution or partition of distributions from the set $\mathcal{F}$. Nonetheless, that does not mean that signals cannot refine uncertainty. It is just that the process requires close attention to what is known and unknown, as I will explore next.

## 3.3 Implications for Bayesian Entrepreneurship

Having set up a formal approach to Knightian uncertainty, I now turn to consider its implications for the Bayesian entrepreneurship research agenda—specifically, the part of that agenda that focuses on how agents learn for the purpose of dealing with uncertainty over their own venture prospects and also as a means of informing others as to a venture's prospects.

### 3.3.1 Experimentation

Experiments and information gathering are at the core of Bayesian decision-making. It is sometimes suggested, however, that when Knightian uncertainty is present, the value of such experiments and information is limited or potentially of no value at all.[13] The logic behind this view is that Bayesian updating procedures cannot be applied to events that have zero probability or where probabilities cannot be established. On the other hand, it seems unlikely that there is no ability to resolve uncertainty, especially given that Knightian uncertainty does, eventually, get resolved to some degree. For instance, an entrepreneur facing Knightian uncertainty might resolve that uncertainty by launching a venture and discovering whether it is successful ex post and why.

This section demonstrates that, based on the formal definition of Knightian uncertainty used here, the role of experimentation and information gathering is undiminished. Indeed, it is shown that many of the predictions regarding experimental choice used in Bayesian entrepreneurship are robust and potentially sharper when Knightian uncertainty is present. The reason for this is that the structure of that uncertainty still posits a (potentially large) set of possible distributions of the state variable, and each of those distributions can be updated and modified based on experiments run. This can have the effect of changing and even reducing that set and so enable more choices by reducing inertia.

Consider again the entrepreneurial launch decision. Suppose that prior to making that decision, the entrepreneur can engage in an experiment that generates a signal, $s$, that takes a value of 1 with probability $\lambda_1$ if the venture is successful and a value of 0 with probability $\lambda_0$ if the venture is unlikely to be successful—that is, $\Pr[s=1|x=V]=\lambda_1$ and $\Pr[s=0|x=0]=\lambda_0$ (see

**Table 3.1**
Experiment maps from signal to state, $\Pr[s|\theta]$

| | (Good news) $s=1$ | (Bad news) $s=0$ |
|---|---|---|
| (Success) $V$ | $\lambda_1$ | $1-\lambda_1$ |
| (Failure) 0 | $1-\lambda_0$ | $\lambda_0$ |

table 3.1). This is the same signal structure that is used by Gans (2023) in a setting without Knightian uncertainty.

As Bewley (1989) noted, a signal will update every distribution in $\mathcal{F}$. This updating can be derived using Bayes' rule. In particular, for a distribution with $\Pr[\text{success}] = p_i$, the posterior distribution characterized by posterior beliefs, $p_i(s)$, will be:

$$p_i(1) = \frac{\lambda_1 p_i}{\lambda_1 p_i + (1-\lambda_0)(1-p_i)}$$

$$p_i(0) = \frac{(1-\lambda_1)p_i}{(1-\lambda_1)p_i + \lambda_0(1-p_i)}$$

for the case of "good news" (i.e., $s=1$) and "bad news" (i.e., $s=0$), respectively. A necessary condition for $p_i(1) \geq p_i \geq p_i(0)$ is that $\lambda_1 + \lambda_0 \geq 1$, which will be assumed to be the case.

Agrawal et al. (2025) examined how the demand for experiments changed as entrepreneurial priors regarding the success of the venture changed. Here, the question of interest is how the demand for experiments changes with the degree of Knightian uncertainty. To examine this, consider first an $N$-default entrepreneur who will, in the absence of an experiment, only launch a venture if $\underline{p}V > C$. In this case, an experiment will be of value if it causes the entrepreneur to launch a venture. In other words, it is the impact of the experiment on $\underline{p}(1)$ that matters.[14]

The realization of $\Pr[s=1]$ depends on the ground truth distribution. However, since this is unknown to the entrepreneur when considering the value of the experiment, they will use $\Pr[s=1]$ evaluated according to the distribution $\underline{p}$. Moreover, since no launch is their default, they will compare the expected payoff if an experiment is deployed with a default payoff of 0. Then, the value of the experiment $\{\lambda_1, \lambda_0\}$ to the entrepreneur is:

$$\underline{p} \leq \tfrac{C}{N} : \Pr[s=1]\left(\underline{p}(1)V - C\right) = \lambda_1 \underline{p}(V-C) - (1-\lambda_0)(1-\underline{p})C$$

$$\underline{p} > \tfrac{C}{N} : \Pr[s=0]\left(C - \underline{p}(0)V\right) = \lambda_0(1-\underline{p})C - (1-\lambda_1)\underline{p}(V-C).$$

This is positive if and only if:

$$\underline{p}(0) \leq \tfrac{C}{V} < \underline{p}(1).$$

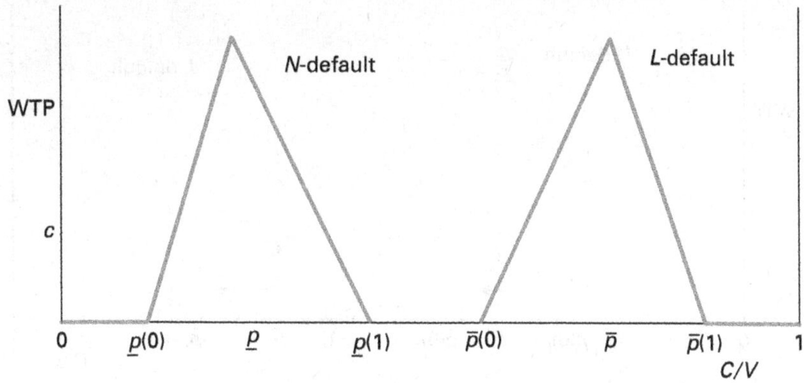


**Figure 3.2**
Willingness to pay (WTP) for experiments.

Intuitively, an experiment is valuable under Knightian uncertainty only where it potentially leads to a change in the entrepreneur's choice. When $\underline{p} \geq C/N$, the experiment shifts the "unknown ranking" set to potentially allow a ranking if $C/V$ is below the new lower bound of the unknown set, $\underline{p}(1)$. Thus, it is in that range where an experiment is valuable. As figure 3.2 shows, the experiment also remains valuable where the entrepreneur would have chosen to launch (at low $C/N$) even under Knightian uncertainty, as it can signal not to launch in that case.

The outcome of the experiment for an $L$-default entrepreneur is the same but for two differences. First, the experiment can resolve uncertainty and provide a ranking where $\bar{p} \geq C/N$ and cause the venture not to launch where it would have launched in the absence of an experiment. Second, the experiment is grounded in the prior, $\bar{p}$, and so its set where the experiment is of positive value is at a correspondingly higher level than for an $N$-default entrepreneur (see figure 3.2).

What impact does the degree of Knightian uncertainty have on the demand for experiments? The analysis, thus far, has demonstrated that experiments can play a role in "resolving uncertainty" in the sense of providing a clear ranking between alternatives where none previously existed. However, depending on the defaults of the entrepreneurs, this resolution is undirectional, with experiments playing their usual role in simply updating priors that, in turn, may change entrepreneurial choices. In this setup, a natural measure of the degree of Knightian uncertainty is the difference, $\Delta \equiv \bar{p} - \underline{p}$. Figure 3.3 shows what happens as $\Delta$ becomes smaller. In that depiction, $\bar{p}$ and $\underline{p}$ are brought closer together. As you can see, this shifts the position of each triangular WTP line toward each other,

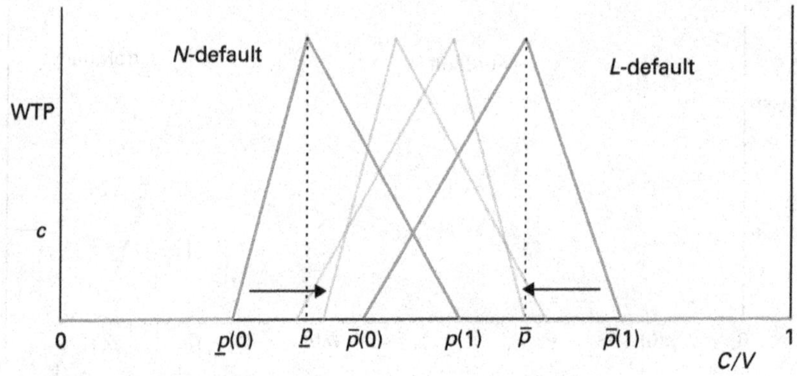


**Figure 3.3**
Impact of a reduction in knightian uncertainty. WTP: willingness to pay.

crossing over when $\bar{p}(0) < \underline{p}(1)$ and eventually fully overlapping when $\Delta = 0$. In that final case, there is no uncertainty, and $\mathcal{F}$ is a singleton yielding the usual single willingness to pay for experiments that arise in the usual Bayesian case (see, e.g., Arora and Fosfuri, 2005).

Although the demand for an experiment appears to be driven by the same fundamentals when Knightian uncertainty is present, it also highlights that Knightian uncertainty means that agents with different defaults in the face of that uncertainty will have positive demand for experiments along distinct ranges of $(V, C)$. $L$-default entrepreneurs will choose to experiment over lower $V$ and higher $C$ levels than $N$-default ones. Those differences could be observed empirically by a researcher in the area.

However, it is also possible that the type of experiments chosen might differ between agents with different defaults. As Gans (2023) demonstrated, agents who are predisposed not to launch a venture (due to low priors or less favourable $[V, C)]$ to their specific situation) will choose experiments with a relatively high $\lambda_0$ and low $\lambda_1$ (termed a "high-bar" or "raising-the-bar" experiment) in order to provide a clearer signal that the venture is worth pursuing. By contrast, agents who are predisposed to launch a venture will choose experiments with a relatively low $\lambda_0$ and high $\lambda_1$ (termed a "low-bar" or "best-foot-forward" experiment) that provides a clear signal when it is not worthwhile to launch a venture. When there is Knightian uncertainty, then it is possible that different agents with different defaults but otherwise facing the same fundamentals, $[V, C]$, would choose to pursue distinct experiment types.

To explore this, suppose that each agent can choose a single experiment from a set $\{\lambda_1, \lambda_0 | \lambda_1 + \lambda_0 \in [1, \Lambda)\}$ with $\Lambda \leq 2$. The following proposition characterizes the outcomes when there is Knightian uncertainty.

**Proposition 3.1.** Suppose that $C/V \in [\underline{p}, \bar{p}]$. Then an $N$-default ($L$-default) entrepreneur will choose an experiment $\{\lambda_1, \lambda_0\} = \{\Lambda - 1, 1\}$ ($= \{1, \Lambda - 1\}$).

The proof of this proposition follows the same logic as proposition 1 in Gans (2023) and is therefore omitted. Intuitively, in the range of Knightian uncertainty, an $N$-default agent will choose not to launch a venture in the absence of an experiment. By setting $\lambda_0 = 1$, if the agent observes $s = 1$, then with probability 1, the venture will be successful—that is, $\Pr[\theta = \text{success}|s = 1] = 1$. The "cost" associated with this choice is that $\Pr[s = 1]$ is low. Thus, a signal pushing that agent to launch a venture will be generated with a relatively low likelihood, but when it arrives, it will be a very clear signal. The opposite holds for an $L$-default entrepreneur who chooses an experiment that generates a clear signal that a venture should not be pursued. What this demonstrates is that (a) if there is Knightian uncertainty, the defaults of different agents might be identified by their experimental choices and (b) the very observation that different experiments are chosen when there are the same or similar fundamentals facing the agents is itself an indicator that Knightian uncertainty is present. Thus, consistent with the implications of the Bayesian entrepreneurship research agenda, experiments themselves explain the broader context associated with entrepreneurial choice.

In summary, Knightian uncertainty does not change the fundamental predictions of Bayesian entrepreneurship regarding the demand for experiments and the type of experiments chosen. It does add nuance to those choices, which may help identify the degree of Knightian uncertainty facing entrepreneurs.

### 3.3.2 Venture Financing

Knight (1921) did not posit what became known as Knightian uncertainty out of mere intellectual curiosity. Instead, he was interested in entrepreneurial risk-bearing—that is, why did they have to bear risks and could not diversify in some way or access some kind of insurance products? He argued that when probabilities could not be established, such markets would be impossible. Bewley (2002) and then Rigotti and Shannon (2005) explored this formally and concluded that uncertainty could prevent what might otherwise be gains to trade in insurance from being realized, and as noted further by Bewley (1989), this could mean that entrepreneurs would not be able to diversify.

To examine this within the context of this model, suppose that the entrepreneur is liquidity constrained and so must raise $C$ from an investor to launch. They will offer an investor an equity share, $1 - \alpha$ with $\alpha \in (0, 1]$, in return for a $C$ investment. If $V$ is realized, the investor earns $(1 - \alpha)V - C$, and the entrepreneur earns $\alpha V$. What $\alpha$ will the entrepreneur offer?

Assume that there are many investors who each have an outside investment opportunity for $C$ normalized to 0. This is the investor's default. Therefore, an uncertainty-averse investor will only invest if:

$$\underline{p}(1-\alpha)V \geq C.$$

This means that the upper bound on $\alpha$ would be:

$$\alpha \leq \hat{\alpha} \equiv \frac{\underline{p}V - C}{\underline{p}V}.$$

Note that if they can secure an investor, an entrepreneur will always want to launch so long as $\hat{\alpha}V \geq 0$. Note that:

$$\hat{\alpha} \geq 0 \Longrightarrow \frac{\underline{p}V - C}{\underline{p}V} \geq 0.$$

That is, an investment contract will be accepted if and only if $\underline{p}(1-\alpha)V \geq C$—that is, an $N$-default entrepreneur would choose to invest.

There are several important things to note here. First, investor preferences drive whether financing occurs. Thus, the existence of Knightian uncertainty for an investor whose alternative options are safe assets with a payoff normalized to 0 will constrain their willingness to participate in funding the venture. Second, the entrepreneur's preferences, optimism, or decisiveness do not play a role. Even if the entrepreneur were $N$ default, the investor's attitude toward Knightian uncertainty would drive the outcome. This reinforces the notion that Knightian uncertainty will constrain entrepreneurial activity.

Perhaps the most striking feature of this analysis is that the broad conclusions that motivate the entrepreneurial finance literature carry over to an environment where there is Knightian uncertainty.

- Staged investment: As Nanda and Rhodes-Kropf (2016) demonstrate, ventures often have staged investment that releases funds condition on reaching certain milestones that themselves involve the resolution of some uncertainty. This is akin to investing a smaller amount than $C$ in return for a signal of venture success. As already demonstrated, this will cause investors to update their priors—specifically, $\underline{p}$—and thereby create conditions that allow them to rank a wider set of possibilities and make corresponding funding choices.
- Experimental choice: Gans (2022) shows that an entrepreneur who places a high prior probability on venture success will find it optimal to choose a high-bar experiment (that sends a clear signal of venture success) to inform potential investors and, should that experiment be successful, allow them to

accept a lower equity share in return for that investment. Here, the value of such experiments would be to allow investors to rank opportunities where, otherwise, $\underline{p}V < C$. Thus, the presence of Knightian uncertainty reinforces the prediction that entrepreneurs looking to secure funds from more cautious investors will choose experiments that set a high bar for success. As noted by Gans (2022), one example of such an experiment would be a minimum viable product (as recommended by Ries, 2011) that has a low probability of success but, if successful, sends a clear signal that the venture would be successful.

- Venture debt: In an unpublished paper, Rigotti (2004) considered an environment where Knightian uncertainty was over whether a venture would generate value of $V$ or some lower amount $v$, that might be sufficient to generate a positive net present value for investors. He demonstrated that, in that case, rather than offering pure equity to an investor, a more debt-like contract with higher investor equity if $v$ is realized than if $V$ is realized. This is because any equity for the outcome $V$ is of little value in persuading investors to invest. This outcome might be achieved either by an entrepreneur using their own wealth as collateral or by offering an equity contract with a put that allows the entrepreneur to repurchase equity at a fixed price should they so choose. Again, this is the type of structure that might arise if the investor and entrepreneur held heterogeneous beliefs regarding venture success, as in Agrawal et al. (2025).

In summary, many of the predictions of models with heterogeneous beliefs apply to a model where there is Knightian uncertainty. Importantly, the theoretical predictions of what happens to financial contracts and outcomes as beliefs become less heterogeneous are identical to the predictions when there is a lower degree of Knightian uncertainty in the sense that this increases $\underline{p}$. Whether there are settings in which the predictions might diverge is left for future research.

### 3.3.3 What Drives Strategic Choice?

Thus far, the uncertainty explored here has been with regard to the value of the underlying idea or entrepreneurial opportunity. However, it is sometimes the case that although the underlying idea of the venture can be evaluated with more certain probability distributions, the strategic options facing the entrepreneur are harder to assess. For instance, an entrepreneur may have an opportunity to offer two product types. One product type may be "safe" with an expected return that is easy to identify, whereas another product type may be "risky" or, in the spirit of this paper, be subject to Knightian uncertainty. This implies that when choosing between distinct strategies, the choice itself will be subject to Knightian uncertainty, as the return on one strategy is generated by a distribution from a large set. The question I examine here is how this impacts entrepreneurial choices.

To formalize this choice, suppose that the strategy space for the entrepreneur is to launch the venture with strategy $A$, launch the venture with strategy $B$, or not launch the venture at all ($N$). The final choice generates a return of 0. For strategy $A$, the probability of success is known at $p_A$ and has a cost of $C_A$ and a potential return of $V_A$. For strategy $B$, the probability of success, $p_B$, comes from a set of probability distributions over the range $p_B \in [\underline{p}_B, \bar{p}_B]$ with cost $C_B$ and potential return of $V_B$. It is assumed that $V_A < V_B$ and $C_A < C_B$. Thus, $B$ is more costly but also has a higher potential upside than $A$. To focus on the interesting case, it will be assumed that $p_A V_A > C_A$ and $p_A V_A - C_A \in (\underline{p}_B V_B - C_B, \bar{p}_B V_B - C_B)$.[15]

Recall that for two options to be ranked, one must have a higher expected payoff than the other for all distributions, $f \in \mathcal{F}$. Therefore, it is possible to specify two better-than sets:

1. $A \succ B$ if $p_A V_A - C_A > \bar{p}_B V_B - C_B$
2. $B \succ A$ if $\underline{p}_B V_B - C_B > p_A V_A - C_A$.

Under the assumptions given here, both of these inequalities can be satisfied. Effectively, neither $A$ nor $B$ can be ranked due to the uncertainty over the outcomes associated with $B$, but each could be preferred if it was the default option.

Interestingly, this does not mean that the model is devoid of empirical predictions. Note that $A$ is more likely to be chosen if $p_A V_A$ is higher and/or if $C_B - C_A$ is higher.[16] Thus, some comparative statics predictions are possible. In addition, as uncertainty is reduced, the prediction of the model becomes clearer in that variations in $V_B$ become more correlated with choice outcomes compared with other "known" parameters. Therefore, there is a sense in which the uncertainty of the underlying environment can be inferred from behavioral observations.

### 3.4 Effectuation

In recent decades, the area of entrepreneurship research that has been built most squarely on a foundation of Knightian uncertainty is effectuation theory. Proposed by Sarasvathy (2001), this theory argues that the entrepreneurial process of decision-making does not proceed according to the approaches grounded in traditional economics. The traditional approach involves the entrepreneur making clear predictions about future variables of interest and then conducting cost–benefit analyses to choose actions with the highest rate of return. By contrast, Sarasvathy argued that entrepreneurs face Knightian uncertainty that makes such prediction and, hence, cost–benefit analysis difficult. In lieu of that, entrepreneurs take actions that build on resources and relationships under their control with the intention of adjusting as uncertainty is resolved. This theory leads to a number of implications that can be distilled by an in-depth analysis of entrepreneurial decision-making.

The methodology used by those examining effectuation processes is distinct from those who use an economically oriented methodology.[17] This makes it difficult to clearly understand and delineate when effectuation-type outcomes will result as compared with other outcomes. Having proposed a formal approach to modeling Knightian uncertainty in entrepreneurship research, my goal here is to examine whether a greater degree of Knightian uncertainty will likely lead entrepreneurs to behave in a manner more consistent with effectuation theory than other approaches.

Although effectuation theory has a number of predictions, I will focus here on two broad implications:

1. Entrepreneurs following an effectuation process will utilize existing resources and networks rather than attract additional resources.[18]
2. Entrepreneurs will focus on affordable loss rather than maximizing the rate of return in selecting options.

Both of these processes contrast with the Bayesian entrepreneurship processes, which focus on maximizing the rate of return for an idea by optimizing all potential resources and capabilities and actively experimenting before choosing actions.

To examine this, consider our earlier model where there are two options for the entrepreneur, $A$ and $B$, should they launch a venture. Several adjustments are made to that model. First, both paths involve Knightian uncertainty, with the range of distributions for option $A$ being $[\underline{p}_A, \bar{p}_A]$ with $\bar{p}_A > \underline{p}_A$. Second, the entrepreneur has their own resources that are sufficient to pursue option $A$; specifically, the entrepreneur has wealth, $W$, with $C_A < W$. Wealth has alternative value to the entrepreneur of $u(W)$. Third, option $B$ requires resources. It is assumed here that these resources cannot come from the entrepreneur's wealth—that is, their wealth is made up of resources and relationships that can only be used in pursuit of option $A$.[19] Consequently, the entrepreneur must secure outside funding to pursue $B$.

In this situation, consuming their own wealth is a natural default option for entrepreneurs, earning them $u(W)$. Importantly, this default payoff can be achieved if the entrepreneur secures outside resources for the venture, but as will be argued below, this comes at the cost of ceding equity to others.

Should the entrepreneur pursue option $A$, given the Knightian uncertainty and the default option, the entrepreneur would only choose $A$ over not launching if:

$$A \succ N \text{ if and only if } \underline{p}_A V_A + u(W - C_A) > u(W).$$

$u(W) - u(W - C_A)$ can be regarded as a measure of the entrepreneur's *affordable loss*. Thus, it can be seen that as Knightian uncertainty is reduced, to the extent this increases $\underline{p}_A$, the entrepreneur's decision calculus shifts from being solely driven by affordable loss to being driven by potential returns, $V_A$, as well.

Now consider option $B$. The entrepreneur must raise $C_B$ from outsiders. As shown earlier, outsiders will base their contributions of resources on $\underline{p}_B$. Thus:

$$\hat{\alpha} = \frac{\underline{p}_B V_B - C_B}{\underline{p}_B V_B}.$$

Note that, so long as $\underline{p}_B > C_B/V_B$, the entrepreneur will prefer option $B$ to $N$—that is:

$$B \succ N \text{ if and only if } \underline{p}_B \frac{\underline{p}_B V_B - C_B}{\underline{p}_B V_B} V_B + u(W) > u(W) \implies \underline{p}_B V_B > C_B.$$

Intuitively, unlike option $A$, the entrepreneur does not forgo the option of consuming their wealth.

This shows, interestingly, that the entrepreneur's affordable loss is *lower* for option $B$ (attracting new resources) than for option $A$ (using existing resources). Nonetheless, there is a sense in which it could be argued that the existing resources are considered too fungible in this instance. Suppose instead that $W$ represents the entrepreneur's time and effort. For option $A$, we assumed that $W$ could be used to mitigate $C_A$, whereas for $B$, suppose that $W$ must be forgone to some degree, leaving $w < W$ for consumption, but it remains that the entirety of $C_B$ must be procured from outsiders. In that case, we have:

$$B \succ N \text{ if and only if } \underline{p}_B \frac{\underline{p}_B V_B - C_B}{\underline{p}_B V_B} V_B + u(w) > u(W)$$

$$\implies \underline{p}_B V_B > C_B > u(W) - u(w).$$

Thus, the affordable loss from $B$ is $u(W) - u(w)$. With this condition, the affordable loss for $A$ can be lower than that for $B$ if $u(W - C_A) > u(w)$. The point here is that affordable loss in of itself does not lead us to conclude that using resources on hand is preferable to procuring resources.

We have shown now that (a) as there is a greater degree of Knightian uncertainty, entrepreneurial choices will be driven more by what might be considered affordable loss (how much of their own utility will they be willing to bet on the venture's success) than other aspects of the rate of return (such as $V_A$ or $V_B$) and (b) affordable loss itself is not a clear driver of a bias toward using resources at hand rather than procuring resources. The question remains whether the degree of Knightian uncertainty can drive such a bias.

One way such a bias could arise is through defaults. The model here starts from the assumption that the entrepreneur's default option is not to launch the venture. This is consistent with the emphasis on affordable loss as a concept guiding decision-making. The entrepreneur faces Knightian uncertainty regarding both options and hence will only choose either option if it is preferred not to launch for all possible distributions.

This was assumed because, as emphasized by Sarasvathy and Kotha (2001), such uncertainty is a constraint for the entrepreneur in considering how to deploy their own resources. But what if we relaxed this assumption and assumed, as in our earlier model, that the entrepreneur does not face Knightian uncertainty on option $A$ and, instead, assesses the probability of success as $p_A$. Assume also that, at this probability, $p_A V_A + u(W - C_A) > u(W)$. Then, $A$ becomes a natural default option for the entrepreneur. In this case, the entrepreneur will stick with that default if:

$$A \succ B \Longrightarrow p_A V_A + u(W - C_A) \geq \underline{p}_B V_B - C_B + u(w).$$

In this case, the impact of Knightian uncertainty reduces to the constraint it places on others in choosing to devote resources to the venture. By contrast, option $A$ does not require the involvement of others to provide resources other than those that already exist. As Knightian uncertainty is reduced, the bias toward using resources at hand is diminished. As Gans (2022) demonstrated, a similar bias can arise in a Bayesian entrepreneurship setting when outsiders hold weaker beliefs on success than the entrepreneur. Ultimately, generating this prediction from effectuation theory using this formal approach to Knightian uncertainty requires additional behavioural assumptions regarding defaults.

In summary, we are left with the conclusion that the formal approach to Knightian uncertainty supports some principles emphasized by effectuation theorists but does not establish a clear relationship between Knightian uncertainty and any bias toward taking advantage of resources at hand rather than procuring resources externality. Indeed, to the extent that Bayesian entrepreneurship assumes that entrepreneurs hold stronger beliefs than others, the very same constraints on procuring external resources and capital arise in each approach with and without Knightian uncertainty.

## 3.5 Conclusion

This paper has demonstrated that incorporating Knightian uncertainty into models of entrepreneurial decision-making need not invalidate the key insights and predictions generated by Bayesian approaches to entrepreneurship. Rather than creating an insurmountable barrier to formal analysis, Knightian uncertainty—when properly modeled using frameworks such as Bewley (2002)—provides additional richness to our understanding of entrepreneurial behavior while preserving important comparative static predictions.

Several key findings emerge from this analysis. First, the value of experimentation remains robust, even in the face of Knightian uncertainty. Although uncertainty may change the specific experiments entrepreneurs choose to run, it does not diminish their fundamental importance in entrepreneurial decision-making. Second, venture financing patterns predicted by Bayesian models continue to hold when accounting for Knightian uncertainty, particularly regarding staged investment and the use of debt-like instruments. Third, strategic choices under Knightian uncertainty exhibit patterns that are empirically distinguishable from choices made under pure risk, providing potential avenues for empirical research.

The analysis also helps bridge some of the methodological divides between formal economic approaches and other entrepreneurship research traditions, particularly effectuation theory. Although the formal model does not perfectly replicate all predictions of effectuation theory, it demonstrates how Knightian uncertainty can generate behavior patterns that align with some effectuation principles while remaining grounded in economic theory.

Here, I have presented one approach to modeling Knightian uncertainty. It was useful because it allowed a way of considering the degree of Knightian uncertainty and its impact on entrepreneurial choices. It is important to note that there are other ways of considering the types of issues arising from non-Bayesian approaches to entrepreneurial decision-making.[20] One fruitful approach would be to consider games with unawareness. These are situations where some players are unaware of states that other players are aware of. This has a natural home in entrepreneurship research, as the entrepreneur can be considered as the agent who is aware, whereas others (including resource holders, competitors, and complementors) are unaware. Bryan et al. (2022) have been instrumental in bringing these considerations into strategy research, and there is a rich vein from that basis that entrepreneurial researchers can surely tap.

Ultimately, this paper suggests that treating Knightian uncertainty as a formal constraint on decision-making rather than an insurmountable barrier to analysis can enrich our understanding of entrepreneurship while maintaining the benefits of formal economic modeling. The challenge for future research is to continue developing frameworks that capture the unique aspects of entrepreneurial uncertainty while generating clear, testable predictions about entrepreneurial behavior.

## Notes

1. Bhidé (2024) argues that Knightian uncertainty has been neglected in formal economic research, although I would contend that this is, in part, caused by the difficulty of integrating it into formal models of any kind. That said, many formal decision theorists have examined these issues, particularly ambiguity; see the review by Gilboa and Marinacci (2013). As Packard et al. (2017) note, by launching a venture, uncertainty is eventually resolved, and thus it is fruitful to examine the changing nature of uncertainty as part of the process of studying entrepreneurial decision-making.

2. See Packard et al. (2021) for a discussion. Recently, Kay and King (2020) took this further and focused on "radical uncertainty," which involves situations where not all outcomes can be described and, consequently, probabilities cannot even be subjectively formed. This is also akin to Sarasvathy and Kotha's (2001) third urn.

3. The reasons Knight gave were arguably related to the standard reasons for the nonexistence of insurance markets grounded in adverse selection and moral hazard; see LeRoy and Singell (1987). That said, others, notably Bhidé (2021), disagree. My purpose here is not to enter the fray as to what Knight meant but rather to use his discussion as a springboard to the examination of uncertainty for entrepreneurial choice.

4. See also Gans (1996).

5. This theory was originally written in 1986 but was unpublished for many years, and its companion piece (Bewley, 1987) was never published.

6. This is paraphrased from Keynes (1937, p. 214): "About these matters there is no scientific basis on which to form any calculable probability whatever. We simply do not know."

7. Bewley (1987) proposed the preference condition in this strict form, but subsequent researchers have shown that it could be relaxed to include weak preferences; see the discussion in Rigotti and Shannon (2005). For example, the decision-maker could be indifferent to two actions if the expected payoff was the same under each distribution. See also Ok et al. (2012) and Gerasimou (2018).

8. Proposition 1 of Rigotti and Shannon (2005).

9. This bears some relationship to the $\epsilon$-contamination approach of Nishimura and Ozaki (2004), although their application is markedly different.

10. Sutton (2006) who adopted a framework based on the unmeasurability of a subset of states used an evolutionary framework that allowed the entry of agents with a fixed action preference so long as it could be justified by one of the possible distributions.

11. Bewley (2002) also considered bounded rationality with the status quo defined by a previous heuristic choice that can be reconsidered in light of a changed decision environment. Other researchers examining this model in light of trade viewed the status quo as involving no trade (Rigotti and Shannon, 2005) or, in a contracting environment, the individual rationality payoff of the decision-maker (Lopomo et al., 2011).

12. This also implies that defaults will play a role in the assessment of other innovation domains. For instance, although start-ups might be viewed as having an $L$ default, incumbents could be viewed in the opposite manner, not simply because of their preferences but also because they earn profits in the status quo; see Gans (2023).

13. This seems implied by, for instance, Shackle (1961) and Wiltbank et al. (2006).

14. Given the setup of the space of distributions in this entrepreneurial context, the distribution of interest that is impacted on by the experiment is the same before and after the experiment. This is because it defines the threshold, and as Bewley (2002) showed, there always is a distribution that defines the "better than set" for the decision-maker.

15. Such settings with strategic choice are explored by Gans et al. (2019), Agrawal et al. (2021), and Chavda et al. (2024). Their focus is on the conflation of signals regarding strategy and the value of the underlying idea and how this impacts the type of search processes in lieu of opportunities for strategic pivots.

16. Ching et al. (2019) informally consider this using what they term "Rosenbergian uncertainty," where the returns from different strategies facing entrepreneurs are not ranked. They then posit that, in such cases, choice will be driven by known factors such as costs. This, however, is imprecise, as it effectively treats the entrepreneur as indifferent with regard to future returns rather than unable to rank them and having incomplete preferences. The formal model here suggests that the empirical predictions associated with incomplete preferences are more subtle.

17. The terminology is also different, and I will set aside there the often quite persuasive terms that have been deployed in effectuation theory to describe its core ideas.

18. This is sometimes referred to as a "bird-in-the-hand" principle but is also related to a "pilot-in-the-plane" principle.

19. This is not an essential assumption, and it can be relaxed to assume that $C_B$ can be mitigated by the entrepreneur's wealth but that this still requires outside funding.

20. This is demonstrated clearly in the work of Ehrig and Schmidt (2022) and Ehrig and Foss (2022). Another approach is to consider ambiguity more directly, as was done by Amarante et al. (2012a) and Amarante et al. (2012b).

## References

Agrawal, A., Camuffo, A., Gambardella, A., Gans, J. S., Scott, E. L., and Stern, S. (2025). The foundations of Bayesian entrepreneurship. *Bayesian entrepreneurship.* MIT Press.

Agrawal, A., Gans, J. S., and Stern, S. (2021). Enabling entrepreneurial choice. *Management Science*, 67(9), pp. 5510–5524.

Allais, M. (1953). Le comportement de l'homme rationnel devant le risque: Critique des postulats et axiomes de l'école américaine. *Econometrica*, 21(4), pp. 503–546.

Amarante, M., Ghossoub, M., and Phelps, E. S. (2012a). *Contracting for innovation under ambiguity.* Center on Capitalism and Society Working Paper (75).

Amarante, M., Ghossoub, M., and Phelps, E. S. (2012b). *Entrepreneurship, ambiguity, and the shape of innovation contracts.* Center on Capitalism and Society Working Paper (76).

Anscombe, F. J., and Aumann, R. J. (1963). A definition of subjective probability. *Annals of Mathematical Statistics*, 34(1), pp. 199–205.

Arora, A., and Fosfuri, A. (2005). Pricing diagnostic information. *Management Science*, 51(7), pp. 1092–1100.

Arrow, K. J., and Hurwicz, L. (1972). An optimality criterion for decision-making under ignorance. *Uncertainty and Expectations in Economics*, 1.

Arrow, K. J., and Hurwicz, L. (1977). An optimality criterion for decision-making under ignorance. In K. J. Arrow and L. Hurwicz (Eds.), *Studies in resource allocation processes* (pp. 463–471). Cambridge University Press.

Aumann, R. J. (1962). Utility theory without the completeness axiom. *Econometrica*, 30(3), pp. 445–462.

Bewley, T. F. (1987). *Knightian decision theory, part II: Intertemporal problems*. Technical Report. Yale University.

Bewley, T. F. (1989). *Market innovation and entrepreneurship: A Knightian view*. Technical Report.

Bewley, T. F. (2002). Knightian decision theory, part I. *Decisions in Economics and Finance*, 25(2), pp. 79–110.

Bhidé, A. (2021). *Renewing Knightian uncertainty: A pragmatic prospectus and demonstration*. Harvard Business School. Working Paper (21–129).

Bhidé, A. (2024). *Uncertainty and enterprise: Venturing beyond the known*. Oxford University Press.

Bryan, K., Ryall, M., and Schipper, B. C. (2022). Value capture in the face of known and unknown unknowns. *Strategy Science*, 7(3), pp. 157–189.

Camuffo, A., Gambardella, A., and Pignataro, A. (2024). Theory-driven strategic management decisions. *Strategy Science*, 9(4), pp. 382–396.

Chavda, A., Gans, J. S., and Stern, S. (2024). Theory-based entrepreneurial search. *Strategy Science*, 9(4), pp. 397–415.

Ching, K., Gans, J. S., and Stern, S. (2019). Control versus execution: Endogenous appropriability and entrepreneurial strategy. *Industrial and Corporate Change*, 28(2), pp. 389–408.

Ehrig, T., and Schmidt, J. (2022). Theory-based learning and experimentation: How strategists can systematically generate knowledge at the edge between the known and the unknown. *Strategic Management Journal*, 43(7), pp. 1287–1318.

Ehrig, T., Felin, T., and Zenger, T. R. (2025). Causal reasoning and the scientific entrepreneur: Beyond Bayes. In A. Agrawal, A. Camuffo, A. Gambardella, J. S. Gans, E. L. Scott, and S. Stern (Eds.), *Bayesian entrepreneurship*. MIT Press.

Ehrig, T., and Foss, N. J. (2022). Why we need normative theories of entrepreneurial learning that go beyond Bayesianism. *Journal of Business Venturing Insights*, 18, e00335.

Ellsberg, D. (1961). Risk, ambiguity, and the savage axioms. *The Quarterly Journal of Economics*, 75(4), pp. 643–669.

Felin, T., and Zenger, T. R. (2017). The theory-based view: Economic actors as theorists. *Strategy Science*, 2(4), pp. 258–271.

Gans, J. S. (1996). On the impossibility of rational choice under incomplete information. *Journal of Economic Behavior & Organization*, 29(2), pp. 287–309.

Gans, J. S. (2022). *A theory of visionary disruption*. Technical Report.

Gans, J. S. (2023). Experimental choice and disruptive technologies. *Management Science*, 69(11), pp. 7044–7058.

Gans, J. S., Scott, E. L., and Stern, S. (2024). *Entrepreneurship: Choice and strategy*. W. W. Norton & Company.

Gans, J. S., Stern, S., and Wu, J. (2019). Foundations of entrepreneurial strategy. *Strategic Management Journal*, 40(5), pp. 736–756.

Gerasimou, G. (2018). On the indifference relation in Bewley preferences. *Economics Letters*, 164, pp. 24–26.

Gilboa, I., and Marinacci, M. (2013). Ambiguity and the bayesian paradigm. In D. Acemoglu, M. Arellano, and E. Dekel (Eds.), *Advances in economics and econometrics* (pp. 179–242). Cambridge University Press.

Gilboa, I., and Schmeidler, D. (1989). Maxmin expected utility with a non-unique prior. *Journal of Mathematical Economics*, 18, pp. 141–153.

Kay, J., and King, M. (2020). *Radical uncertainty: Decision-making beyond the numbers*. W. W. Norton & Company.

Keynes, J. M. (1921). *A Treatise on probability*. Macmillan.

Keynes, J. M. (1937). The general theory of employment. *Quarterly Journal of Economics*, 51(2), pp. 209–223.

Knight, F. H. (1921). *Risk, uncertainty, and profit*. Houghton Mifflin.

LeRoy, S. F., and Singell, L. D., Jr. (1987). Knight on risk and uncertainty. *Journal of Political Economy*, 95(2), pp. 394–406.

Lopomo, G., Rigotti, L., and Shannon, C. (2011). Knightian uncertainty and moral hazard. *Journal of Economic Theory*, 146(3), pp. 1148–1172.

Montiel Olea, J. L., Ortoleva, P., Pai, M. M., and Prat, A. (2022). Competing models. *The Quarterly Journal of Economics*, 137(4), pp. 2419–2457.

Nanda, R., and Rhodes-Kropf, M. (2016). Financing entrepreneurial experimentation. *Innovation Policy and the Economy*, 16(1), pp. 1–23.

Nishimura, K. G., and Ozaki, H. (2004). Search and Knightian uncertainty. *Journal of Economic Theory*, 119(2), pp. 299–333.

Ok, E. A., Ortoleva, P., and Riella, G. (2012). Incomplete preferences under uncertainty: Indecisiveness in beliefs versus tastes. *Econometrica*, 80(4), pp. 1791–1808.

Packard, M. D., Bylund, P. L., and Clark, B. B. (2021). Keynes and Knight on uncertainty: Peas in a pod or chalk and cheese? *Cambridge Journal of Economics*, 45(5), pp. 1099–1125.

Packard, M. D., Clark, B. B., and Klein, P. G. (2017). Uncertainty types and transitions in the entrepreneurial process. *Organization Science*, 28(5), pp. 840–856.

Ries, E. (2011). *The lean startup: How today's entrepreneurs use continuous innovation to create radically successful businesses*. Currency.

Rigotti, L. (2004). *Decisive entrepreneurs and cautious investors*. Duke University Working Paper.

Rigotti, L., and Shannon, C. (2005). Uncertainty and risk in financial markets. *Econometrica*, 73(1), pp. 203–243.

Ryall, M. D. (2009). Causal ambiguity, complexity, and capability-based advantage. *Management Science*, 55(3), pp. 389–403.

Sarasvathy, S. D. (2001). Causation and effectuation: Toward a theoretical shift from economic inevitability to entrepreneurial contingency. *Academy of Management Review*, 26(2), pp. 243–263.

Sarasvathy, S. D., and Kotha, S. (2001). Managing Knightian uncertainty in the new economy. In B. Hunt and S. Kotha (Eds.), *Entrepreneurship and the Internet era: Cases and contexts* (pp. 31–50). Elsevier Science.

Savage, L. J. (1954). *The foundations of statistics*. John Wiley & Sons.

Shackle, G. L. S. (1961). *Decision, order and time in human affairs*. Cambridge University Press.

Sutton, J. (2006). *Flexibility, profitability and survival in an (objective) model of Knightian uncertainty*. Technical Report London School of Economics.

Vaswani, A., Shazeer, N., Parmar, N., Uszkoreit, J., Jones, L., Gomez, A. N., Kaiser, L., and Polosukhin, I. (2017). Attention is all you need. *Advances in Neural Information Processing Systems (NeurIPS)*, 30, pp. 5998–6008.

Von Neumann, J., and Morgenstern, O. (1944). *Theory of games and economic behavior*. Princeton University Press.

Wiltbank, R. N., Dew, N., Read, S., and Sarasvathy, S. D. (2006). What to do next? The case for non-predictive strategy. *Strategic Management Journal*, 27(10), pp. 981–998.

# 4

## Bayesian Entrepreneurship: A User's Perspective

Ashish Arora

### 4.1 Introduction

Entrepreneurship scholars have long searched for a framework that is useful to guide empirical work, especially with observational data, but still descriptively close enough to how entrepreneurs work. There is a vast literature on how decision-makers deal with uncertainty (where probabilities cannot be calculated or estimated) and how one ought to do so, which clearly bears on entrepreneurship research. As an outsider to this literature, it appears to me that there is no clear consensus on how best to proceed if one abandons the expected utility maximization framework, which forms the basis of the Bayesian entrepreneurship approach, particularly if one wants to go beyond verbal descriptions.

Accordingly, I shall say little on the matter other than I find it acceptable, although not comfortable, to stick to the assumption that somehow entrepreneurs form estimates of the quantities of interest—the payoff from a successful venture, the cost of launching the venture, and the prior likelihood of success. But although Bayesian entrepreneurship does not break much new ground on these fundamentals of decision-making, it is not simply old wine in new bottles. To think so is to miss out on some important differences, noteworthy for the promise they hold out.

Perhaps the most important improvement is to do away with common priors. Although many have paid lip service to the importance of differences in beliefs across individuals, when it comes to analytical models, we (at least those of us who do write models) typically seek the safe harbor of common priors. The recourse to common priors involves more than simply a hankering for the familiar. Common priors may be unrealistic, but abandoning them can open up too many degrees of freedom for the researcher. How should one proceed without common priors? Fortunately, it appears that there are ways forward, as described in this volume. Although the technical niceties escape me, I take comfort that experts are engaged in fleshing them out, so that entrepreneurship scholars can have recourse to a principled way of modeling differences in priors.

An immediate implication of allowing for differences in priors is that it leads to the reasonable assumption that entrepreneurs are more optimistic about the prospects of their venture, for otherwise they would not have started the venture in the first instance. As the introductory chapter correctly points out, this is not a personality trait but instead specific to the prospects of the venture.

Another important improvement is the focus on experimentation. The process of entrepreneurship is about resolving uncertainty, and prior beliefs affect not only whether one experiments but also how one experiments. Insofar as entrepreneurs are optimistic about the prospects of their venture (else they would not have started it), they ought to favor low-bar experiments. This should not come as a surprise. Information is valuable only insofar as it causes one to do differently (with some probability) what one would have done in the absence of information. Gans (2023) shows, more directly, optimists should favor experiments that would reliably result in them abandoning the venture (low-bar experiments), whereas skeptics should favor high-bar experiments.[1]

And so they should, but do they? In asking this question, we are traversing from a normative theory into a positive one. Let $V$ be the payoff from a successful venture and $C$ the cost of launching it. Suppose a potential entrepreneur is more optimistic than others about the prospect of her venture, but correctly so. The entrepreneur's prior is $p$ and that of the skeptic is $q$, with $p > q$. Further, suppose, before launch, one can experiment. The probability of a positive signal (good news) if the venture is good is $m_1$ and bad news if the venture is bad is $m_0$. For simplicity, suppose there are two types of experiments, one where $m_1 = 1, m_0 = 1/2$, and another where $m_1 = 1/2, m_0 = 1$. For a low-bar experiment $\{m_1 = 1, m_0 = 1/2\}$, the probability the venture is launched is $p + 1/2(1 - p)$, whereas for the high bar experiment it is smaller, $1/2p$. Moreover, low-bar experiments ought to result in lower performance conditional on not abandoning the venture compared with high-bar experiments. Put slightly differently, Bayesian entrepreneurs, if they do choose low-bar experiments, ought to abandon their venture (or quit) at lower rates, but conditional on not abandoning the venture, they ought to have lower success rates than if they choose high-bar experiments.

There is little direct evidence of which I am aware. The evidence that is available is inconclusive. Camuffo, Gambardella, Messinese, et al. (2024) find that scientific entrepreneurs—Bayesian entrepreneurs trained to form a theory of their venture—are more likely to quit than the control group of entrepreneurs, and do so sooner than quitters from the control group. More encouragingly, scientific entrepreneurs who persist outperform the control group entrepreneurs who persist. Sandy Yu's (2020) analysis using observational data on accelerators similarly shows start-ups in accelerators are more likely to quit than comparable firms not admitted to accelerators. A recent randomized controlled trial by Lee et al. (2024) also indicates that training founders of science- and technology-based start-ups leads to higher quit rates and higher performance conditional on not quitting.

One might wonder if perhaps even entrepreneurs have low enough priors that the optimal experiment is a high-bar experiment. However, for this to be consistent with the empirical evidence, one would have to believe that untrained entrepreneurs (the control group) choose the wrong type of experiment—that is, low bar instead of high bar. Overall, the verdict on Bayesian entrepreneurship's performance as a positive theory is, as my high school report card often read, "Could do better."

To me, the evidence suggests some combination of the following possibilities:

1. Bayesian entrepreneurs are not commonly found in the wild.
2. Entrepreneurial experiments are not optimally chosen.
3. Entrepreneurship training is more about the precision of signals than the choice of experiments.

It may also be that entrepreneurs over estimate payoffs and under estimate costs. An earlier paper by Arora and Nandkumar (2011) on cyber security start-ups found that higher opportunity cost was related to higher quit rates and faster failure as well. Bolton et al. (2024) reach a similar conclusion, albeit where entrepreneurs with higher private value for the start-up are less likely to quit, but consequently also less likely to attract investment on favorable terms. In this view, training helps entrepreneurs form more realistic estimates of costs and benefits, in effect raising the threshold for persisting with the venture.

Clearly, there is more work to be done to sort out experimentation and information gathering, on the one hand, and costs and benefits, on the other. Heterogeneity in costs and benefits can lead to many of the same patterns as differences in priors. For instance, selection into entrepreneurship would imply that, on average, entrepreneurs have lower opportunity cost. Further, this would also affect choice of experiments: An entrepreneur with a low opportunity cost is similar to an "optimist" in terms of choosing low-bar experiments. This is, of course, evident in the introductory chapter itself but perhaps is less salient than one might have wanted.

## 4.2 The Road Ahead

Insofar as the process of entrepreneurship is about experimenting to learn about the venture's prospects, Bayesian entrepreneurship usefully reminds us of the key factors that determine the value of experiments. Whether to experiment and what hypotheses to test depend on one's priors, the importance or criticality of the hypotheses that can be tested, the precision and specificities of experiments available (e.g., what if skeptics can only get experiments with $x' < x$?), and, of course, the opportunity cost of experiments. In particular, if a particular type of experiment forecloses some options, the choice between different types of experiments may not simply depend on priors.

Some further observations follow. Specific and meaningful hypotheses make experiments more helpful. Put differently, a theory of the venture, which both generates these hypotheses and provides a sense of how critical the hypothesis is to the venture's eventual success, may be very useful. The Bocconi group (e.g., Camuffo, Gambardella and Pignataro, 2024) has made some important advances in this regard. Understanding how entrepreneurs develop (and ought to develop) such theories, test them, and modify the theories is a promising area for further work. For instance, experiments are useful not just to inform the entrepreneur but also to persuade stakeholders such as investors and customers. The priors of stakeholders, if they differ from those of the entrepreneur, will affect the choice of experiment, but also perhaps so will differences between the entrepreneur and stakeholders in their theory of the venture.

Second, entrepreneurs are not just scientists—they are also engineers. They seek not only to learn but also to do. Although the framework explicitly disavows "effectuation," I believe this is a missed opportunity. "Grind it out," the advice often given to entrepreneurs, is not just about efficiently acquiring information about the venture's prospects; it suggests that entrepreneurs can (or think they can) sometimes spin gold from straw, especially when a lot is unknown or unknowable (at that time). It would be worthwhile to extend the model where perhaps entrepreneurs can, by expending effort, increase the likelihood of success (or, equivalently, reduce the cost of launching the venture).[2]

More generally, it would be very useful to further develop what is now a normative, decision-theoretic approach into a positive, equilibrium theory. This may require specializing the approach to deal with types of entrepreneurship of interest. For instance, there is growing recognition that the current start-up-based innovation system is markedly lopsided. Digital and biotech start-ups dominate compared with deep-tech start-ups focusing on new materials, climate, and energy (Lerner and Nanda, 2020). Current explanations blame short sighted investors or other types of biases in decision-making. But these are not entirely satisfactory, and I wonder what the Bayesian approach has to offer.

Similarly, many high-growth, innovation-based ventures exit through acquisition. Thus, what also matters for the venture's success is attracting potential acquirers. To this end, as Arora et al. (2021) show, the entrepreneur may choose to provide additional information through experimentation, even when such experimentation may destroy some of the joint surplus shared with an acquirer, as long as it increases the entrepreneur's bargaining power sufficiently. It would be helpful to develop the implications for how experimentation differs (if at all) when the start-up is being developed for acquisition or not.

More generally, a tractable theory that allows researchers to model interactions between entrepreneurs and potential acquirers, or between entrepreneurs and investors, while allowing for differences in priors (or even in theories of the venture) would be very useful. The output of such a theory would include analysis of venture formation and valuation, and how it is conditioned by differences in priors. Insofar as these differences vary with advances in scientific knowledge, or vary with differences in the relative importance of market versus scientific uncertainty, a Bayesian entrepreneurship approach could make a valuable contribution.

## Notes

1. Arora and Fosfuri (2005) also showed, in a different context, optimists, predisposed to persist with the venture, value signals that, when unfavorable, reliably mean that the venture should be abandoned. Skeptics, on the other hand, value signals that, when favorable, reliably mean that the venture is worth pursuing.
2. Persuasion is incorporated into the framework, and persuading stakeholders is like modifying the environment.

## References

Arora, A., and Fosfuri, A. (2005). Pricing diagnostic information. *Management Science*, 51(7), pp. 1092–1100.

Arora, A., Fosfuri, A., and Rønde, T. (2021). Waiting for the payday? The market for startups and the timing of entrepreneurial exit. *Management Science*, 67(3), pp. 1453–1467.

Arora, A., and Nandkumar, A. (2011). Cash-out or flameout! Opportunity cost and entrepreneurial strategy: Theory, and evidence from the information security industry. *Management Science*, 57(10), pp. 1844–1860.

Bolton, P., Liu, S., Nanda, R., and Sundaresan, S. (2024). *Moral hazard in experiment design: Implications for financing innovation*. Working Paper.

Camuffo, A., Gambardella, A., Messinese, D., Novelli, E., Paolucci, E., and Spina, C. (2024). A scientific approach to entrepreneurial decision-making: Large-scale replication and extension. *Strategic Management Journal*, 45(6), pp. 1209–1237.

Camuffo, A., Gambardella, A., and Pignataro, A. (2024). Theory-driven strategic management decisions. *Strategy Science*, 9(4), pp. 382–396.

Gans, J. S. (2023). Experimental choice and disruptive technologies. *Management Science*, 69(11), pp. 7044–7058.

Lee, D., Floyd, E., Hochberg, Y. V., Fehder, D. C., and Bailey-Rihawi, E. (2024). *Learning to quit? A multi-year field experiment with innovation driven entrepreneurs*. Available at SSRN 4865251.

Lerner, J., and Nanda, R. (2020). Venture capital's role in financing innovation: What we know and how much we still need to learn. *Journal of Economic Perspectives*, 34(3), pp. 237–261.

Yu, S. (2020). How do accelerators impact the performance of high-technology ventures? *Management Science*, 66(2), pp. 530–552.

# 5

# Entrepreneurs: Clueless, Biased, Poor Heuristics, or Bayesian Machines?

Thomas Åstebro, Frank M. Fossen, and Cédric Gutierrez

## 5.1 Introduction

Entrepreneurship scholars are interested in understanding and describing how entrepreneurs make decisions under uncertainty, where the probabilities of outcomes are not known but perceived, resulting in ambiguous probabilities. In this context, ambiguity refers to the lack of precise and objective probability assessments and the presence of subjective judgments regarding potential outcomes. In this chapter, we discuss the development of thought on how entrepreneurs perceive and react to uncertainty from Knight (1921) to the present day. Recognizing that entrepreneurs face uncertainty rather than risk and are unlikely to have estimates of all probabilities for all potential outcomes, it becomes difficult to accept expected utility theory (EUT), developed by Savage (1951) and Von Neumann and Morgenstern (1953), as a relevant model for entrepreneurial decision making. We examine a range of decision theories, ranking them in an order starting from EUT and proceeding to the most structure-free models of entrepreneurial choice, allowing for comparisons and contrasts of the main components and underlying concepts as they apply to entrepreneurial decision making.

First order of business is to introduce some definitions of terms. There are three core elements of decision-making (Fischhoff and Broomell, 2020). The first element is judgment, or how people predict the outcomes that will follow possible choices. This can be evaluated in terms of the accuracy of judged probabilities or by their consistency. The most familiar consistency standard is Bayesian inference. Biases distort judgment. The second element is preference, or how people weigh those outcomes. Decision theorists allow individuals to have any preferences but they have a consistency standard: Preferences must satisfy the utility theory axioms. The third element is choice, or how people combine judgments and preferences to reach a decision. This is about the model used to put everything together, and we review a few of these models in this chapter, among which several do not satisfy the utility theory axioms. Heuristics mainly distort choices.[1]

Starting from the left in figure 5.1, one could use EUT to describe how entrepreneurs should choose which action to take. EUT is not a descriptive theory but rather a normative theory, prescribing how an entrepreneur should use information to rank order all options to choose the optimal option under risk. EUT contains a number of axioms that, if fulfilled, allow the decision-maker—the entrepreneur in this case—to rationally choose the best action that, on average (in expectation), delivers the most favored outcome. However, this model's ability to describe decision-making has faced various objections (see, e.g., Starmer, 2000), too numerous to list here. The most problematic aspect of its application in entrepreneurship is that the entrepreneurial decision situation does not meet EUT's data requirements, such as having access to the probabilities of outcomes associated with all alternative options. Entrepreneurial decision-making fundamentally revolves around uncertainty (McMullen and Shepherd, 2006; Townsend et al., 2018), where probabilities of future states, such as market demand or technological opportunities, are unknown (Knight, 1921). Consequently, entrepreneurial scholars have rejected the EUT framework.[2]

However, if this model is rejected, there remains a need for frameworks that help entrepreneurship scholars better understand how entrepreneurs make decisions and assist entrepreneurs in choosing the best possible action among a set of alternatives. Hence, to meet the demands for realism, decision theory researchers have developed alternative theories that address some of the violations of EUT. In addition, these theories describe situations where probabilities are characterized as unknown but subjectively assessable (i.e., ambiguous) or, even more radically, where potential future states and outcomes are not knowable by the decision-maker. In the former case, one can still compare the subjective values of different actions if one can assign self-assessed subjective probabilities for various future states. In the latter case, there are two types of theories to discuss: those in which the entrepreneur is unable to rank order any sets of actions and stands in front of complete (also called absolute) uncertainty, and those in which the entrepreneur is partially able to rank order some but not all potential choices. Finally, an (extreme) alternative would be to completely reject the notion that entrepreneurs evaluate the chances and values of outcomes when making decisions. This is the case of effectuation theory, developed by Sarasvathy (2001), which stands at the opposite end of the spectrum of the theories we will review. We have chosen to review a few prominent theories and a few forgotten but, in our minds, highly relevant alternatives to help understand entrepreneurial choices.

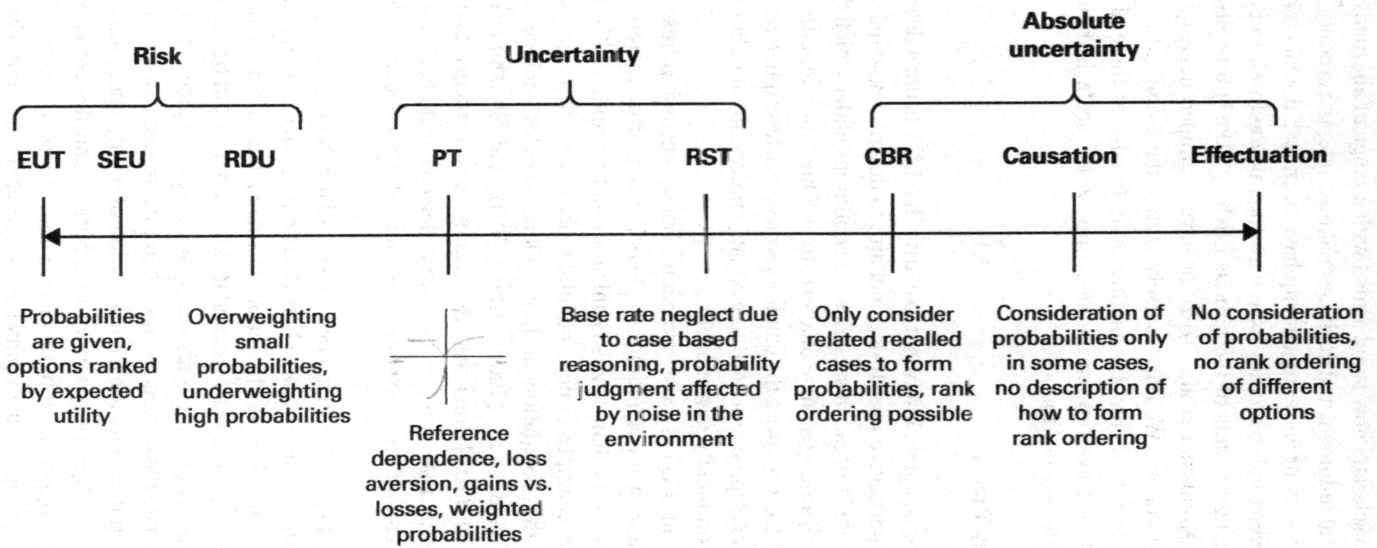


**Figure 5.1**
Theories of (entrepreneurial) probability judgment and decision under risk, uncertainty, and absolute uncertainty. Notes: Picture credits to Daniel Chen. EUT: expected utility theory; SEU: subjective expected utility; RDU: rank-dependent utility. PT: prospect theory; RST: random support theory; CBR: case-based reasoning.

Entrepreneurship scholars now stand armed with a range of alternative theories of entrepreneurial judgment and decision-making under uncertainty. Few of these alternative theories of entrepreneurship have been tested. Only a handful of empirical entrepreneurship studies have expanded beyond risk to behavior under ambiguity, using variants of the famous Ellsberg (1961) urns task (e.g., Holm et al., 2013; Koudstaal et al., 2016), or have examined decision biases associated with uncertainty. We only sparsely review the empirical testing of these theories, saving that review for another time. Finally, at the end of this chapter, we will place Bayesian entrepreneurship on the scale in figure 5.1, which summarizes the various theories.

## 5.2 Expected Utility Theory

Starting with EUT, it is, again, important to clarify that EUT is not a descriptive theory but rather a normative one. It does not aim to describe how individuals usually make decisions. Instead, it specifies the conditions under which the best action can be selected among a range of alternatives. Other theories either modify the conditions of EUT to move the conditions closer to descriptive realism or reject the idea that entrepreneurs need to evaluate choices among alternatives, instead proposing completely new models.

A central tenet in EUT is risk aversion. Entrepreneurship researchers showed an early interest in examining the hypothesis that entrepreneurs are more willing to take risks than others (Kihlstrom and Laffont, 1979). Brockhaus (1980) is one example, where no differences were detected between entrepreneurs and the population at large. Other articles include Cramer et al. (2002), Caliendo et al. (2009), Ahn (2010), and Skriabikova et al. (2014). Åstebro et al. (2014) reviewed the accumulated evidence and found that the differences in risk aversion that were detected between employees and entrepreneurs cannot fully explain why individuals choose to become entrepreneurs.[3]

Knight (1921) told us so. He stated that we should not think of entrepreneurship as simply investment under risk, where decisions are made with respect to an objectively known distribution of returns. He argued that entrepreneurship in such a world would not require any particular skill, and it would be inconceivable that entrepreneurs could earn rents simply for bearing objective risk, as the market should eliminate those rents. Knight therefore argued that the prerequisites for entrepreneurial activity are a combination of highly uncertain returns, which do not have an objectively known distribution, and the entrepreneur's skill in perceiving an opportunity more clearly than others. His work focused attention on the specific individuals pursuing entrepreneurship and what made them distinct, as well as the fact that, in entrepreneurship, the probabilities of outcomes are unknown.

In terms of technical (formal) models of uncertainty versus risk, several recent articles have shown that it really does make a difference how uncertainty is interpreted. Suppose for a moment that "uncertainty" about labor market conditions has increased. Does this induce an individual to search for an entrepreneurial opportunity longer or shorter? In the traditional way to represent uncertainty as risk, where probabilities of outcomes are known, an increase in uncertainty means an increase in the variance of the distribution of entrepreneurial returns around a given mean. A person will then search longer for a higher return because the upside opportunity just increased. However, a person who interprets the news to mean that there is a decrease in her confidence about which earnings distribution among a set of distributions is the true one will instead search for a shorter duration and accept a lower entrepreneurial opportunity earlier (Nishimura and Ozaki, 2004).

Savage (1954) extended EUT to situations where no objective probabilities are available. Under subjective expected Utility (SEU), objective probabilities are replaced with "subjective probabilities," meaning that individuals can form opinions, consciously or not, about the likelihood of events. Although offering a more flexible model adapted to situations of ambiguity, SEU maintains the normative foundation of EUT.

## 5.3 Rank-Dependent Utility

Decision analysts started to recognize a discrepancy between the normative EUT model and the empirical findings on human decision-making behavior. This search for deviations developed into a storm, fueled most prominently by the work of Daniel Kahneman and Amos Tversky and their followers, who took it upon themselves to show, through numerous small-scale experiments, that various inconsistencies with EUT occurred. These inconsistencies, exemplified by the famous Allais paradox (Allais, 1953), were not merely noisy decisions but consistent deviations from EUT. We will describe only some of these deviations below as they pertain to their discussion in entrepreneurship. In this subsection, we describe attempts by theorists to incorporate some of these violations of EUT into a formal model that still allows for the rank ordering of alternative options.

Rank-dependent utility (RDU) was developed to explain why people choose to invest in insurance while also buying lottery tickets. These joint decisions are inconsistent with EUT, as they imply risk aversion in the former case and risk seeking in the latter. To reconcile these choices within a utility framework,

Quiggin (1982) developed a first version of RDU, which was a major breakthrough in decision theory. The main idea of this model is that "the attention given to an outcome depends not only on the probability of the outcome but also on the favorability of the outcome in comparison to the other possible outcomes" (Diecidue and Wakker, 2001, p. 284). In RDU, outcomes are first rank ordered from highest to lowest value. Instead of being weighted by their objective or subjective probabilities, as in EUT or SEU, outcomes are weighted by decision weights that depend on both the probabilities of the events and their ranks among all possible alternatives. The weighting function, which maps probabilities to decision weights, captures the degree of local (i.e., for a certain probability level) risk aversion or risk seeking and is specific to a decision-maker.[4] A common finding in the empirical literature is that this weighting function often has an inverse-S shape: Individuals tend to overweight unlikely extreme outcomes, such as disasters and winning the lottery, but overweight less, or underweight, more common events. This idea later formed one of the central features of cumulative prospect theory (often referred to simply as "prospect theory" or PT) in addition to the incorporation of decision-making under uncertainty (Tversky and Kahneman, 1992).

Why is RDU relevant to entrepreneurship? It turns out that RDU can rationalize why entrepreneurs are insensitive to changes in likelihood over a large range of probabilities of outcomes, focusing instead on extremely lucrative or, alternatively, disastrous outcomes. Verbal representations of this behavior are legion in entrepreneurship, and such narratives have also crept into entrepreneurship theories of various kinds. We may, for example, refer here to the case of Zappos, recounted in Agrawal et al. (2025). In a lab experiment, Laferrière et al. (2023) found that the inverse-S shape of the weighting function can partially explain excess entry in winner-take-all markets, which leads to the skewed distribution of returns in entrepreneurship (Åstebro et al., 2014).

## 5.4 Prospect Theory

Prospect theory, developed by Daniel Kahneman and Amos Tversky, was created to explain systematic deviations from EUT observed in a series of experiments. In 1979, Kahneman and Tversky introduced the first version of PT that they revised in 1992 to extend and refine the original theory, reflect additional research findings, and relabel some of the concepts (Tversky and Kahneman, 1992).

The main components of PT (1979 and 1992 versions combined) are:

1. Reference dependence: This implies that what matters is changes to an earnings position from a reference point, not changes in final wealth. Reference dependence also implies that context matters, as the reference point could change across context.[5]
2. Loss aversion: The disutility of losses is higher than the utility of gains. In other words, losses loom larger than gains.
3. Utility of gains and losses: The utility function is concave for gains and convex for losses.
4. Probability weighting: Probabilities are transformed into decision weights (as in RDU) but can be weighted differently for gains and losses.

The end goal was to construct a model that describes how the decision-maker rank orders different choices. In parallel to constructing the theory, Kahneman and Tversky discussed a list of biases related to probability judgment, defined as the formation of beliefs about the likelihood of events, but did not formally include them in the theory that focuses on the evaluation of uncertain options. The discussed biases include base-rate neglect (i.e., ignoring population-level information) and ordering effects (i.e., taking into account when information is presented). The biases in belief formation were left to a pre-decision "framing" state before the evaluation and decision. In this framing state, decision-makers might decide to discard very unlikely outcomes, focus exclusively on recent decisions or information, make reference to decision situations that the case "represents," or invoke case-based reasoning or base-rate neglect.[6]

Two important updates were made in the 1992 version of PT. First, to resolve a technical issue present in the initial version, the authors incorporated rank dependence. Second, they extended PT to incorporate decision-making under uncertainty. Although Tversky and Kahneman were not the first to develop a model for decisions under uncertainty, they advanced the understanding of how uncertainty may be treated differently from risk. The key insight was that decision weights became "source dependent," meaning they depended on the context of the decision-making process. Hence, in situations of uncertainty, the decision weights could differ from those in contexts of risk and also vary across uncertain decision contexts. Each decision-maker would still have their own weighting function of probabilities, but the weights would be adjusted based on the uncertainty of the data generation process.

Tversky and Kahneman had little empirical evidence when proposing this idea and suggested that "the investigation of decision weights for uncertain events emerges as a promising domain for future research" (Tversky and Kahneman, 1992, p. 317). Abdellaoui et al. (2011) later developed the source method to quantitatively analyze source dependence and found empirical evidence supporting it. In the field of entrepreneurship, Gutierrez et al. (2020) applied the source method to analyze excess entry in markets characterized by uncertain payoffs and skill-based competition, highlighting the critical role of attitudes toward uncertainty beyond mere beliefs. These results align with Wu and Knott's (2006) finding of "apparent risk seeking" in the case of ability uncertainty. However, Wu and Knott attributed this pattern to entrepreneurs' overconfidence in their own skills.

Even though there is empirical evidence of violations of PT (see, e.g., Fox et al., 2015), prospect theory remains one of the most popular theories of decision under uncertainty in terms of its descriptive power, with applications to a broad range of fields (e.g., Barberis, 2013).

## 5.5 Biases, Preferences, and Heuristics in Entrepreneurship

The three theories previously introduced (EUT, RDU, and PT) are theories of decision-making that focus on choices but do not address the formation of beliefs. In other words, they take the probabilities (objective or subjective) as input to the decision model. Hence, they can help explain the behavior of an entrepreneur *conditional* on holding certain beliefs but do not explain how these probabilities are derived or formed in the first place.

However, in the 1980s, interest in the field of entrepreneurship shifted toward concepts associated with probability judgment, such as optimism and overconfidence—an interest that we briefly review below. These concepts align with Knight's (1921) view of entrepreneurs as having unique skills in assessing venture opportunities.

Optimism reflects that all perceived probabilities of positive outcomes are more favorable than objectively measured probabilities. Overconfidence is the exaggerated belief in one's own skills and abilities, often in relation to others, or an excessive faith that you know the truth (see, e.g., Moore and Cain, 2007; Moore and Healy, 2008). Optimism is considered a relatively stable personality trait, whereas overconfidence is more situational. Both have been found associated with entrepreneurship (see Åstebro et al., 2014). More optimistic or overconfident individuals would tend to overvalue the future returns of an entrepreneurial venture. This overvaluation will only affect those who are at the margin of switching—that is, those who are almost indifferent between becoming an entrepreneur or not. Therefore, the overvaluation will bring in a group of marginal performers (see, e.g., Dell'Era et al., 2023).

Surveys illustrate the prevalence of inaccurate beliefs among entrepreneurs. For instance, Cooper et al. (1988) reported that 33 percent of surveyed entrepreneurs rated their odds of success at 10 out of 10, despite giving much lower odds to similar businesses. Shane (2009) showed US entrepreneurs' estimated odds of achieving at least $10 million in sales to be five times higher than empirical data suggest. Whether these survey results reflect overconfidence or optimism is unclear. De Meza et al. (2019) showed that optimism caused lower entrepreneurial earnings due to an increased rate of erroneous entrepreneurial entry. Holm et al. (2013) found that Chinese entrepreneurs were more willing to enter competitive environments reliant on their own skills compared with a control group, suggesting a possible link between overconfidence and entrepreneurial entry. However, the entrepreneurs did not consistently overestimate their performance, indicating that a preference for competition might drive their actions more than biased self-beliefs. Åstebro et al. (2014) highlighted the need for more precise measurement of probability assessment in entrepreneurial contexts to better understand the role of overconfidence and optimism in driving the relationship with entrepreneurship.

Another stream of literature shifted from examining tolerance for risk to exploring tolerance for ambiguity, a concept more closely associated with Knight's (1921) perspective. Already in 1951, Frenkel-Brunswick proposed that "tolerance of ambiguity" is a stable and broad predictive variable across a number of behavioral settings. Among its many instantiations, she proposed a resistance to the reversal of responses to apparent fluctuating stimuli and the early selection and maintenance of one solution in an ambiguous situation. Instead, a person with a high tolerance for ambiguity would perceive ambiguous situations as desirable, challenging, and interesting. Sexton and Bowman (1985) extended these ideas by comparing entrepreneurship students with other types of students, finding that entrepreneurship students had a greater tolerance for ambiguity. However, their results have not held up well when studying entrepreneurs. Furnham and Marks (2013, p. 718) note that "despite work on these subtly different and related concepts there is still no very clear operational definition of Tolerance for Ambiguity at the facet level or a clear differentiation between the manifestations and correlates of Tolerance for Ambiguity." Maybe due to its conceptual unclarity, the construct has not persisted in being used in the field of entrepreneurship. Nevertheless, the idea that an entrepreneur's actions may not respond much to differences in perceived probabilities has survived and has even had a resurgence, as will be described.

Although biases and preferences have received some coverage in entrepreneurship, the use of heuristics by entrepreneurs has obtained very little attention. For two examples, see Åstebro and Elhedhli (2006) and Busenitz and Barney (1997).

## 5.6 Random Support Theory

In contrast to EUT, RDU, and PT, random support theory (RST; Brenner, 1995, 2003) is a theory of probability judgment rather than decision-making. It has been used in conjunction with decision theories that instead focus on choices but do not address the formation of beliefs (e.g., Fox and Tversky, 1998). This approach of separating the probability judgment and action phases is similar to Foss and Klein's beliefs-actions-results framework (Foss and Klein, 2012, 2020).

RST predicts how people assess probabilities in systematically biased ways based on two ideas: (a) humans typically apply case-based reasoning and (b) the noisiness of the decision environment affects how individuals assign subjective probabilities. Using examples and small experiments, Kahneman and Tversky (1973, 1979); and Tversky and Kahneman (1974, 1983) came to believe that intuitive judgments and predictions tend to be driven primarily by characteristics of the specific case at hand (i.e., the entrepreneurial project) and tend to neglect characteristics of the broader class or category to which the specific case belongs (e.g., Business-to-Consumer chatbots). The focus on case-specific characteristics and neglect of class-based aggregate properties then leads to predictable judgmental biases, including base-rate neglect (Tversky and Kahneman, 1974).

Calibration plots, which compare subjective probabilities to objective ones, illustrate these biases (see figure 5.2). Perfect calibration would align on the identity line, but overestimation results in a curve below the identity line, whereas overly extreme judgments yield a flatter curve. These plots have become the workhorse of decision analysts' attempts at trying to understand how and why individuals perceive probabilities in certain ways.

For example, a perfectly flat line would mean that the entrepreneur does not take probabilities into consideration. RST uses the notion of case-based intuitive judgment to predict systematic miscalibration (Brenner, 1995, 2003). It has parameters that are interpretable as reflecting (in)sensitivity to important class-based characteristics (Koehler et al., 2002; Brenner et al., 2006). RST predicts a flatter calibration curve when an entrepreneur is less able to make causal links between characteristics of the venture and its success. This is how many scholars in entrepreneurship think about uncertainty. For example, it may be more difficult to know exactly how to design the business model to be successful when environmental (or technological) uncertainty is high, such as for start-ups using large language models or hydrogen-driven automobiles. Entrepreneurs pursuing these ventures are then predicted to tend to ignore information about the class as a whole, such as the base rate of success,

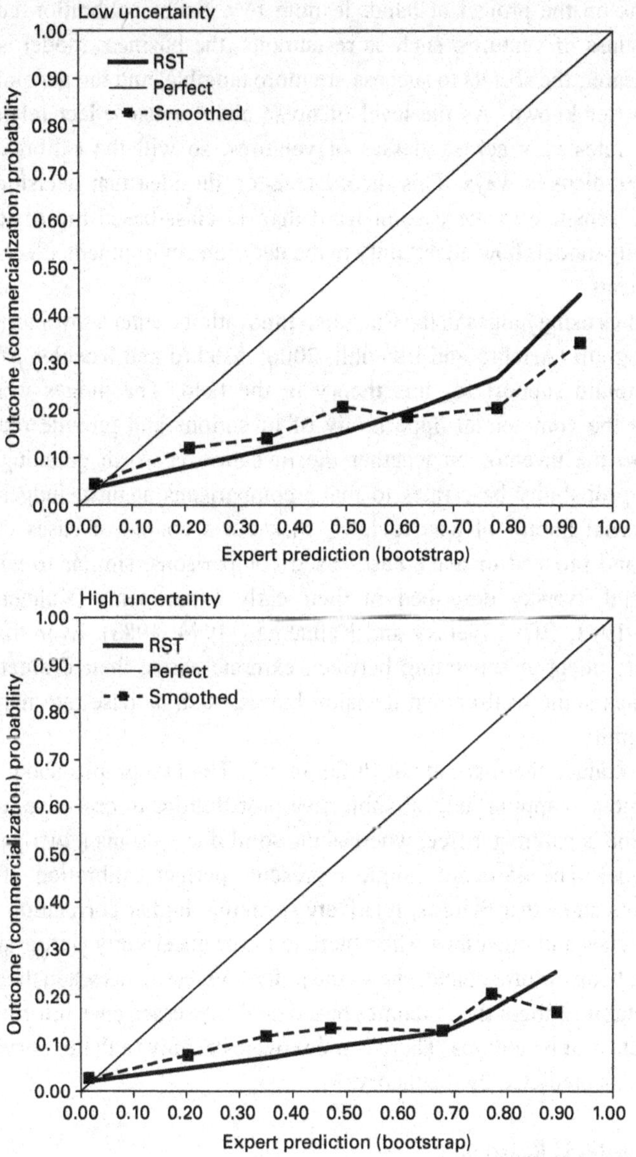


**Figure 5.2**

Calibration of probabilities under low and high uncertainty in entrepreneurship. *Source*: Åstebro and Koehler (2007). *Copyright*: John Wiley & Sons Ltd. Reproduced with permission. *Note*: RST: random support theory. The calibration plots show correspondence between an expert prediction of commercialization probability derived from a model correlating decision criteria with judgment and the actual probability of commercialization ("smoothed") for low and high uncertainty inventions. Fit of RST model to each calibration curve is also shown.

and focus on the project at hand, leading to a flatter calibration curve. For another class of ventures, such as restaurants, the business model is considerably clearer, the secrets to success are more tangible, and the general success rate is better known. As the level of noise about cause–effect relations and the base rates vary across classes of ventures, so will the calibration plots vary in predictable ways. This theory rests on the idea that decision-makers are more sensitive to the case at hand than to class-based information, and specifically models how uncertainty in the decision environment affects the bias in judgments.

In studies using judges in the Canadian Innovation Center's Inventors Assessment Program (Åstebro and Elhedhli, 2006; Åstebro and Koehler, 2007), the authors found support for this theory in the field. The judges were asked to assess the commercial opportunity of inventions and provide recommendations to the inventor on whether the invention is worth pursuing further. Lacking probability base rates to make comparisons against, judges rely on their internal library of past reviews, pick out a couple of cases that seem similar, and proceed to make case-based comparisons, similar to what Kahneman and Tversky described in their early experiments (Kahneman and Tversky, 1973, 1979; Tversky and Kahneman, 1974, 1983). Even though the judges are adept at separating between extreme cases, their evaluation process invites some well-known decision biases, such as base-rate neglect and overextremity.

We reproduce the main result in figure 5.2. The two graphs show the calibration plots, mapping judges' subjective probabilities to objective ones. The dashed line is parameter free, whereas the solid line presents a two-parametric RST model. The 45-degree angle represents perfect calibration of beliefs. The results show that there is, relatively speaking, higher correlation between assessed cues and outcomes when there is lower uncertainty (left graph). The right graph, on the other hand, shows the judges' assessments when there is high unpredictability about the outcomes based on the assessed cues information for another class of inventions. There is more overextremity (a flatter curve) in this situation, as predicted by the theory.

## 5.7 Case-Based Reasoning

Case-based reasoning (CBR) is a formal theory of decision-making under uncertainty, where probabilities are unknown or belong to a set of probability distributions or where only some probabilities can be deduced. It therefore belongs to the classes of models under absolute uncertainty. In CBR, the decision-maker does not have knowledge of future "states of the world," their probabilities, or outcomes and can only vaguely attribute cause-and-effect relationships, since future states are unknown (Gilboa and Schmeidler, 1995).[7]

A key idea in CBR is that the decision-maker recalls a few past cases and makes inferences about what might happen if a similar action is taken in the current situation. This is therefore a model that tries to define the origin of beliefs of decision-makers who do not have complete information. It is an alternative to EUT and SEU, where both states of the world and probabilities are neither given nor can be easily constructed, and therefore belongs to theories of absolute uncertainty. This modeling approach accommodates a number of biases related to belief formation highlighted by Tversky and Kahneman (1974), while, at the same time, the model also prescribes the best mode of action among a limited set of options, but, crucially, these options are not derived from "looking into the future." CBR can further explain behavioral patterns such as satisficing or incorporate concepts such as aspiration levels, both of which have attracted considerable attention in theories and descriptions of decision-making (e.g., Simon, 2013; Cyert and March, 2015).

Further model developments and clarifying arguments followed (e.g., Gilboa and Schmeidler, 2000), explaining that CBR suggests that people make decisions by analogies to past cases: They tend to choose acts that performed well in the past in similar situations and to avoid acts that performed poorly. The decision-maker only needs to recall a few cases, and which cases are recalled may be influenced by biases such as an over-reliance on salient or recent cases (see, e.g., Tversky and Kahneman, 1974).

From this limited set of recalled cases, the decision-maker infers potential outcomes if she takes a similar action for the case at hand—that is, the current entrepreneurial project—as for those recalled from experience. Not all recalled cases have the same weight on the decision. A "similarity function" describes how related the recalled cases are to the current case, and decision-makers weigh the outcomes of each recalled case by their similarity with the current case. Recalled cases do not have to be "real" but can also be hypothetical or counterfactual. CBR retains the formalism to evaluate and rank order different actions but may be considered closer to the realities of decision-making for entrepreneurs and their advisers. It has similarities to the verbal model of effectuation (Sarasvathy, 2001), which will be described later, as both assume that the decision-maker is not able to discern the future states of the world and instead build up perceptions based on past experiences. CBR is an astute formal model of entrepreneurial reasoning that deserves more attention from the scholarly field of entrepreneurship.

## 5.8 Decision Under Absolute Uncertainty

There is not yet any agreement in the entrepreneurship literature on how to describe probability judgment and decision-making under absolute uncertainty, where possible states of the world and events are unknown and where there are no or very weak causal links between actions and outcomes (see, e.g., Packard et al., 2017). For some theorists, it seems implausible that entrepreneurs can come up with a rank order of actions under such conditions. Decision-making in this world may therefore be characterized as making random choices or at least choices unaffected by any form of evaluation of alternatives. Or, as suggested by CBR, choices rely on past experiences plus analogous thinking. To make things more concrete, consider the theory of effectuation. In this theory, the potential entrepreneur behaves like a cook rummaging through the kitchen cabinet to create a meal from whatever ingredients are available. The focus of this theory is not the analysis of the decision situation but rather how to take action using limited means (Sarasvathy, 2001). It is therefore a boundary case where there is no (or sparse) structure in the evaluation of alternative actions. The key limiting factor in this theory is resource constraints, where the chef, in this example, is fully constrained and cannot go on a shopping spree to create the most delicious (and computed to be the most profitable) meal to sell. Instead, the chef must settle for a meal that is sellable at some lower profit, given the resources at hand. This does not mean that the decision-maker does not evaluate alternative meals given what is in the cupboard but rather that she operates under binding constraints.

Other authors, such as Packard et al. (2017), have tried to relate effectuation and other verbal descriptions of entrepreneurial decision-making to Knight's characterization of entrepreneurs making judgments and decisions under absolute uncertainty. Packard et al. (2017) distinguish between open and closed sets of decision options, and between open and closed sets of outcomes, creating a two-by-two typology of different decision environments. Under absolute uncertainty, both sets are open, and the prospective entrepreneurs have to decide on what set to close first to make a decision. The authors propose that effectuation and causation are two paths to reduce the absolute uncertainty to what they call a "perceived opportunity." In their view, effectuation involves taking the set of options as given and immutable (thus closed), being a function of a fixed set of endowed resources, and then trying to figure out what the outcomes will be for each option, without (much) evaluation of their chances of occurrence. In contrast, causation starts by closing the set of possible outcomes, without necessarily considering the available options, before evaluating what options to take. How the outcomes are evaluated in relation to their chances of occurrence is not made precise. It seems that this theory of causation under absolute uncertainty primarily describes the editing stage of PT where information is collated or where there is a selection of cases to refer to as sources of relevant comparisons, as in RST or CBR. None of the theories of PT, RST, or CBR are very precise in how these cases are conceived. PT only obliquely encapsulates how information is collated, although there is certainly a long list of decision biases that affect how information is collated and edited.[8] CBR addresses how past experiences affect future decisions head on but does not explain why certain experiences are recalled and others are not. The ideas presented in Packard et al. (2017) in a verbal way then give more flavor to this process of imagination and unpacking of options while not paying much attention to how probabilities or values of outcomes are later mentally estimated and combined to form a decision.

Yet other authors focus on the process of belief formation in the context of uncertainty and how that leads to action (e.g., McMullen and Shepherd, 2006; Shepherd et al., 2007). These authors make a distinction between a generally observable third-person entrepreneurial opportunity and a first-person opportunity that a specific person may observe and act upon. The actionable opportunity is a function of, for example, individual-specific knowledge. The framework in these theories is not dissimilar to that of PT, RST, or CBR, as it starts with the supposition that there are potential links between cause and effect that can be observed by some and that these decision-making parameters are different for different people based on their past experiences and their ability to recall them.

## 5.9 Bayesian Entrepreneurship

In addition to the different models presented earlier, a recent literature advocates for training entrepreneurs to think more systematically and scientifically about information gathering, making choices, and taking actions. This literature is summarized in this book as reflecting the entrepreneur as a Bayesian updater.

A theory has begun to be formed about Bayesian entrepreneurs. In this framework, the entrepreneur starts with a prior belief about the probability of success, performs a well-structured experiment based on a theory-driven hypothesis, collects the data, and forms a posterior belief about the probability of success (Agrawal et al., 2025, this volume). With this process, these entrepreneurs are said to progress from ex ante opaqueness to ex post clarity about the value of business opportunities to ultimately form a subjective probability distribution treating uncertainty in a way similar to risk (Zellweger and Zenger, 2023, p. 362). Entrepreneurs can then rank order options and choose the most preferred one. Other chapters in this volume describe this approach in more detail. A series of randomized controlled experiments have demonstrated that entrepreneurs can be taught to form a theory about an entrepreneurial opportunity, design an experiment to test it, and gather relevant information to identify the most promising path forward Camuffo et al. (2020); Camuffo, Gambardella Messinese, (2024); Camuffo, Gambardella and Pignataro (2024); Coali et al. (2024).

From the perspective of this chapter, the theory of Bayesian entrepreneurship is positioned as a version of subjective expected utility. It emphasizes the formation and updating of beliefs through subjective probabilities. Over time, these subjective probabilities become more accurate as the entrepreneur gains experience and improves calibration. Essentially, the theory explains how entrepreneurs should form and adjust their beliefs, thereby making more informed decisions based on evolving subjective probabilities. The authors of this chapter would then expect the entrepreneur to still have perceived probabilities that do not exactly fall on the identity line of the calibration plot. More generally, based on our current understanding, we would like to clarify Bayesian entrepreneurship, to the best of our ability, from the perspective of decision theory. We understand the theory to have the following main components: (1) subjective probabilities, (2) optimistic beliefs, (3) initially faulty heuristics, (4) decisions that are malleable through scientific training, and (5) training that offers better heuristics.

We now discuss these components of the theory. Bayesian entrepreneurship theory provides a valuable lens through which to understand entrepreneurial decision-making. However, like every theory, it has some limitations. One limitation of Bayesian entrepreneurship theory, derived from the assumptions of SEU (point 1 above), is its reliance on the independence axiom. A substantial body of empirical literature, beginning with the well-known Allais paradox (Allais, 1953), demonstrates that this axiom is often violated. This has given rise to several behavioral decision theories, many of which we have discussed in various sections of this chapter. Another critical assumption of SEU is probabilistic sophistication across sources of uncertainty, which posits that only subjective probabilities matter when evaluating ambiguous events, irrespective of the source of uncertainty. The Ellsberg paradox and subsequent literature highlight violations of this assumption. People generally exhibit non-neutral attitudes toward ambiguity, treating risk and ambiguity differently.

Bayesian entrepreneurship theory also takes a positive view of the role that strong priors play in entrepreneurial decision-making (point 2 above). For example, in the case of the online shoe retail start-up Zappos, the founder is said to be specifically optimistic about the success chances of the business idea, as he held beliefs that were contrary to the mainstream, driving him to experiment with a business model that others dismissed (Agrawal et al., 2025). However, this focus of Bayesian entrepreneurship theory on the founder's evaluation of likelihood of success may miss alternative explanations for the entrepreneur's actions—for example, different attitudes to ambiguity, different mental rules and strategies, or different repositories of experiences.

The impact of these limitations depends on whether the theory aims to be descriptive or normative. From a descriptive standpoint, failing to account for violations of the independence axiom or probabilistic sophistication can reduce the theory's explanatory power. Failing to account for heuristics and decision biases also ignores a long tradition of research showing their power in affecting judgment and decisions. These could maybe be assumed to fall into the error term as well, but this is perhaps a strong assumption, since biases and heuristics typically lead not to random deviations from the true mean but rather to predictable deviations. Normatively, the issue is more complex. For instance, there remains an ongoing debate about whether ambiguity neutrality should be considered normative. Some authors advocate for a normative approach to ambiguity neutrality (e.g., Li et al., 2018), whereas others argue that, in certain situations of uncertainty, it may be rational to reject ambiguity neutrality (e.g., Gilboa et al., 2009).

The degree of deviations from perfect rationality would then depend on how well Bayesian entrepreneurship training allows the entrepreneur to form well-calibrated probability assessments along the probability distribution. This is yet unclear from the experiments conducted in Bayesian entrepreneurship. What has been discovered so far is that entrepreneurs taking the training make better decisions, which of course is comforting. But it is unclear if the theory of Bayesian entrepreneurship affects only calibration accuracy and consistency or how people combine judgments and preferences to reach a decision (i.e., their decision model, point 3 above) or both. Future work will need to become more precise and decide how to address this challenge. For example, if we presume for a moment that the entrepreneur prior to participating in the experiment was behaving as governed by RST, then obtaining stronger support for the correct hypothesis would both move the expectation closer to the population base rate and make the calibration curve less flat. If, on the other hand, the entrepreneur prior to participating in the experiment was behaving as governed by CBR, then the key impact would be to move the entrepreneur away from the influence of experienced cases and toward the sampling mean from the experiment. Given different theoretical starting points, the training may show vastly different treatment effects. In other words, knowing which biases and heuristics affect poor decisions by entrepreneurs (to start with) helps our understanding of how training in Bayesian entrepreneurship might or might not improve decision accuracy (point 4 above).

Training decision-makers to have better-calibrated beliefs, have better judgment, and make better choices has been the focus of many studies (see, e.g., Fischhoff and Broomell, 2020). Two alternative approaches to training emerge, which address the subject of erroneous decisions differently: debiasing, and changing the decision rules (Larrick, 2004). Debiasing focuses on well-known human biases and designs training methods for a decision-maker to realize they have biased beliefs and to make corrections. The most general way to describe these efforts is to get the decision-maker to switch from System 1 thinking to System 2 thinking, from unconscious and automatic choices to conscious and deliberate choices. This has met with some success (Larrick, 2004; Lawson et al., 2020). The second approach is to make the decision-maker switch from relying on inferior strategies or decision rules to superior, more normative rules. An example of the former decision rules might be "take the first option that appears," whereas the second strategy would be "collect more information, and take the best option among a set of options." In our reading of Bayesian entrepreneurship, it seems that it primarily touches on training the entrepreneur to learn superior decision rules (point 5 above), but this remains unclear, and it is unclear what those superior decision rules are This should be clarified in the future.

To take an example of superior training and calibration, weather forecasters have historically been extremely good at probabilistic forecasting, where ex ante probability forecasts land almost perfectly on the ex post objective probabilities (see, e.g., Murphy and Winkler, 1977). The reader should compare figures 5.2 and 5.3 to get a sense of the mile-wide difference between weather forecasters' and (highly experienced) entrepreneurial venture judges' calibration accuracy!

Three reasons for these differences are the deliberate decision process, the reliance on technical decision support, and the readily available feedback for weather forecasters. Entrepreneurship scholars may consider weather forecasting and entrepreneurship to be two completely different things that cannot be compared. However, for those who doubt that probabilistic estimation is possible even under extreme uncertainty, one need look no further than the many articles produced by the Good Judgment Project. In this project, forecasters were asked to make predictions on questions such as "Will any country officially announce its intention to withdraw from the Eurozone before April 1, 2013?" Forecasters expressed their beliefs by answering the question "How likely is this event?" Probability judgments were later validated against actual outcomes.

A number of results came out of this project, most of which are summarized in Atanasov et al. (2017) and Moore and Cain (2007). Some individuals turn out to be consistently better than others at predicting highly uncertain events;

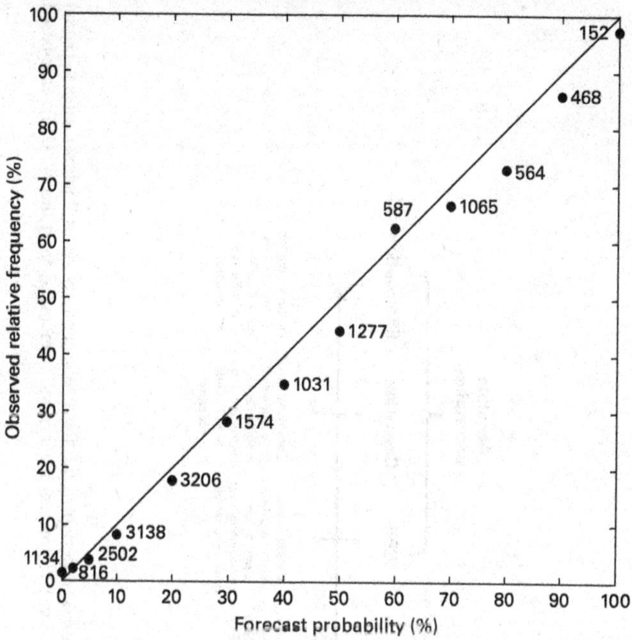


**Figure 5.3**
Calibration curve for US weather forecasters. *Notes:* The numbers in the figure represent the number of forecasts. Weather forecasters show some overprediction of most probabilities, except for very low- and high-probability events. Forecasters could choose among 2, 5, 10, 20, . . ., 90 and 100 percent probability of the event "precipitation."

probabilistic training, incentives, and calibrated aggregation models improve predictions; and both competition and teamwork help in making better forecasts. Training appeared to reduce overconfidence, even many months after the training completed. This is partially an effect of obtaining feedback on the accuracy of their forecasts. The project illustrates that, with tools driving individuals to invoke System 2 thinking, training improves both the ability of judging future events and individuals' confidence in making their forecasts.

We round off the description of the various theories for entrepreneurial decision-making by reverting back to figure 5.1, updating it with the relative position of Bayesian entrepreneurship to form figure 5.4. We have traversed from the normative theory of EUT, which states how a decision-maker should use information to choose the most favorable option in the expectation, to theories that take into account that probabilities may not be commonly known but can be subjectively perceived, to models where part of the probabilities as well as choices and outcomes are unavailable as they depend on a limited set of imagined or recalled cases, and finally to verbal models of decision-making

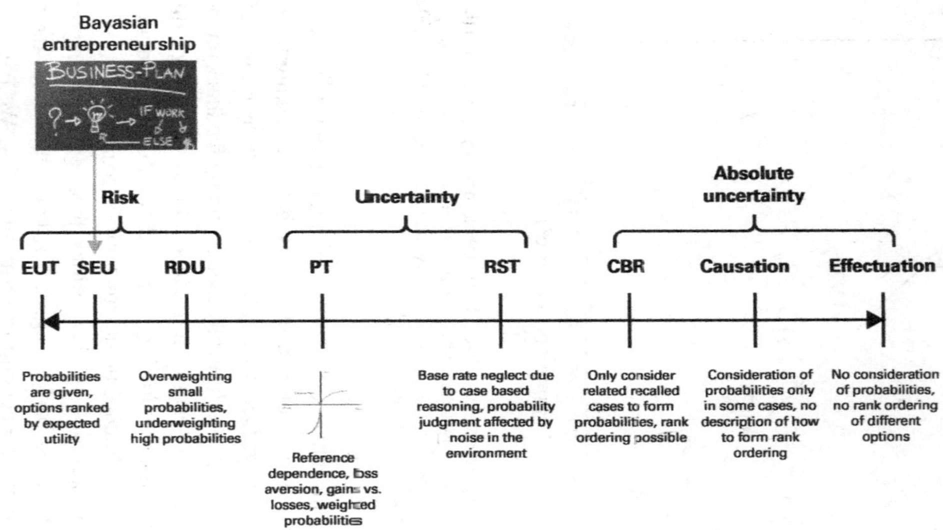


**Figure 5.4**
Bayesian entrepreneurship's position among theories of entrepreneurial judgment under risk, uncertainty, and absolute uncertainty. *Notes:* Picture credits to Daniel Chen and Jonathan Houston. EUT: expected utility theory; RDU: rank-dependent utility; PT: prospect theory; RST: random support theory; CBR: case-based reasoning.

where there is even more uncertainty, characterized, for example, by a lack of knowledge about the relations between causes and effects.

## 5.10 Conclusion

This chapter concludes that there is still a void between economists and decision theorists, on the one hand, and entrepreneurship scholars, on the other, in terms of characterizing judgment and decision-making under uncertainty. Decision theorists have made considerable strides in moving away from expected utility theory to more "realistic" models, thus meeting entrepreneurship scholars' understanding of the decision environment. They have, for example, analyzed the effects of incomplete information, unclear causal effects, and a general degree of ambiguity of the decision space. But, as opposed to the theories of entrepreneurship scholars, this ambiguity is still parametrized. The work of decision theorists has led to new insights—for example, explaining satisficing behavior and aspiration levels using a specific modeling framework. Maybe the most fundamental understanding emanating from this review that impacts the theory of Bayesian formation is that entrepreneurs may come to strikingly different valuations of an opportunity by, for example, having different probability weighting functions or by having different recalled cases to form an estimate from, to have different evaluation models, or by being biased in various ways. Entrepreneurs may, but certainly need not, be more optimistic than other people to perceive an opportunity differently. This calls for theorizing on Bayesian entrepreneurship to clarify the first stage of the theory—the process of forming a "contrarian" belief—even further.[9]

A recent literature has started a more pragmatic approach' trying to help entrepreneurs reduce uncertainty so that they can move as much as possible to a more scientific and rational approach to making decisions. This literature has been supported by a plethora of practical tools aimed at helping entrepreneurs make better decisions. Some of these may work better than others. Bayesian entrepreneurship theory is proposing one practical tool, which involves formulating a hypothesis and conducting an experiment to test it. It is unclear how this nascent theory maps into received theories of decision-making. We have made an attempt at initiating this mapping. It remains unclear to us how the theory maps into the standard view of decision-making in terms of affecting beliefs, heuristics, or both, and how it explains the improvements in entrepreneurial behavior that have been demonstrated in experiments. We do not dispute the experimental results but rather seek clarification of how the theory operates, both beliefs and heuristics. Understanding this would allow the design of more precise experiments and allow for testing the proposed mechanisms of human behavior. As it is a theory that involves training of entrepreneurs, it also seems useful to clarify how it maps to other training methods that either try to debias beliefs or change poor decision rules. We have indicated where to start this clarification as well.

## Acknowledgments

We thank Per Davidsson, Dimo Dimov, Dan Elfenbein, Michael Frese, Itzhak Gilboa, Peter Klein, Don Moore, Holger Patzelt, Scott Shane, and Dean Shepherd for their valuable comments.

## Notes

1. In this chapter, we use the term "heuristic" as a cognitively less demanding and faster alternative to EUT, which may, however, lead to suboptimal decisions. The term "heuristic" is sometimes also used to describe a shortcut in the formation of beliefs. A deeper review of the foundations and the state of the art of the literature of judgment and decision-making can be found in Fischhoff and Broomell (2020).

2. Other optimization models, such as Minimax, where a pessimist chooses the option that minimizes the possible loss for the worst-case scenario, or Maximax, where an optimist chooses the option that makes possible the maximum payoff, do not suffer from the same deficit as EUT being that the probabilities of outcomes associated with all alternative options must be known.

3. More subtle definitions of risk aversion such as in the "need for achievement" (nACH) construct, where entrepreneurs strive for situations with a moderate achievement goal and take calculated risks, are then likely even more difficult to distinguish, as they involve even more complex functional forms of risk aversion to estimate. For reviews of the nACH construct in entrepreneurship, see Johnson (1990) and Stewart and Roth (2007). Judging by these reviews, after being quite popular in research between 1965 and 1985, the construct seems to have waned considerably in popularity, probably partly because of its impreciseness.

4. The literature often interprets the shape of the weighting function in terms of optimism and pessimism.

5. Kahneman and Tversky wrote: "More generally, the preference order between prospects need not be invariant across contexts, because the same offered prospect could be edited in different ways depending on the context in which it appears" (1979, p. 275).

6. Note the similarities here to theories of entrepreneurial decision-making presented by Packard et al. (2017) and Sarasvathy (2001).

7. Preceding this model were models of decision-making under uncertainty that retain the classical state-space approach as in Gilboa (1987), Gilboa and Schmeidler (1989) and Schmeidler (1989). These models retained generalizations of subjective EUT to deal with unknown probabilities (but potentially known and at least explicitly modeled states of the world). None of these models deal with the question of the origin of preferences, which is the key point of CBR.

8. The reader might be amused scrolling through (two of many) lists at https://en.wikipedia.org/wiki/List_of_cognitive_biases and https://thedecisionlab.com/biases.

9. See, e.g., Felin et al. (2024) for an initial attempt.

## References

Abdellaoui, M., Baillon, A., Placido, L., and Wakker, P. P. (2011). The rich domain of uncertainty: Source functions and their experimental implementation. *American Economic Review*, 101(2), pp. 695–723.

Agrawal, A., Camuffo, A., Gambardella, A., Gans, J. S., Scott, E. L., and Stern, S. (2025). *Bayesian entrepreneurship*. MIT Press.

Ahn, T. (2010). Attitudes toward risk and self-employment of young workers. *Labour Economics*, 17(2), pp. 434–442.

Allais, M. (1953). Le comportement de l'homme rationnel devant le risque: Critique des postulats et axiomes de l'école américaine. *Econometrica: Journal of the Econometric Society*, 21(4), pp. 503–546.

Åstebro T., and Elhedhli, S. (2006). The effectiveness of simple decision heuristics: Forecasting commercial success for early-stage ventures. *Management Science*, 52(3), pp. 395–409.

Åstebro, T., Herz, H., Nanda, R., and Weber, R. A. (2014). Seeking the roots of entrepreneurship: Insights from behavioral economics. *Journal of Economic Perspectives*, 28(3), pp. 49–70.

Åstebro T., and Koehler, D. (2007). Calibration accuracy of a judgmental process that predicts the commercial success of new product ideas. *Journal of Behavioral Decision Making*, 20(4), pp. 381–403.

Atanasov, P., Rescober, P., Stone, E., Swift, S. A., Servan-Schreiber, E., Tetlock, P., Ungar, L., and Mellers, B. (2017). Distilling the wisdom of crowds: Prediction markets versus prediction polls. *Management Science*, 63(3), pp. 691–706.

Barberis, N. C. (2013). Thirty years of prospect theory in economics: A review and assessment. *Journal of Economic Perspectives*, 27(1), pp. 173–196.

Brenner, L. A. (1995). *A stochastic model of the calibration of subjective probabilities*. Stanford University.

Brenner, L. A. (2003). A random support model of the calibration of subjective probabilities. *Organizational Behavior and Human Decision Processes*, 90(1), pp. 87–110.

Brenner, L., Griffin, D., and Koehler, D. J. (2006). Modeling patterns of probability calibration with random support theory: Diagnosing case-based judgment. *Organizational Behavior and Human Decision Processes*, 97(1), pp. 64–81.

Brockhaus, R. H., Sr. (1980). Risk taking propensity of entrepreneurs. *Academy of Management Journal*, 23(3), pp. 509–520.

Busenitz, L. W., and Barney, J. B. (1997). Differences between entrepreneurs and managers in large organizations: Biases and heuristics in strategic decision-making. *Journal of Business Venturing*, 12(1), pp. 9–30.

Caliendo, M., Fossen, F. M., and Kritikos, A. S. (2009). Risk attitudes of nascent entrepreneurs—New evidence from an experimentally validated survey. *Small Business Economics*, 32, pp. 153–167.

Camuffo, A., Cordova, A., Gambardella, A., and Spina, C. (2020). A scientific approach to entrepreneurial decision making: Evidence from a randomized control trial. *Management Science*, 66(2), pp. 564–586.

Camuffo, A., Gambardella, A., Messinese, D., Novelli, E., Paolucci, E., and Spina, C. (2024). A scientific approach to entrepreneurial decision-making: Large-scale replication and extension. *Strategic Management Journal*, 45(6), pp. 1209–1237.

Camuffo, A., Gambardella, A., and Pignataro, A. (2024). Theory-driven strategic management decisions. *Strategy Science*, 9(4), pp. 382–396.

Coali, A., Gambardella, A., and Novelli, E. (2024). Scientific decision-making, project selection and longer-term outcomes. *Research Policy*, 53(6), p. 105022.

Cooper, A. C., Woo, C. Y., and Dunkelberg, W. C. (1988). Entrepreneurs' perceived chances for success. *Journal of Business Venturing*, 3(2), pp. 97–108.

Cramer, J. S., Hartog, J., Jonker, N., and Van Praag, C. M. (2002). Low risk aversion encourages the choice for entrepreneurship: An empirical test of a truism. *Journal of Economic Behavior & Organization*, 48(1), pp. 29–36.

Cyert, R., and March, J. (2015). Behavioral theory of the firm. *Organizational Behavior 2*. Routledge, pp 60–77.

De Meza, D., Dawson, C., Henley, A., and Arabsheibani, G. R. (2019). Curb your enthusiasm: Optimistic entrepreneurs earn less. *European Economic Review*, 111, pp. 53–69.

Dell'Era, M., Opromolla, L. D., and Santos-Pinto, L. (2023). Can optimism solve the entrepreneurial earnings puzzle? *The Scandinavian Journal of Economics*, 125(1), pp. 139–169.

Diecidue, E., and Wakker, P. P. (2001). On the intuition of rank-dependent utility. *Journal of Risk and Uncertainty*, 23, pp. 281–298.

Ellsberg, D. (1961). Risk, ambiguity, and the Savage axioms. *The Quarterly Journal of Economics*, 75(4), pp. 643–669.

Felin, T., Gambardella, A., Novelli, E., and Zenger, T. (2024). A scientific method for startups. *Journal of Management*, 50(8), pp. 3030–3104.

Fischhoff, B., and Broomell, S. B. (2020). Judgment and decision making. *Annual Review of Psychology*, 71(1), pp. 331–355.

Foss, N. J., and Klein, P. G. (2012). *Organizing entrepreneurial judgment*. Cambridge University Press.

Foss, N. J., and Klein, P. G. (2020). Entrepreneurial opportunities: Who needs them? *Academy of Management Perspectives*, 34(3), pp. 366–377.

Fox, C. R., Erner, C., and Walters, D. J. (2015). Decision under risk: From the field to the laboratory and back. In C. R. Fox, C. Erner, and D. J. Walters (Eds.), *The Wiley Blackwell handbook of judgment and decision making* (pp. 41–88). Wiley Blackwell.

Fox, C. R., and Tversky, A. (1998). A belief-based account of decision under uncertainty. *Management Science*, 44(7), pp. 879–895.

Frenkel-Brunswick, E. (1951). Personality theory and perception. In R. Blake, & E. Ramsey (Eds.), *Perception: An approach to personality*. Ronald. doi:10.1037/11505-013

Furnham, A., and Marks, J. (2013). Tolerance of ambiguity: A review of the recent literature. *Psychology*, 4(9), pp. 717–728.

Gilboa, I. (1987). Expected utility with purely subjective non-additive probabilities. *Journal of Mathematical Economics*, 16(1), pp. 65–88.

Gilboa, I., Postlewaite, A., and Schmeidler, D. (2009). Is it always rational to satisfy Savage's axioms? *Economics & Philosophy*, 25(3), pp. 285–296.

Gilboa, I., and Schmeidler, D. (1989). Maxmin expected utility with non-unique prior. *Journal of Mathematical Economics*, 18(2), pp. 141–153.

Gilboa, I., and Schmeidler, D. (1995). Case-based decision theory. *The Quarterly Journal of Economics*, 110(3), pp. 605–639.

Gilboa, I., and Schmeidler, D. (2000). Case-based knowledge and induction. *IEEE Transactions on Systems, Man, and Cybernetics-PART A: Systems and Humans*, 30(2), pp. 85–95.

Gutierrez, C., Åstebro, T., and Obloj, T. (2020). The impact of overconfidence and ambiguity attitude on market entry. *Organization Science*, 31(2), pp. 308–329.

Holm, H. J., Opper, S., and Nee, V. (2013). Entrepreneurs under uncertainty: An economic experiment in China. *Management Science*, 59(7), pp. 1671–1687.

Johnson, B. R. (1990). Toward a multidimensional model of entrepreneurship: The case of achievement motivation and the entrepreneur. *Entrepreneurship Theory and Practice*, 14(3), pp. 39–54.

Kahneman, D., and Tversky, A. (1973). On the psychology of prediction. *Psychological Review*, 80(4), p. 237.

Kahneman, D., and Tversky, A. (1979). Prospect theory: An analysis of decisions under risk. *Econometrica*, 47(2), pp. 263–292.

Kihlstrom, R. E., and Laffont, J. J. (1979). A general equilibrium entrepreneurial theory of firm formation based on risk aversion. *Journal of Political Economy*, 87(4), pp. 719–748.

Knight, F. H. (1921). *Risk, uncertainty and profit*, Vol. 31. Houghton Mifflin.

Koehler, D. J., Brenner, L., and Griffin, D. (2002). The calibration of expert judgment: Heuristics and biases beyond the laboratory. In T. Gilovich, D. Griffin, and D. Kahneman (Eds.), *Heuristics and biases: The psychology of intuitive judgment* (pp. 686–715). Cambridge University Press.

Koudstaal, M., Sloof, R., and Van Praag, M. (2016). Risk, uncertainty, and entrepreneurship: Evidence from a lab-in-the-field experiment. *Management Science*, 62(10), pp. 2897–2915.

Laferrière, V., Staubli, D., and Thöni, C. (2023). Explaining excess entry in winner-take-all markets. *Management Science*, 69(2), pp. 1050–1069.

Larrick, R. P. (2004). Debiasing. In D. J. Koehler and N. Harvey (Eds.), *Blackwell handbook of judgment and decision making* (pp. 316–337). Blackwell.

Lawson, M. A., Larrick, R. P., and Soll, J. B. (2020). Comparing fast thinking and slow thinking: The relative benefits of interventions, individual differences, and inferential rules. *Judgment and Decision Making*, 15(5), pp. 660–684.

Li, Z., Müller, J., Wakker, P. P., and Wang, T. V. (2018). The rich domain of ambiguity explored. *Management Science*, 64(7), pp. 3227–3240.

McMullen, J. S., and Shepherd, D. A. (2006). Entrepreneurial action and the role of uncertainty in the theory of the entrepreneur. *Academy of Management Review*, 31(1), pp. 132–152.

Moore, D. A., and Cain, D. M. (2007). Overconfidence and underconfidence: When and why people underestimate (and overestimate) the competition. *Organizational Behavior and Human Decision Processes*, 103(2), pp. 197–213.

Moore, D. A., and Healy, P. J. (2008). The trouble with overconfidence. *Psychological Review*, 115(2), pp. 502–517.

Murphy, A. H., and Winkler, R. L. (1977). Reliability of subjective probability forecasts of precipitation and temperature. *Journal of the Royal Statistical Society. Series C (Applied Statistics)*, 26(1), pp. 41–47.

Nishimura, K. G., and Ozaki, H. (2004). Search and knightian uncertainty. *Journal of Economic Theory*, 119(2), pp. 299–333.

Packard, M. D., Clark, B. B., and Klein, P. G. (2017). Uncertainty types and transitions in the entrepreneurial process. *Organization Science*, 28(5), pp. 840–856.

Quiggin, J. (1982). A theory of anticipated utility. *Journal of Economic Behavior & Organization*, 3(4), pp. 323–343.

Sarasvathy, S. D. (2001). Causation and effectuation: Toward a theoretical shift from economic inevitability to entrepreneurial contingency. *Academy of Management Review*, 26(2), pp. 243–263.

Savage, L. J. (1951). The theory of statistical decision. *Journal of the American Statistical Association*, 46(253), pp. 55–67.

Savage, L. J. (1954). *The foundations of statistics*. Dover.

Schmeidler, D. (1989). Subjective probability and expected utility without additivity. *Econometrica*, 57(3), pp. 571–587.

Sexton, D. L., and Bowman, N. (1985). The entrepreneur: A capable executive and more. *Journal of Business Venturing*, 1(1), pp. 129–140.

Shane, S. (2009). Why encouraging more people to become entrepreneurs is bad public policy. *Small Business Economics*, 33, pp. 141–149.

Shepherd, D. A., McMullen, J. S., and Jennings, P. D. (2007). The formation of opportunity beliefs: Overcoming ignorance and reducing doubt. *Strategic Entrepreneurship Journal*, 1(1–2), pp. 75–95.

Simon, H. A. (2013). *Administrative behavior*. Simon and Schuster.

Skriabikova, O. J., Dohmen, T., and Kriechel, B. (2014). New evidence on the relationship between risk attitudes and self-employment. *Labour Economics*, 30, pp. 176–184.

Starmer, C. (2000). Developments in non-expected utility theory: The hunt for a descriptive theory of choice under risk. *Journal of Economic Literature*, 38(2), pp. 332–382.

Stewart, W. H., and Roth, P. L. (2007). A meta-analysis of achievement motivation differences between entrepreneurs and managers. *Journal of Small Business Management*, 45(4), pp. 401–421.

Townsend, D. M., Hunt, R. A., McMullen, J. S., and Sarasvathy, S. D. (2018). Uncertainty, knowledge problems, and entrepreneurial action. *Academy of Management Annals*, 12(2), pp. 659–687.

Tversky, A., and Kahneman, D. (1974). Judgment under uncertainty: Heuristics and biases: Biases in judgments reveal some heuristics of thinking under uncertainty. *Science*, 185(4157), pp. 1124–1131.

Tversky, A., and Kahneman, D. (1983). Extensional versus intuitive reasoning: The conjunction fallacy in probability judgment. *Psychological Review*, 90(4), p. 293.

Tversky, A., and Kahneman, D. (1992). Advances in prospect theory: Cumulative representation of uncertainty. *Journal of Risk and Uncertainty*, 5, pp. 297–323.

Von Neumann, J., and Morgenstern, O. (1953). *Theory of games and economic behavior* (3rd ed.) Princeton University Press.

Wu, B., and Knott, A. M. (2006). Entrepreneurial risk and market entry. *Management Science*, 52(9), pp. 1315–1330.

Zellweger, T., and Zenger, T. (2023). Entrepreneurs as scientists: A pragmatist approach to producing value out of uncertainty. *Academy of Management Review*, 48(3), pp. 379–408.

# II

# Priors in Entrepreneurial Decision-Making

# 6

## Causal Reasoning and the Scientific Entrepreneur: Beyond Bayes

Timo Ehrig, Teppo Felin, and Todd Zenger

### 6.1 Introduction

The domain of entrepreneurship features varied approaches that seek to make the decision-making process surrounding venture formation under uncertainty more scientific. These include approaches such as lean start-up (Blank, 2013; Blank and Eckhardt, 2023; Leatherbee and Katila, 2020), the theory-based/scientific approach to entrepreneurial decision-making (Camuffo et al., 2020; Ehrig and Schmidt, 2022; Felin and Zenger, 2009), different forms of experimentation (Contigiani, 2023; Koning et al., 2022; Adner and Levinthal, 2024; Shelef et al., 2025), and frameworks of entrepreneurial strategy and choice (Agrawal et al., 2021; Gans et al., 2019).

In their introductory chapter, Agrawal et al. (2025) propose Bayesian entrepreneurship as an umbrella term and approach that integrates this literature, providing a unifying logic, organizing principle, and formal framework for entrepreneurship and decision-making under uncertainty. The core claims of Bayesian entrepreneurship, Agrawal et al. (2025) argue, are a focus on (a) heterogeneous prior beliefs, (b) belief strength or optimism, and (c) beliefs that are updated from evidence using Bayes' rule. On one level, the concepts of heterogeneous beliefs, entrepreneurial optimism, and learning from data are uncontroversial when applied to entrepreneurs (Ehrig and Schmidt, 2022; Felin and Zenger, 2009, 2017; Zellweger and Zenger, 2023).

In this paper, we question whether the Bayesian label—and the predominant focus on updating priors through experimentation and data gathering—captures the central features of the entrepreneur's efforts to generate value. Although we broadly endorse the effort to integrate these various approaches under a common banner, we argue that the most central mechanisms of learning for the entrepreneur—including the causal formation and revision of a theory—are underdeveloped in this Bayesian framing. The central processes of scientific entrepreneurship are not merely about updating priors. Rather, they involve writing and rewriting the structural assumptions encoded in them. As we will discuss, this operation is key to the entrepreneurial process, and it is strictly non-Bayesian. It involves using causal reasoning to build theories of value that often run counter to current facts and evidence and then revising that logic based on feedback surrounding assumptions embedded in that logic. We argue that these two central activities of theory-based scientific entrepreneurship go well beyond the Bayesian approach. Although Bayesian updating based on evidence or data is part of the entrepreneurial process, Agrawal et al.'s 2025 effort to shoehorn it to cover all of a scientific approach to entrepreneurship overlooks the boundaries and limitations of Bayesianism, and focuses excessive attention on a somewhat narrow part of the entrepreneurial learning process. In short, we argue that Bayesian entrepreneurship does not capture the richness of the idea of entrepreneurs as theorists and scientists—namely, the role of causal reasoning in articulating a path to value creation.

We advocate for an expanded perspective in which entrepreneurs behave like scientists or pragmatists who select problems, frame and formulate them, develop causal theories about how to solve them, and then set about testing these theories, revising them as needed. The fundamental difference in emphasis here surrounds the development and revision of causal theories that envision, define, and revise state spaces through logic and reasoning with the aid of very small samples of data. In an important sense, data are an outcome of this entrepreneurial process rather than input. The learning process is thus far more than revising beliefs based on evidence. It is about building theories and causal models that enable an entrepreneur to become aware of the correct interventions and informative experiments to generate new data and revise theories to enable the process of start-up-specific value creation. Admittedly, Agrawal et al. (2025) reference many of these processes (e.g., theory revision, experimentation), including theory formation. But, as we discuss, most of these activities are non-Bayesian, and labeling them as Bayesian undersells their centrality and significance in the process of entrepreneurial learning. From a normative perspective, Agrawal et al.'s 2025 articulation of Bayesian entrepreneurship offers an inadvertently thin vision of what start-ups should focus on, whereas the ambition of scientific approach to entrepreneurship is to offer practical and normative guidance well beyond prior beliefs and belief updating, including centrally important issues such as governance, team formation, funding, and stakeholder persuasion (Wuebker et al., 2023).

## 6.2 Bayesian Entrepreneurship Versus a Scientific Approach to Entrepreneurship

To highlight our concern with the Bayesian label for the entrepreneur's process in seeking to create value, we begin by first articulating the theory-building process that we believe merits greater attention and should feature more prominently in a theory-based, scientific approach to entrepreneurship. We then review the three tenets or core claims of Agrawal et al.'s 2025 conception of Bayesian entrepreneurship and highlight how a focus on theory formation and revision adds depth and nuance to these core claims. Notice that we do not have concerns with the core claims themselves per se if the word "Bayesian" were to be replaced with the broader label of "theory-based" or "scientific" and much greater emphasis placed on theories, theorizing, and the process of logical reasoning and revision. Our concern is with how the application of Bayesian logic leads us to treat beliefs, updating, and learning in relatively narrow and normatively limited fashion.

Notice that in using the term "theory-based entrepreneurial scientists"—and contrasting it with the term "Bayesian entrepreneurship"—we use a composite term that features intuition published by us and many others (Camuffo et al., 2020; Coali et al., 2024; Ehrig and Schmidt, 2022; Felin and Zenger, 2009; Felin et al., 2024; Zellweger and Zenger, 2023). We recognize that this represents our (perhaps biased) interpretation of the literature. We certainly do not mean to imply that our arguments about entrepreneurs as scientists have the endorsement of the scholars we cite. We want to recognize this at the outset, as the introductory chapter by Agrawal et al. (2025) on Bayesian entrepreneurship includes scholars who are coauthors, as well as those whom we consider to be close fellow travelers in our joint effort to understand entrepreneurship and decision-making under uncertainty. We focus on differences in the details and prescriptions suggested by the idea of Bayesian entrepreneurship versus entrepreneurs as scientists, but we recognize and applaud the broad-based similarities in overall ethos and ambition of these approaches.

We start with highlighting two fundamental features of a scientific entrepreneurial process that are non-Bayesian: developing and revising counterfactual causal theories and "seeing" different state spaces or possibilities.

### 6.2.1 Developing and Revising Counterfactual Causal Logic Is Non-Bayesian

The fundamental differences between the approach we advocate and what Agrawal et al. (2025) term "Bayesian entrepreneurship" all connect to the role of theory and causal logic that are the foundation of the theory-based view (Felin et al., 2024). This process is initiated by some form of contrarian belief about a future state of the world, often delineated by a problem that, if solved, realizes that future state. Within the perspective we advocate, causal theories are systems of causal beliefs or assumptions that are jointly sufficient to solve a problem. These causal beliefs and logic frequently take the form of subproblems that, if solved—or assumptions that, if rendered true—combine to enable the future state in which the larger problem is solved.

The non-Bayesian aspects of the theory-based view of entrepreneurship—and the idea of causal logic—are perhaps best illustrated with a stylized example. Consider Airbnb. The venture founders had a contrarian belief about the plausible use of vacant apartments as hotel accommodation. Although there was no seeming evidence for their belief at the time—in fact, hotel experts and prominent venture capitalists were extremely skeptical—the founders formulated specific problems that needed to be solved to enable them to realize the value they envisioned. Airbnb's founding theory might be summarized as follows: "We can realize our envisioned future state if we can (a) efficiently match idle capacity to demand, (b) facilitate informationally efficient professional listings, (c) facilitate secure payment, and (d) facilitate trust in transacting with strangers." The articulation of these problems provides the concrete set of actions that Airbnb believed it needed to take to broker safe, reliable, and efficient access to "idle hotel capacity" (see Felin et al., 2021). Note that, in this illustration, causal logic defines beliefs that need to be made true for the realization of value or some counterfactual future state (Ehrig and Schmidt, 2022). It is thus through causal logic that entrepreneurs build their beliefs that then shape actions.

Judea Pearl's seminal work on causality offers a useful framework for understanding the need to extend and move beyond Bayesian reasoning (Pearl, 2000). Notice that although Pearl's emphasis on causal reasoning is closely linked to the theory-based view of entrepreneurship and strategy, we also emphasize that Pearl's causal logic focuses mostly on *backward-looking* explanation, whereas our emphasis is on *forward-looking* decision-making and interventions.

Perhaps the best way to summarize Pearl's work on causal reasoning is to briefly revisit the example he most frequently discusses in his papers—namely, the relationship between smoking and cancer. To put this in Bayesian terms, we might compute conditional probabilities for the relationship—$P(Y \mid X)$—where $X$ represents smoking status (smoker or nonsmoker) and $Y$ lung cancer (yes or no).

Pearl's work extends the notion of conditional beliefs to causal beliefs, where one considers the likelihood not only of an event given another (as in traditional conditional probability) but also of an event given that a particular intervention has occurred. This is captured by Pearl's "do" calculus and the concept of "do" operators (e.g., $P(Y \mid do(X))$), which allow for reasoning about the effects of potential interventions on outcomes. We might ask, if we "forced" a particular individual to smoke, would they develop cancer? Or what would be the probability of lung cancer if we could somehow intervene and make a person smoke, independent of their natural tendencies or genetic predispositions? In this framework, conditional beliefs become crucial when determining how likely a certain outcome is, given not just that some condition is observed but because an intervention has taken place. This is a more refined and powerful type of belief that explicitly considers causality. And beyond the broad correlation between smoking and cancer, varied interventions might enable scientists to isolate causal mechanisms more carefully to tease out alternative explanations such as genetic disposition, environmental factors, or various person-specific effects. This is, of course, the central logic behind randomized control trials, where the treatment represents an intended intervention for a hoped-for outcome.

Now, although Pearl's focus is largely on *backward*-looking causal reasoning in science (in terms of understanding existing correlations and causal relationships), the theory-based view extends this logic into the domain of *forward*-looking decision-making by entrepreneurs. The theory-based view of start-ups represents a normative agenda where entrepreneurs are essentially asked to counterfactually reason about future states and to consider the "interventions"—the actions they need to take, types of experimentation they need to engage in, or problems they need to solve—to create the conditions for the realization of these future states. In the context of entrepreneurship (which is our focus here), Pearl's all-important "why" question is not so much about why something has happened in the past (Pearl and Mackenzie, 2018) but rather about why an entrepreneur thinks—given their causal logic and associated interventions—something *could* happen. Notice that belief revisions and learning based on this logic can also happen through argumentation (before real-world interventions or experimentation). From a theory-based perspective, an entrepreneur is asked to delineate a causal path for realizing a future state that presently lacks evidence. This normative agenda cannot be captured by focusing on Bayesian priors, belief updating, and learning.

Bayesian approaches unfortunately cannot fully capture the process of counterfactual reasoning and the development of causal logic. Thus, like Pearl (2001), we are "half-Bayesian" in that we think priors and belief updating are important for certain settings—but more is needed for the context of entrepreneurship. Pearl's earlier work building on Bayes is often conflated with his work on causal logic and inference. But the latter work is not Bayesian (or, put differently, extends beyond it), given the distinction between the identification of causality and counterfactuals (Pearl, 2000; Pearl and Mackenzie, 2018). A key insight—relevant to empirical research in the sciences, including management and strategy science—is that identifying causality requires *stable and known settings, and repetition*. Entrepreneurial decision-making, however, is characterized by uncertainty. This places a premium on counterfactuals and interventions or, put differently, on counterfactually reasoning about causal paths (and associated actions and experiments) to realize theorized states of the world. The central point is that traditional Bayesian learning mechanisms focus on belief updating rather than generating and revising hypothetical causal structures, which we argue is critically important as entrepreneurs craft novel theories and strategies.

The Airbnb example illustrates the importance of causal logic in the context of entrepreneurship. The settings in which entrepreneurs operate are uncertain (as they want to change the world), and unknown (as, if they were known, there would be little space for profits; Knight, 1921), and interventions can typically be done only once. Entrepreneurs postulate causal logic. What entrepreneurs do when they craft strategies and formulate and rewrite causal logic cannot fully be captured with Bayesian reasoning. This is because entrepreneurs engage in counterfactual causality, and revising counterfactual causality under simulated conditions requires some form of technology for revising beliefs. Indeed, Pearl was one of the early proponents of belief revision technology (Darwiche and Pearl, 1997). And in the domain of strategy and entrepreneurship, Ehrig and Schmidt (2022) show how strategists and entrepreneurs can learn and improve postulated causal logic using belief revision, and how this process is rational but not Bayesian. The critical point here is also that, in the absence of evidence (or if evidence is contested), beliefs and causal reasoning can be developed and updated on the basis of argumentation (Ehrig and Schmidt, 2022).

### 6.2.2 Seeing and Awareness Are Non-Bayesian

A second concern we have with the Bayesian umbrella concept proposed by Agrawal et al. (2025) is that Bayesian logic presumes that actors agree on some kind of state space along which they assign beliefs and probabilities. Put differently, the implicit assumption is that the "things" about which one has beliefs are (somehow) known. But how the something that one has priors about is identified in the first place is central for entrepreneurship but inherently lies outside Bayesian logic.

The problems with a Bayesian approach, specifically when dealing with novelty, are perhaps best illustrated in the context of perception. Bayesian models work well for understanding routine perception and familiar domains, where actors are engaged in repeated encounters with known objects or visual scenes that generate an underlying, known probability distribution (Kersten et al., 2004; see also Chater et al., 2010). But as discussed by Koenderink, Bayesian "priors apply to *known* domains" (Koenderink, 2015, p. 252, emphasis added; see also Felin and Koenderink, 2022). Bayesian logic works well in domains where the relevant variables are known and where evidence is incontrovertible (Pinker, 2022). It is in these situations where we can perhaps assess whether humans "update rationally" or whether they are "rational Bayesians" (cf. Augenblick and Rabin, 2021; Pilgrim et al., 2024). But not everything we encounter is familiar, and we certainly do not have priors about what might be possible. If we are dealing with unknown or unfamiliar domains—as is the case with entrepreneurs—where these priors come from requires non-Bayesian logic.

Whether we are talking entrepreneurship or perception, the problem with Bayesian approaches is that they assume an all-seeing eye (Felin et al., 2016). That is, the domains of belief or knowledge along which we might assign probabilities are assumed to be known. But the theory-based approach to entrepreneurship is more interested in *hidden* and *possible* domains, including the (aforementioned) interventions that might be taken for the realization of novel or contrarian beliefs. Theories provide salience for things about which we have no prior (Felin and Zenger, 2017). A theory tells us what to look for—it unlocks awareness of novelty beyond what might be seemingly obvious (Felin and Koenderink, 2022). This goes not just for perception but also for awareness of new sources of economic value and rents (Bryan et al., 2022; Ehrig and Zenger, 2024a).

Another way to clarify the all-seeing-eye problem we have ascribed to Bayesian approaches is to specifically look at the underlying models that use this approach and point to how they operationalize beliefs and updating. In large part, these models start with a delineation of options (or a full state space) and then see whether boundedly rational agents engage in rational Bayesian updating and learning (e.g., Agrawal et al., 2021; Gans et al., 2019). The problem with this approach is aptly summarized by Ehrig and Schmidt. They argue that, in Bayesian (and related) approaches, "the space of possible signals is assumed to be known to the strategist and—importantly—unchanging, that is, strategists face a signal extraction problem rather than the problem of evaluating and possibly changing the mapping from possible signals to outcomes" (Ehrig and Schmidt, 2022, p. 1289). Put in these terms, we argue that entrepreneurial activity is not a signal extraction problem but rather a problem of specifying a theory and causal path for realizing some form of contrarian belief. The problem is that signals—like customer validation by a minimum viable product (MVP; Felin et al., 2024)—can lead entrepreneurs astray. Therefore, we argue that a theory is needed to guide any form of learning and search by entrepreneurs.

## 6.3 Revisiting Agrawal et al.'s Core Claims

With the above as background, we are equipped to carefully revisit Agrawal et al.'s 2025 three core claims. Note that our effort is not to dispute these claims per se but rather to highlight how a more complete embrace of theories and causal reasoning brings new meaning to these claims and provides a far richer understanding of the entrepreneurial process and the tools (or the scientific method) an entrepreneur can use to generate economic value.

### 6.3.1 Heterogeneous Beliefs

Agrawal et al.'s 2025 first core claim for Bayesian entrepreneurship focuses on the recognition or acknowledgment that entrepreneurs lack common priors and instead have heterogeneous beliefs. We concur that heterogeneous beliefs are important if economic agents are to create significant value (Felin et al., 2021). Beliefs that are in some form heterogeneous—contrarian, discrepant, or different—are central, as they can enable start-ups and firms to see value in places that are not obvious to others (Felin and Zenger, 2017).

Although belief heterogeneity is a premise shared by us, it is in the Bayesian elaboration of what belief heterogeneity means for entrepreneurship that we start to see differences. Namely, Agrawal et al. (2025) argue that belief heterogeneity represents the idea that "agents do not share a common prior belief about important elements of the state space—that is, they effectively agree to disagree." This Bayesian articulation of belief heterogeneity places emphasis on disagreement about the distribution over a state space, as manifest by the different, subjective probabilities that agents might assign to states, but not disagreement about the state space itself. Through experiments and associated Bayesian updating, actors learn of the accuracy of their priors—the accuracy of their models—and accordingly update their probabilities.

The problem with this approach, as discussed by us in the previous section, is the lack of attention to where prior beliefs come from in the first place. The belief about "something" (a state, core element, or space as a whole)—that is, what a belief is *about*—is an important starting point for entrepreneurs and scientists more broadly. An entrepreneur-as-scientists conception of heterogeneous beliefs is *not* about the fact that agents might assign different probabilities to a *particular* known belief (or particular set of beliefs or states of the world)—although it might include this—but is more about *what* the belief is about in the first place. That is, the plausible belief states and spaces—the things that one might "bet" on—are not given. This requires a theory. Furthermore, bets about future states are entangled with an agent's willingness to take action (Ramsey, 1931).

In short, the Bayesian apparatus is focused mostly on belief updating based on evidence but not on novel beliefs and the articulation of a causal path to enable their realization. This is evident, for example, in the lack of attention on the idea of beliefs and conflicting evidence, where agents might disagree about what a data point or piece of evidence means. Furthermore, Bayesian logic provides no mechanism for outlining the causal path that an entrepreneur needs to take to enable the realization of a contrarian belief. From the perspective of theory-based scientific entrepreneurship, that causal path is outlined by the articulation of problems that need to be solved and experiments that need to be made to enable the creation of new value (Felin et al., 2021). Although Bayesian logic might provide a mechanism through which specific models can be calibrated and tested, it says little about how to build new models, how to select new problems, or how to generate or respond to evidence.

That said, Agrawal et al. (2025) do acknowledge that actors may build different theories and state spaces, which we applaud. But the process of selecting problems, forming and revising theories, and responding to evidence falls outside a strictly Bayesian rubric and merits much greater attention and focus. Entrepreneurs create value by imagining future states, composing an architecture of necessary and sufficient conditions—subproblems that must be solved or assumptions that must be true or rendered true to solve some larger problem. Such expressions of conditional causal logic do not lend themselves to one-off tests or experiments. That is, there are important intervening steps for realizing a future state, which fall outside the remit of Bayesian entrepreneurship. A theory-based scientific approach to entrepreneurship is focused on the hard work of articulating an underlying *causal path* for how to reach an imagined future state, including framing the problems that need to be solved and the experiments that need to be taken to realize value. Thus, the exercise of entrepreneurship is one of discriminating judgment, where formulated problems are matched with the right types of experiments and governance forms (Wuebker et al., 2023). This also offers highly practical, start-up-specific guidance for decisions that are faced by entrepreneurs, including decisions about the type of financing they need to secure, the protection of intellectual property, and so forth. The start-up serves an integrating and coordinating function that looks more expansive than the central aspects of start-up activity articulated by Agrawal et al. (2025).

### 6.3.2 Strength of Belief

A second core claim of Agrawal et al. (2025) is that Bayesian entrepreneurship rests on an assumption that entrepreneurs are optimistic—that the strength of their belief or optimism in their priors is simply greater than that of other individuals. In other words, entrepreneurs believe in their beliefs more than others believe in their beliefs. Of course, remedying this gap—seeking to elevate the confidence of others in a belief—is a central task of the entrepreneur. What motivates experimentation and evidence gathering by the entrepreneur is a desire to validate or confirm their beliefs and generate evidence by which to persuade others. As outlined in Agrawal et al. (2025), evidence gathering and the additional information that results causes posterior probabilities held by the entrepreneur and others to converge. Of course, this convergence may either confirm the Bayesian entrepreneurs' priors or disconfirm them, but either way, convergence occurs. In Bayesian entrepreneurship, therefore, effective entrepreneurs are simply better Bayesian learners.

From the perspective of theory-based entrepreneurship, the focus is on learning with theories. Here, the concepts of belief strength and optimism perform a different and more robust role in guiding the learning process. Namely, belief strength and confidence are more directly linked to the causal logic itself rather than some form of optimism. Belief strength relates directly to confidence in the underlying causal structure of an entrepreneur's theory: the interventions, experiments, and overall path outlined for generating some future state. Put differently, strength of belief is confidence in the underlying assumptions that are made and the identification of weakest premises (Ehrig and Schmidt, 2022) or confidence in the causal logic that connects premises (Ehrig and Zenger, 2024a). If, as previously argued, theories are composed of sets of necessary and sufficient conditions or assumptions that must be true to enable a future state, then belief strength is a critical metric by which to assess the soundness or efficacy of a theory, long before a business is launched and the larger belief about achieving a future state can be directly tested.

The process of argumentation (of proposing theories and counter-theories) and debating alternative paths to value (which experiments will prove a belief or solve a problem) can serve as an ex ante mechanism for theory revision and learning. As Ehrig and Schmidt (2022) argue, belief strength surrounding constituent assumptions inherent to a theory is a metric that can be used to establish when and how theories need revision. From a theory-based perspective, building the logically driven (and, eventually, data-supported) case for an entrepreneur's optimism is the object of the entrepreneur's learning process. It is about building confidence. The process here is more than simply updating priors on the larger belief or theory about a future state. It is about elevating confidence in assumptions and ultimately the causal logic that connects them (Ehrig and Zenger, 2024a). In this manner, belief strength is both the object of and an impetus for learning. Ultimately, entrepreneurs seek to gain confidence in their beliefs sufficient to justify and enable launching a venture that they believe will validate their beliefs.

We wholeheartedly embrace the call in Agrawal et al. (2025) for more formal foundations to our understanding of entrepreneurship. But, in addition to Bayesianism, these foundations must comprise unawareness and belief revision modeling. There are valuable linkages among these new concepts that need to be explored. For instance, causal logic can be used to directly define state spaces (Ehrig and Zenger, 2024a,b). State spaces associated with a theory can be modeled as a vector of states defined by combinations of assumptions being true or false, or subproblems being solved or not solved. A confident or optimistic entrepreneur may assign a high probability and payoff to the state in which all assumptions are true or subproblems resolved, and lower probabilities to other states in which one or more assumptions are false or subproblems unsolved. Through data gathering, entrepreneurs can then update their priors as defined by the causal logic and conditional beliefs inherent to a theory. To illustrate, the Airbnb founders may place a high initial probability on their capacity to solve the aforementioned four key subproblems and place a high probability on the conditional belief that, with these subproblems solved, the larger problem will be resolved as well. Note that although evidence gathering may update probabilities about assumptions and subproblems and therefore state spaces in a Bayesian fashion, the test of the capacity to solve the larger problem—the test of what happens when all assumptions are rendered true—often requires launching the business (Ehrig and Zenger, 2024a). But given the high cost of launching the venture, as well as the largely unrepeatable nature of such experimentation, the resulting learning is not of a simple Bayesian form.

Of course, frequently, the process of learning about assumptions and thus updating probabilities on states generates updates that undermine the theory's causal logic. Subproblems are found to be unsolvable or assumptions likely untrue, generating a need to revise causal logic. Entrepreneurs must find an alternate set of assumptions and causal logic to solve the selected problem, or entrepreneurs must simply pivot to another problem and theory (Ehrig and Zenger, 2024a). Importantly, these processes of theory revision and problem selection, central to the entrepreneur's learning and exploration, are distinctly non-Bayesian.

### 6.3.3 Belief Updating

A third core claim is that beliefs are updated via Bayes' rule through evidence gathering. We, of course, concur and see the process of evidence gathering to update beliefs as central to the entrepreneur's value creation efforts. But by placing theories, their construction, and revision more centrally in a theory of the entrepreneurial process, belief updating takes on greater complexity and meaning. In particular, causal theories expand our understanding of belief updating in four ways.

First, emphasizing causal theories highlights that belief updating from evidence involves more than Bayesian updating about priors on the overarching belief. Evidence may also inform the validity of assumptions and thus highlight how the process of belief or theory revision should proceed, or precisely when the overarching belief itself should be abandoned and a wholesale new problem or new theory should be taken up. As highlighted in our discussion of Pearl's work, this process of theory writing, rewriting, or abandonment is non-Bayesian in nature.

Second, an emphasis on causal theories highlights how entrepreneurs learn with a paucity of data. Entrepreneurs are not simply looking at (or for) evidence that validates the possibility of achieving an end state in which a problem is solved. Rather, they are looking at evidence through the lens of their causal logic, observing evidence that supports assumptions or enables the resolution of subproblems. Importantly, an entrepreneur's causal logic enables them to generate interventions or experiments that create *new* evidence. Belief updating therefore is theory guided. It entails entrepreneurs rationally learning from arguments (i.e., hypothetical causal logic advocated by others). As Ehrig and Schmidt (2022) demonstrate, this non-Bayesian process can be rational and follow the axioms of belief revision.

Third, causal theories provide guidance on precisely what forms of evidence to seek out and what experiments to run. Evidence gathering is more than composing minimum viable products by which to update estimates that a theory is valid. Customer-related validation tools (like MVPs) might provide a certain type of "evidence." But in many cases, updating beliefs based on this form of evidence would merely lead entrepreneurs astray (Felin et al., 2024). Entrepreneurial theories provide crucial guidance as to which evidence to take seriously and which not. And importantly, theories provide guidance for delineating the types of experiments that are needed to validate a theory (or its subcomponent). The theory itself enables entrepreneurs to define critical experiments to test the weakest premise (Ehrig and Schmidt, 2022) or to search out solutions to the most difficult to solve subproblems. In short, theories provide a roadmap for experimentation and data gathering.

Fourth, the possession of a well-formulated theory is essential to learning from the inevitable surprise that data gathering generates. In fact, it is only through the lens of a theory that surprises are recognized in the first place. As Herbert Simon observed in his article titled 'The Scientist As Problem Solver': "One good reason for running an experiment or for spending one's time just observing phenomena closely is that you may be surprised. The best things that come out of experiments are things that we did not expect to come out—especially those that we couldn't even have imagined in advance as possibilities" (Simon, 1989, p. 389). Of course, such learning from surprises is a distinctly non-Bayesian activity.

## 6.4 Practical Implications and Conclusion

The perspective outlined in this paper has important practical implications for both entrepreneurs and those who seek to support them. The guidance, interventions, and treatments that are given, prescribed, or used by entrepreneurs are critically important in shaping not just start-up performance but also broader outcomes related to innovation and public policy.

For entrepreneurs, we underscore the critical importance of developing and testing causal theories through belief revision rather than simply relying on Bayesian learning. We argue that the most effective entrepreneurs are those who can articulate clear causal theories and identify key assumptions to test and problems to solve through design-revealing experiments while remaining open to surprising results. Entrepreneurs must use not only Bayesian-powered decision tools but also other tools that focus on theory formation and problem solving (Felin et al., 2021), belief revision (Ehrig and Schmidt, 2022), and unawareness (Bryan et al., 2022).

For educators, incubators, and policymakers, the key takeaway is that the most helpful interventions may be those that push entrepreneurs to refine their causal logic. Rather than just encouraging aspiring founders to envision a future state, we should strive to help them craft robust theories of how to achieve that state. Furthermore, our arguments question the foundations of popular approaches such as lean start-up, which form the intellectual basis of advice given in many incubators, university entrepreneurship classes, and even the National Science Foundation's Innovation Corps program. Although lean start-up claims to be "an application of the scientific method to entrepreneurship" (Blank and Eckhardt, 2023, p. 2), its central mechanism of validation—customers and MVPs—is not a panacea (Felin et al., 2024). The Bayesian approach to entrepreneurship, as outlined by Agrawal et al. (2025), thankfully does not succumb to this form of simplicity. But for educators, incubators, and policymakers alike, it is imperative to teach not only Bayesian decision-aid tools but also tools that emphasize how theory-based causal logic can direct entrepreneurs to utilize *various* forms of experimentation in start-up-specific ways. We need educational interventions that teach entrepreneurs how to rationally compose and revise theories to balance an unrelenting push in our teaching for entrepreneurs to engage in experimentation and testing—important as these are. And rather than prescribe one-size-fits-all solutions that focus on customer validation through tools such as MVPs, start-up-specific entrepreneurial theories need to guide the process of experimentation and validation.

Ultimately, the goal should not be to determine whether entrepreneurs are (or should be) strict Bayesians but rather to develop pragmatic theories and practices that actually enhance value creation and enable better decision-making under uncertainty. By recognizing the central role of causal theories in the entrepreneurial process and by embracing the messy, non-Bayesian nature of entrepreneurial learning, we can take important steps in that direction.

In conclusion, although Bayesian learning is undoubtedly an important part of what entrepreneurs do, it is not the whole story. The formation and testing of causal theories—essentially, a scientific entrepreneur's cheat code for navigating uncertainty—deserves at least equal billing. Thus, we as educators must invest in technologies and modeling approaches that support entrepreneurs in their theory formation, problem solving, belief revision, and understanding of unawareness. By understanding the interplay between various modes of learning (and their contingent uses) and by supporting entrepreneurs in developing and refining their theories, we can help unlock the full potential of entrepreneurial value creation.

## References

Adner, R., and Levinthal, D. A. (2024). Strategy experiments in nonexperimental settings: Challenges of theory, inference, and persuasion in business strategy. *Strategy Science*, 9(4), pp. 311–321.

Agrawal, A., Camuffo, A., Gambardella, A., Gans, J. S., Scott, E. L., and Stern, S. (2025). The foundations of Bayesian entrepreneurship. *Bayesian entrepreneurship*. MIT Press.

Agrawal, A., Gans, J. S., and Stern, S. (2021). Enabling entrepreneurial choice. *Management Science*, 67(9), pp. 5510–5524.

Augenblick, N., and Rabin, M. (2021). Belief movement, uncertainty reduction, and rational updating. *The Quarterly Journal of Economics*, 136(2), pp. 933–985.

Blank, S. (2013). Why the lean start-up changes everything. *Harvard Business Review*, 91(5), pp. 63–72.

Blank, S., and Eckhardt, J. T. (2023). The lean startup as an actionable theory of entrepreneurship. *Journal of Management*, 50(8), pp. 3012–3034.

Bryan, K., Ryall, M., and Schipper, B. C. (2022). Value capture in the face of known and unknown unknowns. *Strategy Science*, 7(3), pp. 157–189.

Camuffo, A., Cordova, A., Gambardella, A., and Spina, C. (2020). A scientific approach to entrepreneurial decision-making: Evidence from a randomized control trial. *Management Science*, 66(2), pp. 564–586.

Chater, N., Oaksford, M., Hahn, U., and Heit, E. (2010). Bayesian models of cognition. *WIREs Cognitive Science*, 1(6), pp. 811–823.

Coali, A., Gambardella, A., and Novelli, E. (2024). Scientific decision-making, project selection and longer-term outcomes. *Research Policy*, 53(6), p. 105022.

Contigiani, A. (2023). Experimentation and appropriability in early-stage ventures: Evidence from the US software industry. *Strategic Management Journal*, 44(9), pp. 2128–2174.

Darwiche, A., and Pearl, J. (1997). On the logic of iterated belief revision. *Artificial Intelligence*, 89(1–2), pp. 1–29.

Ehrig, T., and Schmidt, J. (2022). Theory-based learning and experimentation: How strategists can systematically generate knowledge at the edge between the known and the unknown. *Strategic Management Journal*, 43(7), pp. 1287–1318.

Ehrig, T., and Zenger, T. (2024a). Rational Confidence. Universite Libre de Bruxelles, Working Paper.

Ehrig, T., and Zenger, T. (2024b). Competing with theories: Using awareness and confidence to secure resources and rents. *Strategy Science*, 9(4), pp. 416–432.

Felin, T., Gambardella, A., Novelli, E., and Zenger, T. (2024). A scientific method for startups. *Journal of Management*, 50(8), pp. 3080–3104.

Felin, T., Gambardella, A., and Zenger, T. (2021). Value lab: A tool for entrepreneurial strategy. *Management and Business Review*, 1(2), pp. 68–76.

Felin, T., and Koenderink, J. (2022). A generative view of rationality and growing awareness†. *Frontiers in Psychology*, 13. https://doi.org/10.3389/fpsyg.2022.807261.

Felin, T., Koenderink, J., and Krueger, J. I. (2016). Rationality, perception, and the all-seeing eye. *Psychonomic Bulletin Review*, 24(4), pp. 1040–1059.

Felin, T., and Zenger, T. R. (2009). Entrepreneurs as theorists: On the origins of collective beliefs and novel strategies. *Strategic Entrepreneurship Journal*, 3(2), pp. 127–146.

Felin, T., and Zenger, T. R. (2017). The theory-based view: Economic actors as theorists. *Strategy Science*, 2(4), pp. 258–271.

Gans, J. S., Stern, S., and Wu, J. (2019). Foundations of entrepreneurial strategy. *Strategic Management Journal*, 40(5), pp. 736–756.

Kersten, D., Mamassian, P., and Yuille, A. (2004). Object perception as Bayesian inference. *Annual Review of Psychology*, 55(1), pp. 271–304.

Knight, F. H. (1921). *Risk, uncertainty, and profit*. Houghton Mifflin.

Koenderink, J. (2015). To Bayes or not to Bayes... *Perception*, 45(3), pp. 251–254.

Koning, R., Hasan, S., and Chatterji, A. (2022). Experimentation and startup performance: Evidence from A/B testing. *Management Science*, 68(9), pp. 6434–6453.

Leatherbee, M., and Katila, R. (2020). The lean startup method: Early-stage teams and hypothesis-based probing of business ideas. *Strategic Entrepreneurship Journal*, 14(4), pp. 570–593.

Pearl, J. (2000). *Causality: Models, reasoning, and inference*. Cambridge University Press.

Pearl, J. (2001). *Bayesianism and causality, or, why I am only a Half-Bayesian*. In D. Corfield and J. Williamson (Eds.), *Foundations of Bayesianism* (pp. 19–36). Springer Netherlands.

Pearl, J., and Mackenzie, D. (2018). *The book of why: The new science of cause and effect*. Basic Books.

Pilgrim, C., Sanborn, A., Malthouse, E., and Hills, T. T. (2024). Confirmation bias emerges from an approximation to Bayesian reasoning. *Cognition*, 245, p. 105693.

Pinker, S. (2022). *Rationality: What it is, why it seems scarce, why it matters*. Penguin.

Ramsey, F. (1931). *Foundations of mathematics and other logical essays*. Cambridge University.

Shelef, O., Wuebker, R., and Barney, J. B. (2025). Heisenberg effects in experiments on business ideas. *Academy of Management Review*, 50(3), 560–580.

Simon, H. A. (1989). The scientist as problem solver. In D. Klahr and K. Kotovsky (Eds.), *Complex information processing: The impact of Herbert A. Simon* (pp. 375–398). Lawrence Erlbaum Associates.

Wuebker, R., Zenger, T., and Felin, T. (2023). The theory-based view: Entrepreneurial microfoundations, resources, and choices. *Strategic Management Journal*, 44(12), pp. 2922–2949.

Zellweger, T. M., and Zenger, T. R. (2023). Entrepreneurs as scientists: A pragmatist approach to producing value out of uncertainty. *Academy of Management Review*, 48(3), pp. 379–408.

# 7

# A Scientific Approach to Entrepreneurial Decision-Making as Bayesian Learning

Elena Novelli

## 7.1 Introduction

Recent academic work has advanced the concept of Bayesian entrepreneurship (Agrawal et al., 2025; Chavda et al., 2024; Ehrig and Schmidt, 2022; Zellweger and Zenger, 2022). This concept refers to how entrepreneurs make decisions. Central to Bayesian entrepreneurship is the idea that entrepreneurs face choices characterized by high degrees of uncertainty. Bayesian entrepreneurs address this uncertainty by learning about their environment and the cause–effect linkages associated with their actions and by acting in ways that maximize their subjective utility. Specifically, they develop subjective beliefs (or prior distributions) regarding the probabilities of various opportunities or scenarios they might encounter and select those they deem most likely to be conducive to the desired outcomes. Aware of the uncertainty in their prior distributions, Bayesian entrepreneurs collect signals from the environment in a deliberate attempt to learn about the entrepreneurial opportunities they envision. These signals function similarly to a likelihood distribution, allowing them to update their prior probability distributions and develop posterior distributions, which then inform subsequent entrepreneurial actions.

The notion that entrepreneurs can learn in a Bayesian manner adds a significant perspective to the lenses through which entrepreneurial decision-making can be interpreted, potentially leading to valuable insights. To advance our understanding of whether, to what extent, and how this approach can offer value to entrepreneurs, three fundamental issues arise. First, although the key elements of a Bayesian learning-based decision-making approach are clear (prior formation, testing, and updating), it is not obvious how such an approach works in practice. Second, there is a lack of evidence on the implications of Bayesian decision-making approaches in entrepreneurship. Finally, a relevant issue concerns the extent to which a Bayesian approach to entrepreneurship can be effectively taught, as this could affect the phenomenon's magnitude under investigation.

This chapter aims to begin addressing these issues by building on insights from recent research that has promoted a scientific approach to decision-making in entrepreneurship (Agarwal et al., 2025; Camuffo, Gambardella, Messinese, et al., 2024; Novelli and Spina, 2024; Zellweger and Zenger, 2022). A scientific approach involves theory development, hypothesis formation, testing, and evaluation, applying the principles of the scientific learning process to the context of entrepreneurship. Notably, a scientific decision-making approach shares important similarities with Bayesian learning, particularly in how decision-makers rely on prior beliefs and engage in a deliberate process of evidence collection to update these beliefs and guide their actions. The purpose of this chapter is to present the most recent empirical results on a scientific approach to decision-making and, by highlighting the similarities with Bayesian entrepreneurship, to offer insights that might inform and support future research in this area. In doing so, we also hope to contribute to filling a broader empirical gap: Existing field evidence on Bayesian learning is limited, with current evidence mostly derived from environments where true probabilities are known or can be reliably estimated, such as gambling, lotteries, and sports (Benjamin, 2019).

To this end, the chapter is structured as follows. First, we provide an overview of the scientific approach to entrepreneurial decision-making, highlighting its similarities to the process of Bayesian learning. Next, we describe the methodology employed in the recent research program on this scientific approach. Following this, we present the key findings from this body of research and discuss their implications for the study of Bayesian entrepreneurship. Finally, we conclude by identifying open questions in this area that could guide future research in Bayesian entrepreneurship.

## 7.2 A Scientific Approach to Decision-Making as Bayesian Learning

Recent research has conceptually advanced and empirically explored the implications of a scientific approach to decision-making (Agarwal et al., 2025; Camuffo, Gambardella, Messinese, et al., 2024; Novelli and Spina, 2024; Zellweger and Zenger, 2022). In contrast with the idea that entrepreneurs make decisions following their gut feelings (Bennett and Chatterji, 2019), the approach advocates that entrepreneurs can benefit when informing their decisions with a deliberate and structured approach to learning, which follows the same structured approach used by scientists when learning about phenomena. Specifically, scientific entrepreneurs follow four fundamental steps:

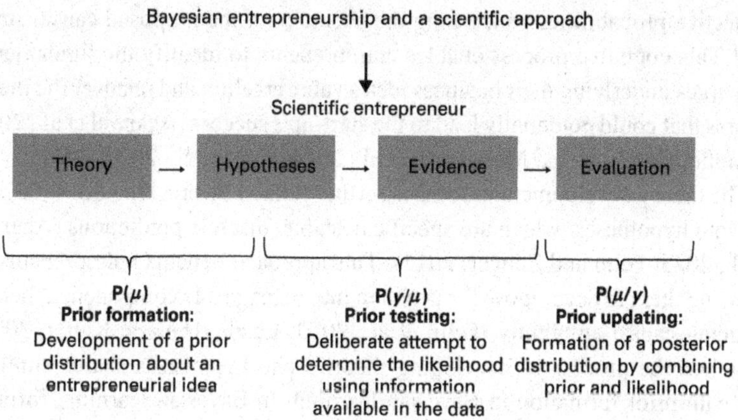


**Figure 7.1**
Bayesian entrepreneurship and a scientific approach to entrepreneurial decision-making.

1. The formulation of a theory of value
2. The derivation of hypotheses from this theory
3. The acquisition of evidence to support or challenge these hypotheses
4. The rigorous evaluation of the collected evidence

The study of a scientific approach to decision-making represents a potential opportunity for the advancement of research on Bayesian entrepreneurship because these four steps essentially reflect important elements of Bayesian learning. The process of theory and hypotheses development can be equated to a process of prior distribution formation; the process of evidence gathering resembles a process of likelihood estimation; the disciplined evaluation of the evidence collected can be reconnected to a process of posterior development (figure 7.1).

### 7.2.1 Theory and Hypotheses Development as Prior Formation

Entrepreneurs using this scientific approach start by defining the problem with a theory of value, which serves as a conceptual framework for understanding how their business generates value (Felin and Zenger, 2017; Zellweger and Zenger, 2022). This framework involves linking the core elements of their business idea with relevant environmental contingencies. Using the language of Camuffo, Gambardella, and Pignataro (2024), these core elements can be referred to as attributes. These attributes are interconnected through causal links. As part of the theory development process, entrepreneurs articulate their prior beliefs or subjective probabilities concerning the accuracy of the proposed causal structure. This cognitive process enables entrepreneurs to identify the fundamental principles underlying their business idea's value creation and uncover the mechanisms that could potentially lead to the start-up's success (Agarwal et al., 2025; Camuffo, Gambardella, Messinese, et al., 2024; Felin et al., 2020, 2024).

The theory development process is further refined by breaking down the theory into hypotheses, which are specific, testable, discrete predictions (Agarwal et al., 2025; Felin and Zenger, 2017). This approach benefits entrepreneurs by allowing them to decompose the problem into manageable components, thereby reducing causal ambiguity (Felin et al., 2020; Leatherbee and Katila, 2020). Together, the process of developing theories and hypotheses shares similarities with prior formation in Bayesian learning. In Bayesian learning, forming priors involves specifying probability distributions that encode initial beliefs about the model parameters based on existing knowledge, thereby creating a baseline understanding. The definition of attributes, causal linkages, and beliefs in the entrepreneurial theory parallels the definition of parameters, structure in a Bayesian model and prior distributions.

### 7.2.2 Evidence Gathering as Prior Testing

In the context of the scientific approach to decision-making, defining a research design and collecting relevant data to test hypotheses provide critical feedback that can be used as a signal on the validity of the theory of value. This process is consistent with Bayesian learning, in that both processes involve a feedback loop in which new data are collected to refine initial assumptions. In both Bayesian learning and the scientific approach, new data provide evidence that strengthens or weakens the plausibility of the initial beliefs or hypotheses. This signal informs whether the entrepreneur's current understanding requires revision.

### 7.2.3 Evidence Evaluation as Prior Updating

Finally, in the scientific approach, the methodical and critical evaluation of the gathered evidence enables entrepreneurs to compare the results against the original theory and determine whether the evidence collected supports it, refines it, or refutes it. This is similar to the process of prior updating in Bayesian learning, where initial beliefs (priors) are assessed against new data. In Bayesian learning, new evidence is combined with the prior distribution via Bayes' rule to produce a posterior distribution. If the evidence is probable under the prior the posterior remains close to the prior; if it refutes it, the original belief is adjusted.

## 7.3 Scientific Decision-Making in Entrepreneurship: A Research Program

The scientific approach to decision-making has been empirically studied through a large-scale research program involving approximately four thousand entrepreneurs. This program began with an initial study in Milan (Camuffo et al., 2020) and has since expanded to include a series of international randomized controlled trials (RCTs) in various locations: London, Turin, Hyderabad, Shanghai, Dar-Es-Salaam, Rotterdam, Bogotá, Barcelona, and Chennai. Although the specific design and treatments varied across the different RCTs, reflecting the unique research questions each aimed to address, the program maintained a common methodology: Entrepreneurs were randomly assigned to different treatment conditions and monitored for at least ten months. This period was used to observe the impact of the treatments on their decisions and performance across a range of locations. The rationale for conducting multiple RCTs was to ensure reproducibility across different settings and to enhance statistical power. This approach is consistent with the research design implemented in other large-scale RCTs across various contexts (Banerjee et al., 2015; Bowers et al., 2017; Davis et al., 2023).

### 7.3.1 Intervention Details

Experimental groups participated in an entrepreneurship business support program consisting of approximately twenty hours of training over a three-month period. The sessions were experiential, with small class sizes ensuring personalized feedback. In the first batch of RCTs (RCTs 1–4), the program aimed to compare the effectiveness of a structured scientific approach against traditional entrepreneurship training methods. Participants were divided into a treatment group and a control group. Both groups received training that included elements of cognitive-based decision-making, such as strategy frameworks and tools (e.g., the Business Model Canvas or Balanced Scorecard), and evidence-based decision-making, incorporating various data collection and testing techniques such as surveys, qualitative interviews, and A/B testing to suit different entrepreneurial contexts. However, only the treatment group was explicitly encouraged to, adopting a scientific approach to decision-making. Specifically, the treatment group was instructed to use the strategy frameworks introduced in the training to develop a theory of value, derive hypotheses from this theory, and subsequently use data-gathering and analysis techniques to test these hypotheses. Coali, Novelli, et al. (2024) provide more insights on the characteristics of the andragogical tools that can support the delivery of these types of interventions.

### 7.3.2 Data-Collection Process

Entrepreneurs provided data on their decision-making processes and business performance throughout the training program via monthly telephone interviews conducted by research assistants (RAs). RAs were rigorously trained to ensure consistency in data collection and coding. Interviews included closed- and open-ended questions on various aspects of business development, with data recorded and stored securely. Several measures were implemented to minimize attrition, such as offering the program free of charge through prestigious universities and organizing monthly events (identical but run separately for treatment and control) to encourage ongoing participation.

## 7.4 Key Findings

The current available empirical evidence on a scientific approach refers to three key entrepreneurial outcomes: project termination, pivoting, and economic performance. Results from this stream of research provide insights that can be used to understand more about the implications of Bayesian entrepreneurship.

### 7.4.1 Project Termination

**Bayesian entrepreneurship and termination** One important entrepreneurial outcome is project termination. As ventures progress, entrepreneurs naturally or deliberately gather data and receive signals about the market, customer interest, competition, and overall feasibility. This new information serves as evidence that can be used to update the prior probability into a "posterior probability"—a refined estimate of the venture's success prospects. For example, if an entrepreneur initially believes that a new product has a high probability of success but then encounters consistent market feedback indicating a lack of interest, Bayesian reasoning would involve adjusting this belief (or probability estimate) downward.

However entrepreneurs might suffer from cognitive biases, such as overconfidence or the sunk cost fallacy, leading them to ignore or misinterpret new evidence (Chen et al., 2018). Instead of adjusting their beliefs in response to negative data, they might irrationally maintain the original high probability of success. This behavior can result in continued investment until external constraints, such as running out of funds, force termination (Artinger and Powell, 2016).

**A scientific approach and termination: Empirical evidence** The research program on the scientific approach provides evidence relevant to the dynamics discussed above. In a large-scale study involving 759 entrepreneurs observed for up to eighteen months following the start of their training, Camuffo, Gambardella, Messinese, et al. (2024) demonstrate that a scientific approach to decision-making is associated with more frequent and timely project terminations. Specifically, entrepreneurs trained in the scientific method are 9.8 percentage points more likely to terminate their projects than traditional entrepreneurs and do so 2.7 weeks earlier. This aligns with the patterns observed in Bayesian entrepreneurship, wherein scientifically trained entrepreneurs are better able to identify unpromising ideas more quickly.

Firms may choose to terminate projects or ventures for a variety of reasons, each with different implications (Wennberg and DeTienne, 2014). To distinguish between these scenarios and understand the underlying mechanisms, Coali, Gambardella, et al. (2024) go beyond merely examining the termination decision itself. They explore the entrepreneurs' own estimates of the value of their projects throughout the observation period, particularly before the decision to terminate. Their findings indicate a general disconnect between termination decisions and entrepreneurs' expectations prior to the termination event, suggesting that entrepreneurs do not always anticipate termination. However, this disconnect is less pronounced among entrepreneurs trained to think and act like scientists. Notably, these trained entrepreneurs tend to lower their expectations about the value of their ideas and anticipate project termination earlier than those in the control group. This behavior supports the notion that scientifically trained entrepreneurs act as Bayesian decision-makers, adjusting their initial beliefs as they gather new data.

This intuition is further supported by the analysis of the longer-term implications of the scientific approach. Although entrepreneurs in the treatment group showed a higher probability of terminating their projects in the short term (up to eighteen months after the start of the program), the long-term results (up to five years after the start of the program) indicate that the proportion of projects terminated by treated and control groups is *not* significantly different. Over time, the initial discrepancy in termination rates between the two groups leveled out. By 2023, the termination rate was 75.8 percent for treated entrepreneurs compared with 81.8 percent for the control group. Interestingly, the results also show that, after termination, treated entrepreneurs initiated more new projects compared with their control group counterparts, with an average of 0.805 new projects per treated entrepreneur versus 0.484 for those in the control group. This suggests the possibility, which should be tested by further research, that treated entrepreneurs who terminated projects earlier were able to redirect resources more effectively in line with Felin et al.'s (2020) intuition. This finding is significant, as it demonstrates that, over time, all entrepreneurs, regardless of whether they adopt a deliberate learning process, adjust their expectations downward, likely due to natural market feedback. However, the deliberate process followed by scientific entrepreneurs, combined with their willingness to base decisions on learned insights, enables them to update their expectations more quickly in line with Bayesian updating.

Coali, Gambardella, et al. (2024) address whether the downward adjustment made by scientific decision-makers accurately also reflects an update of their priors based on environmental signals, or if they might react too negatively, potentially increasing false negatives while reducing false positives. To determine which of these two possible interpretations is more likely, they make the following assumption: If the second interpretation is correct and the treatment made entrepreneurs excessively critical, one should observe that the projects that are discarded by treated entrepreneurs are of higher quality compared with those discarded by the control group. To investigate this, they used two relatively "objective" indicators of the value of entrepreneurial projects: (1) the successful acquisition of external funding (blind to the treatment) and (2) expert assessments of projects at the baseline. Their examination of the terminated projects reveals that the proportion of terminated projects securing external funding did not significantly differ between treated and control groups, nor were the expert assessments at the baseline superior for projects terminated by treated entrepreneurs. This suggests that the scientific approach does not lead to an overcorrection in the idea selection process.

### 7.4.2 Pivoting

**Bayesian entrepreneurship and pivoting** Another important entrepreneurial outcome is pivoting, which refers to a change in a firm's strategy that reorients its strategic direction (Kirtley and O'Mahony, 2023, p. 199; see also Gans et al., 2019; Pillai et al., 2020). Given the uncertainty faced by entrepreneurs, their ideas are subject to change over time. For entrepreneurs using a Bayesian approach, pivots are a natural outcome of the learning process. Their ideas and specific configurations of their value propositions are based on prior beliefs (Felin and Zenger, 2017; Zellweger and Zenger, 2022). As they update these priors with new information from the environment, their value propositions also evolve, leading to pivots. However, Bayesian entrepreneurs, guided by a theoretical framework and structured prior beliefs, are more likely to conduct a highly targeted search process (Felin et al., 2023). This suggests that although Bayesian entrepreneurs may pivot more frequently, their pivots are more focused: When they pivot, they do so based on signals obtained through their structured learning process, with a clear and well-reasoned configuration in mind, rather than pivoting indefinitely in a trial-and-error manner.

In the context of a scientific approach to decision-making, Camuffo, Gambardella, Messinese, et al. (2024) identify two core mechanisms: methodic doubt and efficient search. They suggest that the mechanism of efficient search underscores the superior efficiency of scientific entrepreneurs in exploring potential solutions to business challenges. This efficiency stems from their enhanced ability to prioritize ideas that are more likely to succeed. Conversely, the mechanism of methodic doubt increases their awareness of factors that could diminish the likelihood of an idea's success. Entrepreneurial ideas are considered valuable only when supported by thorough reasoning, strong arguments, and consistent evidence.

**A scientific approach and pivoting: Empirical evidence** In line with the above, Camuffo, Gambardella, Messinese, et al.'s (2024) findings reveal that the relationship between scientific intensity (i.e., the extent to which an entrepreneur's decision-making adheres to the principles of a scientific approach to decision-making) and pivoting is not linear. Instead, scientific decision-makers are more likely to engage in fewer but more focused pivots rather than not pivoting at all or pivoting multiple times. Specifically, the scientific treatment increases the probability of executing a radical pivot once (versus not pivoting or pivoting many times) by 8.3 percentage points.

Valentine et al. (2024) expand on this by exploring the synergies between theorization and evidence evaluation—the two key components of a scientific approach to decision-making, which closely mirror the processes of prior formation, testing, and updating. Their qualitative analysis, provides abundant evidence of the way in which these processes work in the practice of entrepreneurs.

Through a series of case studies, they illustrate how when the process of theorization guides subsequent evidence gathering, pivot become more focused and targeted. The process might also uncover unexpected "surprises"—that is, new relevant attributes that were not initially considered in their priors. Recent work by Gagliardi and Novelli (2025) provides further support for the disciplined-pivot argument. Drawing on three randomised controlled trials that tracked female entrepreneurs and ethnic entrepreneurs, the authors show that using a scientific approach can help user entrepreneurs escape narrow niches, redirect their strategies toward broader and more promising markets, and ultimately improve venture performance.

### 7.4.3 Economic Performance

**Bayesian entrepreneurship and economic performance** An important question in research on Bayesian entrepreneurship concerns whether making decisions based on this approach ultimately leads to superior economic performance. In principle, if Bayesian learning results in a more accurate assessment of the key attributes in the entrepreneur's environment and the causal linkages between them, it could be argued that this should enhance the entrepreneur's ability to optimize actions and, consequently, maximize utility. Therefore, it is rational to assume that improved economic performance would be a desirable outcome of this approach.

**A scientific approach and economic performance: Empirical evidence** Camuffo, Gambardella, Messinese, et al. (2024) demonstrate that entrepreneurs who adopt a scientific approach generally achieve superior economic performance, generating more revenue compared with control firms, conditional on remaining active. In the top 25 percent of cases, start-ups using the scientific method generated an average of €28,000 more than control companies over the course of the experiment. For the top 5 percent, these start-ups earned an average of €492,000 more.

Additional insights are provided by a study conducted by Agarwal et al. (2025), which underscores the importance of theory-based prior formation over an approach focused solely on evidence gathering. In their RCT, they compared a treatment that combined theorization and evidence evaluation as the basis for decision-making with a treatment that emphasized evidence evaluation without theorization. The results indicate that entrepreneurs who developed their priors through a comprehensive articulation of their theory were able to make coordinated changes in both the core and operational elements of their business models. This comprehensive approach led to superior performance.

Novelli and Spina (2024) examine the contingencies that influence prior formation and theory development. They particularly note that the utility of a scientific approach in supporting decisions varies significantly throughout the entrepreneurial journey. Early in the process of business model development, entrepreneurs have made only a low degree of strategic commitment to key business model choices. Radical changes are still possible, and the decision-making process has many degrees of freedom. In contrast, later in the process of business model development, strategic commitments have been made, and decisions involve more granular refinements.

Their study reveals that entrepreneurs exposed to a scientific approach early in their business model development experience a *"back to the drawing board"* effect. They begin to question their macro choices, increasing their epistemic uncertainty about the value of the business model. As a result, they delay action to reconsider their choices with the intention of making changes that will position them on a better trajectory later. Consequently, their economic performance is lower compared with nonscientific entrepreneurs who continue with business as usual. This finding aligns with Gans et al. (2019), who show that feedback early in the search process can induce further exploration.

Conversely, when entrepreneurs are exposed to a scientific approach later in their business model development, after making strategic commitments on core business model choices, they undergo a process of fine-tuning the existing model and quickly refine and improve it. As a result, their economic performance increases rapidly. Treated firms that were at an average degree of business model development at the program's start experienced revenues that were 13 percent lower than those of the control group by the end of the program. However, treated firms that were one standard deviation above the mean in their degree of business model development at the program's start reported revenues that were 38 percent higher than the control group at the end of the program.

## 7.5 Discussion and Open Questions for Future Research on Bayesian Entrepreneurship

Overall, the insights from research on a scientific approach suggest that entrepreneurs who base their decision-making on a clear articulation of their prior beliefs, and actively update those beliefs with evidence, tend to engage in earlier project selection and more focused pivots, and achieve superior economic performance. The findings also indicate that such decision-making approaches can be effectively taught to entrepreneurs. These insights raise important questions that could inform future research on Bayesian entrepreneurship.

One key area for further exploration is the process of forming prior beliefs. The research on scientific decision-making shows that entrepreneurs can be encouraged to articulate their prior beliefs into a coherent theory of value through dedicated interventions. However, evidence also indicates that some entrepreneurs in the control group demonstrated high levels of theorization without such interventions (Valentine et al., 2024). Therefore, it would be valuable to investigate alternative mechanisms that enable entrepreneurs to develop articulated priors. Valentine et al. (2024) provide qualitative evidence suggesting that entrepreneurial experience is a significant factor for high theorization among control entrepreneurs. Similarly, Gagliardi and Novelli (2025) and Frosi et al. (2023) highlight that an entrepreneur's educational background and domain expertise also influence the development of priors. Jannace and

Camuffo (2023) further show that the priors developed by scientific decision-makers differ in their level of parsimony compared with others. Future research should delve deeper into these aspects to understand the role of environmental and behavioural contingencies in shaping these important cognitive processes.

A second line of inquiry concerns the process of evidence gathering. Qualitative evidence on the scientific approaches to decision-making has revealed that entrepreneurs employ a wide range of methods to collect evidence supporting their prior beliefs. These methods include conducting surveys, detailed customer interviews, A/B testing, as well as conceptual experiments or the observation of comparative case studies of competitors (Camuffo, Gambardella, and Pignataro, 2024; Novelli and Spina, 2024; Valentine et al., 2024). Future research could benefit from developing a comprehensive rubric of all approaches used by entrepreneurs in the process of prior testing under different contingencies and analyzing the implications of different approaches on the development of effective signals.

A third important area for future research involves a closer investigation into the process of updating prior beliefs. Qualitative data collected in the context of a scientific approach to decision-making has revealed that some entrepreneurs exhibit resistance to updating their beliefs. This finding aligns with existing experimental work on Bayesian learning, which has documented various systematic learning biases, such as under- or overreaction to information, overconfidence, and correlation neglect (Benjamin, 2019). Understanding the characteristics that make entrepreneurs more receptive to Bayesian decision-making would be crucial. Finally, future research would benefit from a closer analysis comparing the Bayesian approach to decision-making with other structured approaches. This comparison could help assess the effectiveness of each approach in guiding entrepreneurial decisions, particularly under various types of environmental contingencies.

In conclusion, the findings from research on a scientific approach to decision-making offer valuable insights, suggesting that by fostering a systematic theory-and-evidence-based methodology, Bayesian entrepreneurship can enhance entrepreneurs' ability to make informed and adaptive decisions. As the field of entrepreneurship continues to evolve in an increasingly complex and uncertain world, adopting Bayesian principles could provide a promising pathway to more effective and successful entrepreneurial endeavors.

## References

Agarwal, R., Bacco, F., Camuffo, A., Coali, A., Gambardella, A., Msangi, H., Sonka, S., Temu, A., Waized, B., and Wormald, A. (2025). Does a theory-of-value add value? Evidence from a randomized control trial with tanzanian entrepreneurs. *Organization Science*, 36(2), pp. 601–625.

Agrawal, A., Camuffo, A., Gambardella, A., Gans, J. S., Scott, E. L., and Stern, S. (2025). The foundations of bayesian entrepreneurship. *Bayesian entrepreneurship*. MIT Press, 2025.

Artinger, S., and Powell, T. C. (2016). Entrepreneurial failure: Statistical and psychological explanations. *Strategic Management Journal*, 37(6), pp. 1047–1064.

Banerjee, A., Duflo, E., Goldberg, N., Karlan, D., Osei, R., Parienté, W., Shapiro, J., Thuysbaert, B., and Udry, C. (2015). A multifaceted program causes lasting progress for the very poor: Evidence from six countries. *Science*, 348(6236), p. 1260799.

Benjamin, D. J. (2019). Errors in probabilistic reasoning and judgment biases. In B. D. Bernheim, S. DellaVigna, and D. Laibson (Eds.), *Handbook of behavioral economics—Foundations and applications 2* (Vol. 2, pp. 69–186). North-Holland / Elsevier.

Bennett, V. M., and Chatterji, A. K. (2019). The entrepreneurial process: Evidence from a nationally representative survey. *Strategic Management Journal*, 44(1), pp. 86–116.

Bowers, J., Higgins, N., Karlan, D., Tulman, S., and Zinman, J. (2017). Challenges to replication and iteration in field experiments: Evidence from two direct mail shots. *American Economic Review*, 107(5), pp. 462 465.

Camuffo, A., Cordova, A., Gambardella, A., and Spina, C. (2020). A scientific approach to entrepreneurial decision-making: Evidence from a randomized control trial. *Management Science*, 66(2), pp. 564–586.

Camuffo, A., Gambardella, A., Messinese, D., Novelli, E., Paolucci, E., and Spina, C. (2024). A scientific approach to entrepreneurial decision-making: Large-scale replication and extension. *Strategic Management Journal*, 45(6), pp. 1209–1237.

Camuffo, A., Gambardella, A., and Pignataro, A. (2024). Theory-driven strategic management decisions. *Strategy Science*, 9(4), pp. 382–396.

Chavda, A., Gans, J. S., and Stern, S. (2024). Theory-based entrepreneurial search. *Strategy Science*, 9(4), pp. 397–415.

Chen, J. S., Croson, D. C., Elfenbein, D. W., and Posen, H. E. (2018). The impact of learning and overconfidence on entrepreneurial entry and exit. *Organization Science*, 29(6), pp. 989–1009.

Coali, A., Gambardella, A., and Novelli, E. (2024). Scientific decision-making, project selection and longer-term outcomes. *Research Policy*, 53(6), p. 105022.

Coali, A., Novelli, E., Sirigiri, A., and Spina, C. (2024). A scientific approach to decision-making: Key tools and design principles. *Journal of Business Venturing Design*, 3, p. 100023.

Davis, A. M., Flicker, B., Hyndman, K. B., Katok, E., Keppler, S., Leider, S., Long, X., and Tong, J. (2023). *A replication study of operations management experiments in management science*. SSRN Scholarly Paper (No. 4135500).

Ehrig, T., and Schmidt, J. (2022). Theory-based learning and experimentation: How strategists can systematically generate knowledge at the edge between the known and the unknown. *Strategic Management Journal*, 43(7), pp. 1287–1318.

Felin, T., Gambardella, A., Novelli, E., and Zenger, T. (2024). A scientific method for startups. *Journal of Management*, 50(8), pp. 3080–3104.

Felin, T., Gambardella, A., Stern, S., and Zenger, T. (2020). Lean startup and the business model: Experimentation revisited. *Long Range Planning*, 53(4), pp. 101889.

Felin, T., Kauffman, S., and Zenger, T. (2023). Resource origins and search. *Strategic Management Journal*, 44(6), pp. 1514–1533.

Felin, T., and Zenger, T. R. (2017). The theory-based view: Economic actors as Theorists. *Strategy Science*, 2(4), pp. 258–271.

Frosi, C., Mariani, M., and Gagliardi, L. (2023). Is it a match? Training entrepreneurs to make decisions favors the prepared mind. In *Academy of Management Proceedings* (Vol. 2023, No. 1, p. 14047). Briarcliff Manor, NY 10510: Academy of Management.

Gagliardi, L., and Novelli, E. (2025). Female entrepreneurs targeting women: Strategic redirection under scientific decision-making. *Organization Science*, 36(2), 718–736.

Gans, J. S., Stern, S., and Wu, J. (2019). Foundations of entrepreneurial strategy. *Strategic Management Journal*, 40(5), pp. 736–756.

Jannace, D., and Camuffo, A. (2023). Curbing the addition bias: Scientific approaches and propensity to subtract in entrepreneurial ideas. In *Academy of Management Proceedings* (Vol. 2023, No. 1, p. 15426). Briarcliff Manor, NY 10510: Academy of Management.

Kirtley, J., and O'Mahony, S. (2023). What is a pivot? Explaining when and how entrepreneurial firms decide to make strategic change and pivot. *Strategic Management Journal*, 44(1), pp. 197–230.

Leatherbee, M., and Katila, R. (2020). The lean startup method: Early-stage teams and hypothesis-based probing of business ideas. *Strategic Entrepreneurship Journal*, 14(4), pp. 570–593.

Novelli, E., and Spina, C. (2024). Making business model decisions like scientists: Strategic commitment, uncertainty, and economic performance. *Strategic Management Journal*, 45(13), pp. 2642–2695.

Pillai, S. D., Goldfarb, B., and Kirsch, D. A. (2020). The origins of firm strategy: Learning by economic experimentation and strategic pivots in the early automobile industry. *Strategic Management Journal*, 41(3), pp. 369–399.

Valentine, J., Novelli, E., and Agarwal, R. (2024). The theory-based view and strategic pivots: The effects of formalization in cognition and action on the type and nature of pivots. *Strategy Science*, 9(4), 433–460.

Wennberg, K., and DeTienne, D. R. (2014). What do we really mean when we talk about "exit"? A critical review of research on entrepreneurial exit. *International Small Business Journal*, 32(1), pp. 4–16.

Zellweger, T., and Zenger, T. (2022). Entrepreneurs as scientists: A pragmatist alternative to the creation-discovery debate. *Academy of Management Review*, 47(4), pp. 696–699.

# 8

## Priors, Experiments, Learning, and Persuasion in (Bayesian) Entrepreneurial Finance

Ramana Nanda

### 8.1 Introduction

At the heart of entrepreneurial finance lies a persuasion challenge: Regardless of the strength of an entrepreneur's belief in the potential of their idea, they typically need to convince investors of its potential in order to raise the financial capital required for its commercialization. Can entrepreneurs take certain steps to increase the chance of raising the required external finance from investors? Why does it appear harder to convince venture capital investors to fund start-up commercialization in certain industries and regions and in certain periods of time? Are there systematic frictions preventing entrepreneurs from effectively persuading investors in certain settings, and if so, what can be done to reduce them?

In this chapter, I discuss answers to these and other questions through a conceptual framework in which investors update their beliefs about a start-up's prospects through a sequence of investments over time. In section 8.2, I first show how this learning perspective to multistage financing is closely related to the ideas discussed in Agrawal et al.'s (2025) chapter on Bayesian entrepreneurship. Next, in section 8.3, I discuss how this learning perspective helps explain patterns of venture capital finance across industries and over time that are hard to understand when considering staged financing exclusively as a means of governance to mitigate agency problems. In section 8.4, I discuss important generalizable and teachable lessons related to the practice of entrepreneurial finance that emerge from this perspective. Finally, in section 8.5, I end with some thoughts on the promise of more closely connecting the entrepreneurship literature on Bayesian entrepreneurship outlined by Agrawal et al. (2025) with the economics and finance literatures on learning through experimentation, particularly in the context of entrepreneurial finance.

## 8.2 A Learning Perspective to Multistage Financing

A key contribution of the finance literature on venture capital has been to document the important role that venture capitalists (VCs) play in supporting the growth of start-ups through monitoring, governance, and value-added support of entrepreneurial teams (Sahlman, 1990; Lerner, 1995; Hellmann, 1998; Hellmann and Puri, 2002; Bernstein et al., 2016). Among the many tools that VCs bring to bear in this process, scholars have also documented the particular role of multistage financing in this governance process by enabling investors to provide funding for the venture only when management teams continue to meet targets and hence mitigating agency problems between investors and management (Gompers, 1995; Cornelli and Yosha, 2003; Tian, 2011).

A conceptually distinct role of staged financing, however, is to help both entrepreneurs and investors *learn* about the prospects of a venture before committing further resources toward its development (Metrick and Yasuda, 2006). Although this idea is perhaps implicit in research considering multistage financing from a monitoring and governance perspective, I outline below why it helps to explicitly consider it as a distinct benefit of multistage financing.

First, many entrepreneurial ventures, particularly those commercializing entirely new products or technologies, face fundamental uncertainty about the venture's prospects. Which of the many novel cancer treatments currently under development will pass multiple clinical trials and be deemed efficacious enough to become commercially available treatments? Which of the several battery technologies under development will be sufficiently cost-effective to be adopted at scale? Which of the hundreds of consumer-facing software applications being developed will gather sufficient traction with consumers to make them worth financing to scale? Although asymmetric information and agency frictions between investors and entrepreneurs are always relevant, a large share of the uncertainty in these contexts is likely to be symmetric—in that neither the entrepreneur nor the investor knows the answers to those questions.

Second, and perhaps more importantly, entrepreneurs and investors do not just passively receive cost-less signals about the ultimate promise of a venture. More information about the venture's potential can only be generated through an investment into the start-up (Bergemann and Hege, 2005; Bergemann et al., 2008; Metrick and Yasuda, 2006). Entrepreneurs and investors can make decisions about what they want to learn, in what sequence, and with what fidelity. This approach to learning builds on a long literature in economics that considers the optimal approach to acquiring or generating economically valuable information in dynamic settings (Grossman et al., 1977; Weitzman, 1979; Aghion et al., 1991; Bolton and Harris, 1999; Bergemann and Hege, 1998, 2005). In this conceptualization, the act of incurring a cost to generate the relevant information is described as conducting an experiment. The experiment generates information that allows decision-makers to update their prior beliefs. Learning stems from priors being updated (which is typically assumed to happen using Bayes' rule). In turn, learning enables economic agents to make better decisions.

Explicitly focusing on the learning perspective to staged financing is therefore valuable because, in this framing, one of the important ways in which value-added investors such as VCs can help entrepreneurs is by designing more effective experiments that allow for better decision-making around the future path of the venture completely independent of agency considerations. It also allows for an understanding of how costs and constraints to the ability to learn and/or persuade investors can have consequential effects on the rate and direction of venture capital in a manner that is distinct from factors related to monitoring and governance.

This learning perspective to multistage financing is closely related to the ideas discussed in Agrawal et al.'s (2025) chapter on Bayesian entrepreneurship. Indeed, they share an intellectual lineage in that they both build on fundamental insights emerging from information and decision theory and can both be seen as trying to understand how to improve entrepreneurial decision-making in the context of uncertainty.

### 8.2.1 A Framework for Learning Through Multistage Financing

To guide the discussion in the next two sections, I outline a simple framework for multistage financing that is focused on learning through experimentation. To focus on the insights emerging from this learning perspective, the framework abstracts away from potential agency frictions between entrepreneurs and investor and focuses attention on the investor's beliefs. Specifically, I consider a start-up that requires $\$K$ to commercialize its technology that may or may not work. The investor believes that it will be successful with probability $p_0$ and, if so, worth $V$, whereas with probability $1-p_0$, it will be worth nothing. Abstracting away from the discount rate for simplicity, the expected value of the project is $p_0 V - K$, and the venture will only be financed if $p_0 V > K$.

Alternatively, the investor can finance an experiment at cost $C$ before committing the full amount $K$. The result of the experiment could involve achieving some milestone such as showing initial customer traction or initial proof of technological success in the lab. The likelihood that the experiment generates positive intermediate information and passes the milestone is $p_M$, whereas the likelihood of generating negative information and failing to achieve the milestone is $(1-p_M)$. On passing the milestone, the investor updates their belief about the chance of ultimate success to $p_P$, whereas their posterior probability

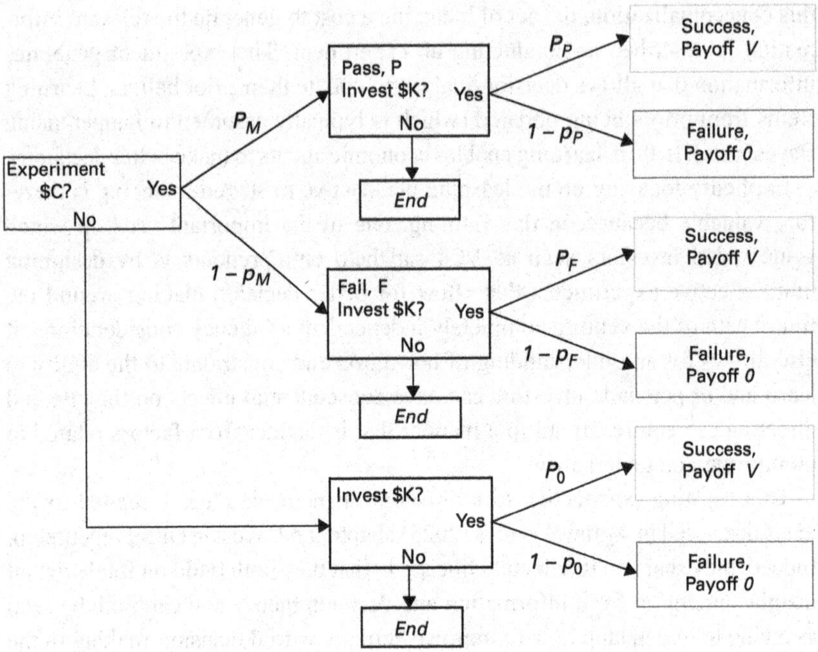


**Figure 8.1**
Investor's decision tree.

of the chance of success if the experiment fails is $p_F$. To be equivalent to the scenario when no experiment is run, it is assumed that $p_P \times p_M \times p_F \times (1 - p_M) = p_0$—that is, the unconditional probability of success is the same whether or not the experiment is run. The investor's decision tree is depicted in figure 8.1.

Looking at the investor's decision tree highlights two important points. First, financing an experiment is not always the dominant strategy for the investor. The investor will only prefer to finance an experiment in cases where it impacts their decision on whether to invest. This cannot hold, for example, for any project that has a positive expected value even after the experiment fails or for a project that has negative expected value even after the experiment succeeds. In these cases, running the experiment is just a waste of resources because the information generated has no impact on the investor's decision to invest or not. However, suppose that $p_0 \times V - K < 0$ and that $p_P \times V - K > 0$ while $p_F \times V - K < 0$. Then, the investor's default position without an experiment is not to invest, but if the investor financed the experiment, they would want to reinvest if the venture meets the milestone and avoid investing if the venture fails to meet the milestone. The investor would therefore be willing to pay to update their beliefs about the likelihood of ultimate success based on whether the venture met the milestone, since the results of the experiment can change their default decision not to invest.[1]

Second, since the value of an experiment stems from its ability to change decisions, the experiment that is most valuable to a decision-maker depends on *their specific prior belief* about the venture's prospects. In this sense, the most relevant experiment in a given context is intricately tied to the priors of the decision-maker. Later, we discuss how the amount the investor is willing to pay to finance the experiment is also related to the degree they are able to update their beliefs (i.e., how much they learn from the experiment's results). For now, we can see that financing the experiment is most valuable to the investor in situations when relatively small investments can reveal information that impacts their investment decision, and that the ideal experiment itself develops on their priors.

## 8.3 Implications for Entrepreneurial Finance

In this section, I discuss a number of implications for entrepreneurial finance that arise from this learning perspective to multistage financing, including ones that are not immediately obvious when considering staged financing as a tool only for monitoring and governance.

### 8.3.1 Value Inflection Is Tied to the Resolution of Uncertainty

The most natural and intuitive implication of the learning perspective to staged financing is that experiments generate information that resolve uncertainty. In the context of the framework above, we can think of $V \times (p_P - p_F)$ as the amount of the information revealed by the experiment or the amount that the decision-maker can learn from the experiment's results.

On one extreme, an experiment might demonstrate nothing (i.e., $V \times p_P = V \times p_F$). That is, the probability of earning $V$ is the same no matter the experiment's outcome, so no uncertainty is resolved by the experiment. Alternatively a perfect experiment would imply $p_P = 1$ and $p_F = 0$, meaning that passing or failing the milestone would completely resolve uncertainty about the prospects of the venture.[2] Holding $K$, $p_0$, $V$, and $C$ constant, one can see that the resolution of uncertainty leads to an increase in the posterior probability of success for ventures that pass the milestone. This, in turn, leads to an increase in expected value, or value inflection.

Figure 8.2, using data on the pre-money valuation of all US start-ups between 2013 and 2022, documents the progression of start-up valuation across rounds of funding. The pattern is consistent with a view where investors update their beliefs about a start-up's potential across multiple rounds of funding, reinvesting only in those where they update positively and abandoning funding in ventures

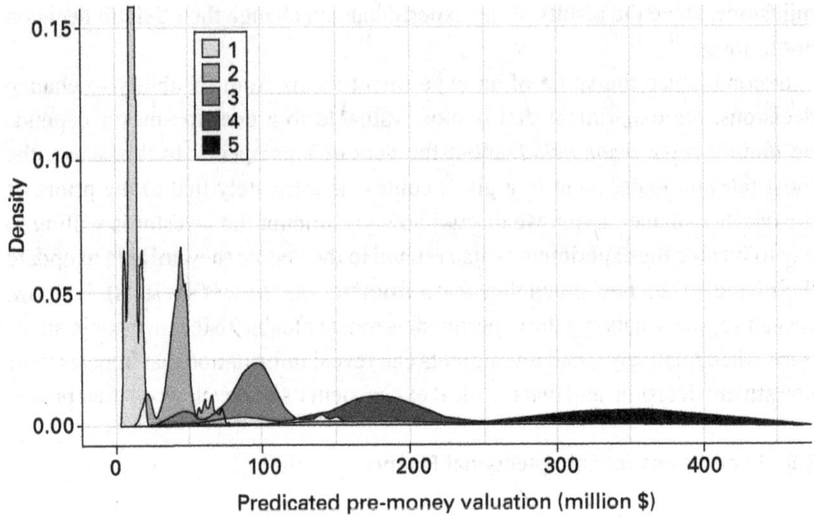


**Figure 8.2**
Evolution of pre-money valuation across rounds of financing. This figure displays density plots of predicted pre-money valuation in levels, taken from the Poisson regressions (Pre-money) = $\beta\text{DealOrder}_i + \gamma\text{FirmAge}_i + \alpha\text{Industry}_i$. FirmAge is defined as the difference in years between the deal year and the year the start-up was founded. Density plots of pre-money valuations are deal orders. To improve readability, firms with greater than 450 million pre-money valuation or density more than 0.15 are not included in the plots despite being in the regression.

where they don't.[3] It is worth emphasizing that, in this framework, failing to meet a milestone is not a reflection of entrepreneur effort, since there is no agency friction in this model. Failure in this model stems simply from the fact that the experiment revealed that the venture is unlikely to work at scale (say, because of weak demand or the inability of the technology to work as initially hypothesized). This is consistent with the notion that failure can sometimes be good and that, in these instances, failing fast is often lauded as the best course of action.

**Designing more informative experiments leads to greater value inflection (if the milestone is achieved)** Having discussed how value inflection is tied to the resolution of uncertainty, I turn to discussing how *experiment design* can impact the degree of information generated by experiments. Similar to the discussion in Agrawal et al. (2025) and Bolton et al. (2024), one can characterize an experiment by its ability to correctly predict success and failure. Similar to language used in medical diagnostic tests, let $s_1 = P(s = P|v = V)$ be the probability the venture passes the milestone in the event the venture will actually be successful and will generate $V$, and $s_2 = P(s = F|v = 0)$ be the probability that the experiment fails to pass the milestone when the venture is a failure. In other words, the parameter $s_1$ captures the sensitivity of the experiment, with $(1-s_1)$ denoting the rate of false-negative outcomes if the venture failures to reach the milestone. Similarly, $s_2$ captures the specificity of the experiment, with $(1-s_2)$ denoting the rate of false-positive outcomes if the venture passes the milestone.

| | Success (Payoff $V$) | Failure (Payoff 0) |
|---|---|---|
| Pass Milestone, $s=P$ | *Sensitivity* $(s_1)$ | *FalsePositive* $(1-s_2)$ |
| Fail Milestone, $s=F$ | *FalseNegative* $(1-s_1)$ | *Specificity* $(s_2)$ |

Given priors that the technology works are $p_0$, the probability of passing milestone $p_M = p_0 s_1 + (1-p_0)(1-s_2)$. The posterior probability of achieving $V$ is:

$$p_S = \frac{p_0 s_1}{p_0 s_1 + (1-p_0)(1-s_2)}.$$

Holding $K, p_0, V$, and $C$ constant, and given that $V > K$, it can be seen that the net gain from the experiment is strictly increasing in both $s_1$ and $s_2$. This implies that the best experiment design in terms of maximizing value gained from information discovery is the one with the greatest sensitivity (the highest $s_1$) and the greatest specificity (the highest $s_2$). Note that a perfect experiment has $s_1 = 1$ and $s_2 = 1$, the greatest specificity and sensitivity possible, and therefore, if feasible, yields the highest gain.

It is also worth nothing that $dP_M/ds_2 < 0$ and $dP_S/ds_2 > 0$—that is, designing experiments with a lower likelihood of generating false positives makes it less likely that the venture will achieve the milestone but, *conditional on meeting the milestone*, will lead to a higher posterior probability of success. This captures the notion that more rigorous experiments are harder to pass but generate more value inflection if achieved.

Investors value more rigorous experiments—equivalent of having met more aggressive milestones at a given level of development—because it reduces the likelihood of them having to commit further funding to a venture due to a false positive. In turn, they are willing to pay a higher price for the information generated by this experiment. In practical terms, this stems from them requiring a lower share of the equity in the venture to provide the same amount of capital to finance the scale-up. In sectors where there are a large number of equivalent firms, VC investors even provide benchmarks for ventures, with firms having met more aggressive benchmarks receiving higher valuations.[4] Indeed, as seen in figure 8.2, firms at equivalent points of their development can have very different valuations based on investor beliefs.

### 8.3.2 Investor Persuasion

One of the important insights emerging from the framework above, which is also emphasized in Agrawal et al. (2025), is that an optimal experiment does not exist in isolation. Rather, the experiment that generates the most amount of information for a given venture depends on the specific priors of the decision-maker involved. This insight is important, as it highlights that, since an entrepreneur needs to persuade an investor to fund their venture, it is the *investor's priors* that the entrepreneur cares about updating. To the extent that their own beliefs about the prospects of the venture are different from those of the investor, the entrepreneur needs to focus on generating the information that will be most important in shifting investor priors.

This also highlights the importance of understanding investor priors and how they are shaped. Indeed, the specific networks, incentives, and prior experiences of individual investors appear to be important in shaping how information is acquired, aggregated, and filtered by investors (Bernstein et al., 2017; Hu and Ma, 2024), which has important implications for what innovations are financed and, in turn, on the direction of innovation (Ewens and Townsend, 2019; Gompers et al., 2014; Howell and Nanda, 2023). Equally important, the ability to frame and interpret experiments may itself be a skill, particularly in the context of power law information distributions (Malenko et al., 2024). This motivates the need to better understand how the backgrounds of decision-makers and the organizational contexts in which they make decisions can have a systematic impact on which ventures tend to receive financing (Lerner and Nanda, 2020).

Bolton et al. (2024) also document how there may sometimes be incentives for entrepreneurs to design experiments in ways that make it more likely to pass an intermediate milestone. Conditional on passing the milestone, however, the venture is less likely to be ultimately successful. They show that this moral hazard in experiment design can undermine investor confidence and lead to a market failure in the provision of funding where investors are unable to fully understand the nature of the experiment from which the results were generated. Approaches to validating the experiment design or reducing such moral hazard can play a role in reducing such market failure.

### 8.3.3 Costs and Constraints to Financing Experimentation

There is growing evidence that costs and constraints to investors' ability to finance experimentation can also have real effects. In this section, I discuss how such costs and constraints can shape variation in what is financed across industries and regions and over periods of time.

**Industry variation in costs and informativeness of early experiments** Recent work discusses the narrow band of innovation that venture capital investors tend to finance (Lerner and Nanda, 2020). Although there are many potential non–mutually exclusive explanations for this, I document here how the cost and information value of each round of financing vary across industries, and show how these patterns are consistent with the disproportionate rise in the financing of software-related applications by venture capital investors in recent years.

Figure 8.3 shows the median cost and the median step-up in valuation associated with the first five rounds of financing across different sectors. It can be seen that software ventures tend to consistently generate higher-value inflections across rounds and also require substantially less capital than hardware and pharmaceutical sectors. This makes software a much more attractive sector to invest in and helps to explain why the more than five fold growth in venture capital–backed start-ups in the past decade have been driven largely by enterprise software, consumer Internet, and other sectors building heavily on software technology.

Ewens et al. (2018) trace this dramatic rise to the substantial drop in the cost of early rounds of funding for software ventures, driven by the advent of cloud computing. They show how the projects that were most likely to benefit from this drop in the cost of early experiments were long-shot bets—start-ups

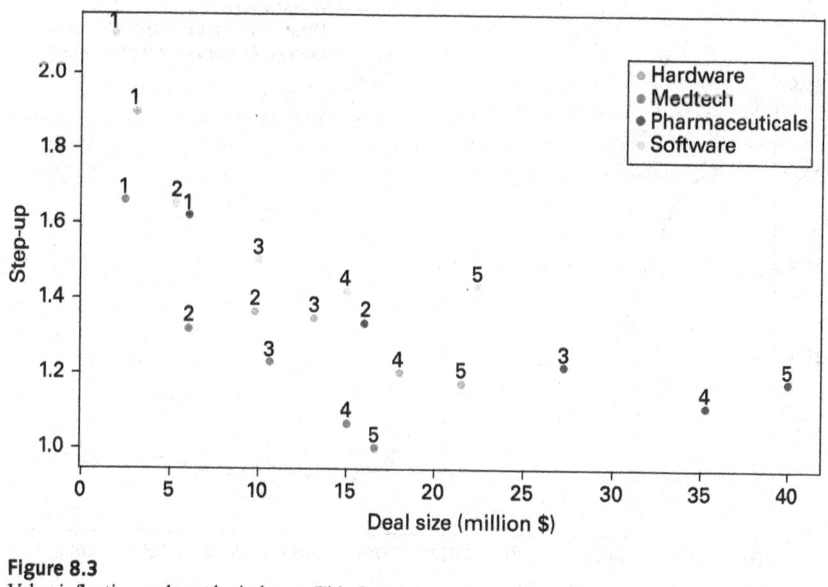


**Figure 8.3**
Value inflection and cost by industry. This figure shows the relationship between deal size and step-up. Step-up at round $t$ is defined as $\frac{\text{Pre-Money}_{t+1}}{\text{Post-Money}_t}$. Industry groups are color coded. The number on each point represents the round number.

with small chances of a large outcome, where an investor can learn a significant amount from an initial experiment.[5] Consistent with the predictions of the model, they find that, relative to sectors that did not benefit from the cloud computing shock, industries that benefited from a decline in the cost of starting due to cloud computing experienced a large increase in the number of start-ups that were more likely to fail. Conditional on not failing, however, these start-ups experienced larger step-ups in value from the first round to the next. Ewens et al. (2018) highlight that by making the real option value of these early experiments greater, the advent of cloud computing led to a shift in the composition of start-ups that VCs funded toward sectors where the value of learning through the initial experiment had increased.[6] Although technological advances have since also led to falls in the cost of early experiments in other sectors (e.g., gene sequencing and gene editing technologies such as CRISPR for drug development and rapid prototyping/3D printing for hardware technologies), the lower costs and higher step-ups associated with the typical software investment continue to make them attractive investments for VCs. As can be seen from figure 8.4, the total investment into IT hardware, BioTech and MedTech, and energy, materials and resources has declined substantially over the past twenty years, from about 50 percent in the early 2000s to about 25 percent in 2023.

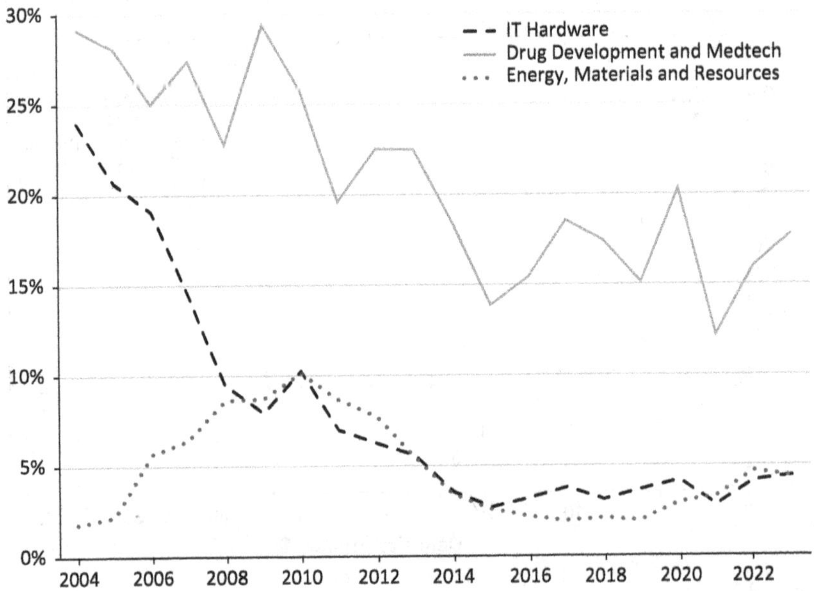


**Figure 8.4**
Share of US venture capital investment across sectors, 2004–2023.

**Venture capital booms and busts** Venture capital funding for start-ups is notoriously volatile (Gompers et al., 2008; Janeway et al., 2021). In fact, even the *risk* of reduced future funding can lead venture capital investors to shift the nature of firms that they back during downturns (Nanda and Rhodes-Kropf, 2013, 2017a; Howell et al., 2020).[7]

When seen through the learning perspective of multistage financing, this is consistent with early-stage investors reacting to what they believe to be shifts in the priors of later-stage investors. Specifically, if early-stage investors forecast a drop in the willingness of later-stage investors to fund start-ups in a given sector at the next round of financing, it is rational for them to respond to this belief by changing what they choose to finance today (Nanda and Rhodes-Kropf, 2017a). A belief that it will be harder to secure a round of financing in the next period also reduces the value of experimentation in the current period (since, even if milestones are hit, firms are less likely to get funded), leading to safer, less novel firms being prioritized in terms of funding. Another important implication of this is that a few deep-pocketed investors—who are the ones involved in later-stage financing—can have a disproportionate role in shaping what is funded because the much larger number of early-stage investors (rationally) focus on trying to predict the priors of these late-stage investors and respond accordingly.

**Institutional and cultural factors** Although many explanations abound for the differences in the rates of high-risk entrepreneurship across regions, one set of arguments has focused on institutional factors such as bankruptcy law (Acharya and Subramanian, 2009; Cerqueiro et al., 2017) as well as on cultural factors such as the stigma of failure (Landier, 2002). Nanda and Rhodes-Kropf (2017b) highlight how including a cost of early failure for the entrepreneur if intermediate information is negative reduces the value of experimentation for investors, even though they do not directly pay the failure costs. This is because the VC and entrepreneur must negotiate over any surplus generated by the project, and stigma from early failure lowers the entrepreneur's expected payoff. If the total expected value of the project does not generate enough to cover the costs borne by both the entrepreneur and investor, then the entrepreneur and investor will not be able to find a deal that will induce them both to participate. The deals where they do agree will be ones with less uncertainty. By committing to be failure tolerant, investors may shift the projects they fund away from ventures where more can be learned through experiments. To the extent that more radical innovations are associated with greater uncertainty, this implies that a commitment to being failure tolerant can change what investors would want to finance. This provides a different perspective on the relationship between failure tolerance and innovation (Manso, 2011; Tian and Wang, 2011) by showing that failure tolerance impacts both the desire to engage in risky innovation by entrepreneurs and the willingness to finance such innovation by investors.

## 8.4 Insights for Entrepreneurs

The learning perspective on multistage financing is also appealing in that it provides clear actionable insights for practitioners, which I have found to resonate considerably with entrepreneurs, investors, and other individuals engaging in the financing ecosystem. I discuss a few of these below.

### 8.4.1 How Much to Raise and How to Structure Financing

The insight that entrepreneurs need to focus on the priors of those who will provide the next round of financing leads to multiple useful insights. First, it focuses attention on the specific information that will need to be generated to convince the next round's investor to engage. Understanding what they need to see to be convinced to invest and what needs to be learnt to help convince them transforms an amorphous fundraising task into something quite concrete.

Second, it allows entrepreneurs to target value inflection milestones. The milestones that entrepreneurs aim for at each financing stage should have the potential to meaningfully increase the value of the venture if achieved. In other words, each milestone should help resolve a meaningful amount of uncertainty about the venture's viability—in the eyes of the next round's investors—and thus justify a significant "step-up" in valuation from one financing round to the next (e.g., from a series A to series B round). By understanding variation in the milestones that are achieved for ventures at a similar stage of progression as the focal venture, it also allows an understanding for the level of "experiment design rigor" that needs to be targeted to achieve a desired level of value inflection (e.g., how variation in the degree of traction with key customers relates to variation in the valuation at a given round of financing).

Third, by understanding what needs to be learnt to persuade investors and the experiments that need to be run to generate this information, it also helps estimate the time and money that will be needed to generate the information needed to achieve the value inflection. This enables entrepreneurs to raise enough to achieve the next value inflection point, plus some extra cushion for unanticipated needs. Raising too much can lead to excess dilution, whereas raising too little creates execution risk.

Fourth, realizing that it is investors' priors that are relevant highlights that there is path dependence in funding strategy. Raising money from certain early investors unlocks certain types of funding in the next stage but can also preclude an entrepreneur from raising funds from certain funds (due to the signal it sends those later-stage investors). Recognizing this allows entrepreneurs to be deliberate about building their investment syndicate.

Fifth, entrepreneurs may need to anticipate market downturns and plan appropriately due to the presence of financing risk. At times, it may make sense to try and raise more than is needed to protect a venture against a market downturn. This is true even if an entrepreneur anticipates that the fundamentals of the venture itself will remain strong but believes that investors will be harder to persuade due to the increased financing risk.

### 8.4.2 Addressing the Challenges of Financing Deep Tech

As discussed above, certain industries appear to face a natural handicap in terms of their attractiveness for venture capital. Although there are many explanations for this, one that is relevant in the context of the framework above pertains to deep tech ventures that build on fundamental science and engineering. Several challenges make these types of ventures harder to commercialize. First, lab testing often occurs under very different conditions than would be seen in an actual commercial deployment at scale. Even when early experiments show promising results, those results are often less predictive of ultimate success than seeing early customer traction from a minimum viable product in a software venture. What constitutes a successful result can be ambiguous due to a lack of well-established benchmarks and testing protocols that would allow for clear interpretation of experimental results and how they map to commercial viability, making it hard to update priors.

Given the importance of these technologies for addressing global challenges faced by society, overcoming these barriers will likely require a multipronged approach. Prior work has documented how policymakers who provide nondilutive financing for such ventures to a point where there is clear value inflection can help unlock venture capital (Howell, 2017). The framework above also sheds light on the promise of developing ways to develop greater confidence in the predictions of early experiments for ultimate success. It also highlights the importance of investors with specialized expertise in understanding and interpreting these signals, as well as programs that can support pre-commercial development of ventures toward addressing the specific concerns that early-stage investors have about the ventures (Lerner, 1999; Howell, 2017, 2020).[8] Doing so can reduce frictions in the persuasion challenge faced by deep tech entrepreneurs.

## 8.5 Conclusion

As the Bayesian entrepreneurship research agenda develops, my hope is that entrepreneurship scholars continue to work on integrating these streams of related work to make the power of this framework clear. Otherwise, it risks remaining niche and fragmented. For example, Agrawal et al. (2025) explicitly define Bayesian entrepreneurship as having heterogeneous priors and, in fact, go to the point of noting that having shared priors is *not* Bayesian entrepreneurship. This seems to me to be an unnecessarily strong exclusion criterion and one that reduces the scope for learning across related literatures. I would suggest that, instead, like many other research streams, including economics, this literature develops a perspective that Bayesian entrepreneurship is not *necessarily* about homogeneous priors. This is because starting with homogeneous priors is a weaker assumption and therefore requires a higher hurdle to generate desired results. It will also allow scholars to understand in which instances heterogeneous priors are critical to generating insights that common priors cannot and heterogeneous beliefs might evolve in the first place (Acemoglu et al., 2006; Nimark and Sundaresan, 2019). Relatedly, it would be valuable to ensure that a shared language is used when defining disagreement between entrepreneurs and other stakeholders, in particular distinguishing between statements such as stronger versus weaker priors and optimism versus pessimism. This will allow for greater communication between literatures and enable greater cross-fertilization of ideas.

In conclusion, I have tried to highlight how the conceptual insights developed by Agrawal et al.'s (2025) chapter on Bayesian entrepreneurship are closely related to a growing entrepreneurial finance literature on learning through multistage financing. I have also highlighted how such a learning perspective to staged financing has strong pedagogical value because it helps provide a clear, actionable approach to navigating the process of financing highly uncertain new ventures. In turn, this enables a promising research agenda that can help both identify and quantify frictions to effective learning and the related solutions to more effectively persuading of investors about the potential of different ventures.

## 8.6 Appendix

| Industry | Amount raised | Dilution | Pre-money | Stepup Multiples | Exit Valuation |
|---|---|---|---|---|---|
| | | Round 1 | | | |
| Hardware | 5.22 | 26.29 | 12.76 | | |
| MedTech | 5.31 | 29.16 | 10.20 | | |
| Pharmaceuticals | 20.37 | 40.93 | 24.10 | | |
| Software | 3.52 | 22.79 | 11.74 | | |
| | | Round 2 | | | |
| Hardware | 14.35 | 24.89 | 52.41 | 4.08 | 104.85 |
| MedTech | 10.24 | 25.30 | 29.17 | 2.91 | 416.58 |
| Pharmaceuticals | 43.49 | 32.65 | 103.36 | 3.18 | 530.69 |
| Software | 10.26 | 22.85 | 43.75 | 5.25 | 44.96 |
| | | Round 3 | | | |
| Hardware | 23.96 | 21.22 | 106.57 | 1.81 | 137.90 |
| MedTech | 19.88 | 25.14 | 60.08 | 2.08 | 174.29 |
| Pharmaceuticals | 59.06 | 26.46 | 225.31 | 1.91 | 457.09 |
| Software | 19.82 | 20.45 | 108.35 | 2.78 | 169.58 |
| | | Round 4 | | | |
| Hardware | 40.16 | 18.85 | 253.92 | 1.80 | 168.25 |
| MedTech | 31.47 | 22.30 | 113.71 | 1.61 | N/A |
| Pharmaceuticals | 60.60 | 23.88 | 226.33 | 1.93 | 468.64 |
| Software | 38.82 | 18.28 | 258.02 | 2.27 | 132.15 |
| | | Round 5 | | | |
| Hardware | 91.88 | 16.22 | 529.18 | 1.61 | 302.86 |
| MedTech | 23.86 | 17.44 | 115.79 | 1.23 | N/A |
| Pharmaceuticals | 60.26 | 20.57 | 306.72 | 1.46 | 1,160.76 |
| Software | 71.04 | 16.25 | 588.24 | 2.08 | 362.44 |

Amounts raised are in USD million for non-exit deals. Dilution represents the percentage of shares acquired in a non-exit deal. Pre-money valuation is in USD million for non-exit deals. Step-up multiples are defined as $Stepup_t = \frac{PreMoney_t}{PreMoney_{t-1}}, t = 2, 3, \ldots$ Exit valuation is the pre-money valuation of exit deals (i.e., MA or IPO) in USD million. Round $t$ refers to the exit valuation of firms that raised $t-1$ round(s) of private financing before exit.

**Figure 8.5**
Mean of amount raised, dilution, and exit valuations across industries. This table reports values for all venture capital deals of firms incorporated in the US from 2013 to 2022. The data source is Pitchbook. The analysis excludes rounds with missing valuation or deal size and rounds that follow a missing round and only consider rounds up to round 5. Hardware companies are classified as those manufacturing computer hardware, manufacturing and distributing communication equipment or providing communication services, manufacturing and designing semiconductors and circuits, or providing energy-related products. MedTech companies are defined as those manufacturing health-care devices and supplies for consumers and other health-care organizations. Pharmaceuticals companies are defined as those engaged in drug discovery and delivery of pharmaceuticals or biotechnology. Software companies are defined as those that design and develop software for both business and consumers.

| Industry | Amount Raised | Dilution | Pre-money | Stepup Multiples | Exit Valuation |
|---|---|---|---|---|---|
| | | | Round 1 | | |
| Hardware | 2.22 | 25.93 | 7.00 | | |
| MedTech | 2.00 | 27.92 | 5.01 | | |
| Pharmaceuticals | 6.50 | 39.14 | 10.00 | | |
| Software | 1.80 | 23.08 | 5.90 | | |
| | | | Round 2 | | |
| Hardware | 8.50 | 24.70 | 25.00 | 2.57 | 11.54 |
| MedTech | 5.00 | 23.95 | 17.38 | 1.71 | 380.81 |
| Pharmaceuticals | 20.05 | 30.77 | 40.00 | 1.94 | 198.49 |
| Software | 5.28 | 22.48 | 19.00 | 2.34 | 20.55 |
| | | | Round 3 | | |
| Hardware | 15.18 | 22.56 | 50.00 | 1.53 | 72.80 |
| MedTech | 8.19 | 23.51 | 30.00 | 1.37 | 120.00 |
| Pharmaceuticals | 38.50 | 23.50 | 100.00 | 1.43 | 288.14 |
| Software | 11.00 | 19.64 | 40.00 | 1.87 | 64.75 |
| | | | Round 4 | | |
| Hardware | 28.48 | 18.36 | 122.93 | 1.66 | 100.00 |
| MedTech | 12.00 | 21.85 | 41.00 | 1.31 | N/A |
| Pharmaceuticals | 37.00 | 21.66 | 149.16 | 1.27 | 374.45 |
| Software | 20.00 | 16.28 | 90.00 | 1.76 | 75.03 |
| | | | Round 5 | | |
| Hardware | 50.00 | 14.37 | 236.00 | 1.24 | 302.86 |
| MedTech | 12.60 | 15.16 | 57.75 | 1.34 | N/A |
| Pharmaceuticals | 34.89 | 15.17 | 207.45 | 1.11 | 460.54 |
| Software | 39.93 | 13.28 | 215.00 | 1.71 | 236.70 |

Amounts raised are in USD million for non-exit deals. Dilution represents the percentage of shares acquired in a non-exit deal. Pre-money valuation is in USD million for non-exit deals. Step-up multiples are defined as $Stepup_t = \frac{PostMoney_t}{PostMoney_{t-1}}, t = 2, 3, \ldots$ Exit valuation is the pre-money valuation of exit deals (i.e., MA or IPO) in USD million. Round $t$ refers to the exit valuation of firms that raised $t-1$ round(s) of private financing before exit.

**Figure 8.6**
Median of amount raised, dilution, and exit valuations across industries. This table reports values for all venture capital deals of firms incorporated in the US from 2013 to 2022. The data source is Pitchbook. The analysis excludes rounds with missing valuation or deal size and rounds that follow a missing round and only consider rounds up to round 5. Hardware companies are classified as those manufacturing computer hardware, manufacturing and distributing communication equipment or providing communication services, manufacturing and designing semiconductors and circuits, or providing energy-related products. MedTech companies are defined as those manufacturing health-care devices and supplies for consumers and other health-care organizations. Pharmaceuticals companies are defined as those engaged in drug discovery and delivery of pharmaceuticals or biotechnology. Software companies are defined as those that design and develop software for both business and consumers.

## Acknowledgments

I thank Shannon Liu, Matthew Rhodes-Kropf, William Sahlman, Savi Sunderasan, and participants at the MIT Bayesian Entrepreneurship Conference for very helpful discussions. Xuelai Li provided exceptional research assistance. This chapter is related to several prior coauthored review articles and, in particular, builds on Kerr et al. (2014), Nanda and Rhodes-Kropf (2016), and Lerner and Nanda (2020). This paper is part of a project that has received funding from the European Research Council (ERC) Horizon 2020 research and innovation programme (grant number 865127).

## Notes

1. Similarly, even if $p_0 * V - K > 0$, an investor may find it worthwhile to finance an experiment if learning the result enables them to exercise their abandonment option, thereby changing their default decision not to invest.

2. Note that the experiment is not more or less important if the project is riskier—a risker project might be one with a larger $V$ and smaller probabilities of success, $p_G$ and $p_B$—but the information revealed by the experiment, $V * p_G - V * p_B$, could be the same. Thus, a project with valuable experiment and the risk of the project are related but not the same.

3. Note that although consistent with the view of updating beliefs about valuation across rounds, the figure itself does not provide definitive evidence on this view.

4. This is common practice for enterprise software companies. For example, see https://www.scalevp.com/scale-studio/.

5. In the context of the framework above, with small probabilities of success ($p_0$) but where the information and hence value of the abandonment option is high ($V * p_P - V * p_F$ is large).

6. Ewens et al. (2024) also point to the ability of VCs to reduce the burden of nondiversifiable risk (Hall and Woodward, 2010) for founders by providing founder CEOs higher levels of cash compensation once the firm has a product. They show that, consistent with theory, sectors in which uncertainty is resolved more slowly and hence have a longer time to product have have a lower likelihood of selection into VC-backed entrepreneurship.

7. See also https://articles.sequoiacap.com/rip-good-times.

8. This can often relate to the ability of the management team to execute, something that entrepreneurial teams under emphasize in their assessment of the risks to the venture's success.

## References


Acemoglu, D., Chernozhukov, V., and Yildiz, M. (2006). *Learning and disagreement in an uncertain world.* NBER Working Paper (No. 12648).

Acharya, V., and Subramanian, K. (2009). Bankruptcy codes and innovation. *Review of Financial Studies*, 22(12), pp. 4949–4988.

Aghion, P., Bolton, P., Harris, C., and Jullien, B. (1991). Optimal learning by experimentation. *The Review of Economic Studies*, 58(4), pp. 621–654.

Agrawal, A., Camuffo, A., Gambardella, A., Gans, J. S., Scott, E. L., and Stern, S. (2025). The foundations of Bayesian entrepreneurship. *Bayesian entrepreneurship*. MIT Press.

Bergemann, D., and Hege, U. (1998). Venture capital financing, moral hazard, and learning. *Journal of Banking & Finance*, 22(6–8), pp. 703–735.

Bergemann, D., and Hege, U. (2005). The financing of innovation: Learning and stopping. *RAND Journal of Economics*, 36, pp. 719–752.

Bergemann, D., Hege, U., and Peng, L. (2008). *Venture capital and sequential investments*. Cowles Foundation Discussion Papers (No. 1682).

Bernstein, S., Giroud, X., and Townsend, R. R. (2016). The impact of venture capital monitoring. *The Journal of Finance*, 71(4), pp. 1591–1622.

Bernstein, S., Korteweg, A., and Laws, K. (2017). Attracting early-stage investors: Evidence from a randomized field experiment. *The Journal of Finance*, 72(2), pp. 509–538.

Bolton, P., and Harris, C. (1999). Strategic experimentation. *Econometrica*, 67(2), pp. 349–374.

Bolton, P., Liu, S., Nanda, R., and Sunderesan, S. (2024). *Moral hazard in experiment design: Implications for financing innovation* innovation. Working Paper.

Cerqueiro, G., Hegde, D., Penas, M. F., and Seamans, R. C. (2017). Debtor rights, credit supply, and innovation. *Management Science*, 63(10), pp. 3311–3327.

Cornelli, F., and Yosha, O. (2003). Stage financing and the role of convertible securities. *Review of Economic Studies*, 70, pp. 1–32.

Ewens, M., Nanda, R., and Rhodes-Kropf, M. (2018). Cost of experimentation and the evolution of venture capital. *Journal of Financial Economics*, 128(3), pp. 422–442.

Ewens, M., Nanda, R., and Stanton, C. (2024). Founder CEO compensation and selection into venture capital-backed entrepreneurship. *The Journal of Finance*, 79(5), pp. 3361–3405.

Ewens, M., and Townsend, R. R. (2019). Are early stage investors biased against women? *Journal of Financial Economics*, 135(3), pp. 653–677.

Gompers, P. (1995). Optimal investment, monitoring, and the staging of venture capital. *Journal of Finance*, 50, pp. 1461–1489.

Gompers, P., Kovner, A., Lerner, J., and Scharfstein, D. (2008). Venture capital investment cycles: The impact of public markets. *Journal of Financial Economics*, 87(1), pp. 1–23.

Gompers, P. A., Mukharlyamov, V., Weisburst, E., and Xuan, Y. (2014). *Gender effects in venture capital*. Working Paper.

Grossman, S. J., Kihlstrom, R. E., and Mirman, L. J. (1977). A Bayesian approach to the production of information and learning by doing. *The Review of Economic Studies*, 44(3), pp. 533–547.

Hall, R. E., and Woodward, S. E. (2010). The burden of the nondiversifiable risk of entrepreneurship. *American Economic Review*, 100(3), pp. 1163–1194.

Hellmann, T. (1998). The allocation of control rights in venture capital contracts. *RAND Journal of Economics*, 29, pp. 57–76.

Hellmann, T., and Puri, M. (2002). Venture capital and the professionalization of start-up firms: Empirical evidence. *Journal of Finance*, 57(1), pp. 169–197.

Howell, S. T. (2017). Financing innovation: Evidence from R&D grants. *American Economic Review*, 107(4), pp. 1136–1164.

Howell, S. T. (2020). Reducing information frictions in venture capital: The role of new venture competitions. *Journal of Financial Economics*, 136(3), pp. 676–694.

Howell, S. T., Lerner, J., Nanda, R., and Townsend, R. R. (2020). *How resilient is venture-backed innovation? Evidence from four decades of U.S. patenting*. NBER Working Paper (No. 27150).

Howell, S. T., and Nanda, R. (2023). Networking frictions in venture capital, and the gender Gap in entrepreneurship. *Journal of Financial and Quantitative Analysis*, 59(6), pp. 2733–2761.

Hu, A., and Ma, S. (2024). Persuading investors: A video-based study. *Journal of Finance, forthcoming.*

Janeway, W., Nanda, R., and Rhodes-Kropf, M. (2021). Venture capital booms and startup financing. *Annual Review of Financial Economics*, 13(1), pp. 111–127.

Kerr, W. R., Nanda, R., and Rhodes-Kropf, M. (2014). Entrepreneurship as experimentation. *The Journal of Economic Perspectives*, 28(3), pp. 25–48.

Landier, A. (2002). *Entrepreneurship and the stigma of failure*. MIT Working Paper.

Lerner, J. (1995). Venture capitalists and the oversight of private firms. *The Journal of Finance*, 50(1), pp. 301–318.

Lerner, J. (1999). The government as venture capitalist: The Long-run impact of the SBIR program. *The Journal of Business*, 72(3), pp. 285–318.

Lerner, J., and Nanda, R. (2020). Venture capital's role in financing innovation: What we know and how much we still need to learn. *Journal of Economic Perspectives*, 34(3), pp. 237–261.

Malenko, A., Nanda, R., Rhodes-Kropf, M., and Sundaresan, S. (2024). Catching outliers: Committee voting and the limits of consensus when financing innovation. *Journal of Finance, forthcoming.*

Manso, G. (2011). Motivating innovation. *Journal of Finance*, 66(5), pp. 1823–1860.

Metrick, A., and Yasuda, A. (2006). *Venture capital and the finance of innovation*. John Wiley & Sons.

Nanda, R., and Rhodes-Kropf, M. (2013). Investment cycles and startup innovation. *Journal of Financial Economics*, 110(2), pp. 403–418.

Nanda, R., and Rhodes-Kropf, M. (2016). Financing entrepreneurial experimentation. *Innovation Policy and the Economy*, 16(1), pp. 1–23.

Nanda, R., and Rhodes-Kropf, M. (2017a). Financing risk and innovation. *Management Science*, 63(4), pp. 901–918.

Nanda, R., and Rhodes-Kropf, M. (2017b). Innovation policies. *Entrepreneurship, Innovation and Platforms*, 37, pp. 37–80.

Nimark, K. P., and Sundaresan, S. (2019). Inattention and belief polarization. *Journal of Economic Theory*, 180, pp. 203–228.

Sahlman, W. A. (1990). The structure and governance of venture-capital organizations. *Journal of Financial Economics*, 27, pp. 473–521.

Tian, X. (2011). The causes and consequences of venture capital stage financing. *Journal of Financial Economics*, 101(1), pp. 132–159.

Tian, X., and Wang, T. Y. (2011). Tolerance for failure and corporate innovation. *Review of Financial Studies*, 27(1), pp. 211–255.

Weitzman, M. L. (1979). Optimal search for the best alternative. *Econometrica*, 47, pp. 641–654.

# III

## Entrepreneurial Experimentation and Communication

# 9

# Learning from Data in Entrepreneurial Experimentation

Hyunjin Kim

## 9.1 Introduction

Experimentation plays a pivotal role in entrepreneurial and strategic decision-making (Camuffo et al., 2020; Gans, 2023; Levinthal, 2017). The Bayesian entrepreneurship framework underscores this importance by positioning experimentation at the heart of entrepreneurial decisions, where the value of experimentation stems from its ability to update entrepreneurs' beliefs, thereby shaping their strategic choices (Agrawal et al., 2025). Through this lens, experiments encompass any activity that generates information for belief updating, be it feedback from customers, product launches, or A/B testing. In essence, the fundamental objective of an experiment is to produce informative data that facilitates learning for the entrepreneur.

Despite the recognized importance of experimentation, there has been less insight on how entrepreneurs design experiments to generate informative data and how they translate this data into learning. Much of the existing research has abstracted the data from experiments—both its design and its interpretation—into simplified signals. Studies on data-driven decision-making also often represent information acquisition as a binary input, emphasizing the value of leveraging information for firm decisions (Nagaraj, 2022; Kim, 2025; Brynjolfsson and McElheran, 2016). In practice, however, entrepreneurs do not receive abstract signals or binary inputs. Instead, they must generate and derive insights from complex, multifaceted data. Yet, this process of data generation and interpretation has been largely overlooked.

The challenges of designing experiments that yield informative data and interpreting them to extract learning are aptly illustrated by Netflix, the online streaming giant. As they noted, "Good businesses pay attention to what their customers have to say. But what customers ask for (as much choice as possible, comprehensive search and navigation tools, and more) and what actually works (a few compelling choices simply presented) are very different"

(Gomez-Uribe and Hunt, 2015, p. 7). This insight underscores that experiments are not alike in the data they generate: In this case, an experiment asking customers what they want may not generate data as informative as launching product changes and observing real reactions. Moreover, the interpretation of experimental data presents its own set of challenges. Although Netflix has run experiments for decades, they emphasize that interpreting results "remains partly art" and requires "more sophisticated experiment design and analysis" to extract learning (Gomez-Uribe and Hunt, 2015, p. 11).

This highlights the importance of understanding the process of how entrepreneurs learn from data in experimentation, which this chapter aims to unpack. By anchoring on the premise that experimenting is fundamentally about purposefully acquiring data, it highlights that learning from experiments is far from straightforward. Entrepreneurs often face three challenges: (1) failing to collect informative data, (2) neglecting to pay sufficient attention to the data collected, and (3) making incorrect inferences even when attention is paid. Building on these observations, I propose that although the quality of hypotheses remains important, the value of experimentation is critically dependent on how entrepreneurs generate, attend to, and interpret data from their experiments. I then discuss the implications of these insights for entrepreneurial strategy and the potential role of artificial intelligence (AI) in the experimentation process.

The chapter is structured as follows. Section 9.2 provides an overview of research on entrepreneurial experimentation, reviewing existing definitions and key findings. It then introduces the Bayesian entrepreneurship framework and its emphasis on information and learning through experimentation. Section 9.3 explores why experiments may not lead to learning, highlighting three key frictions. Section 9.4 discusses the increasing value of the ability to learn from data in the context of recent advancements in AI. Finally, section 9.5 concludes by summarizing insights and identifying open questions for future research.

## 9.2 Entrepreneurial Experimentation

### 9.2.1 What Is Experimentation?

Experimentation has been a growing topic in entrepreneurship and strategy research. A significant literature emphasizes experimentation as a key part of the entrepreneurial learning process, helping early-stage ventures purposefully learn about the value of their idea or strategy (Murray and Tripsas, 2004; Eisenmann et al., 2012; Kerr et al., 2014). The widely adopted lean start-up approach highlights the importance of rapid iteration to test hypotheses, focusing on customer feedback to guide the experimentation process (Ries, 2011a; Eisenmann et al., 2012; Blank, 2013). Camuffo et al. (2020) emphasize the role of theory-driven hypotheses in improving entrepreneurial strategy and performance, whereas Gans et al. (2019) stress the importance of experimenting with different but equally viable alternatives to make a choice. This process extends beyond entrepreneurial firms to strategic decisions more broadly: The real options approach to strategy emphasizes how initial investments allow firms to collect signals about possible options (Adner and Levinthal, 2004). Moreover, Levinthal (2017) underlines an experimental approach to strategic decision-making, emphasizing the role of a Mendelian executive who explores different alternatives, tests them through experiments, and selects the best path.

A defining attribute of experimentation is that it is purposeful rather than simply a by-product of trial-and-error actions (Murray and Tripsas, 2004). Experimentation allows entrepreneurs to learn about the value of their idea or strategy (Agrawal et al., 2021), exploring aspects of their technology, market, or business model (Murray and Tripsas, 2004). This learning may not necessarily be low cost: Although the lean start-up approach emphasizes small and low-cost experiments, Gans et al. (2019) note that what might appear to be a low-cost test may involve partial commitment to one plan over other equally viable alternatives. A growing body of work empirically evaluates the consequences of experimentation, finding it to result in more pivots and terminations (Camuffo et al., 2020) and positively correlated with product changes and higher performance measures conditional on survival (Koning et al., 2022).

The process of experimentation involves three key stages: (1) hypothesis generation, (2) testing, and (3) choice selection. Research has provided much insight on the first step of hypothesis generation, which serves an important purpose: Experiments require ideas to test, and not all possibilities can be explored. This work advocates for articulating scientific, falsifiable hypotheses that can be validated through experiments (Camuffo et al., 2020; Blank, 2013) and emphasizes the value of experimenting with more uncertain hypotheses that inform the entrepreneur's broader theory (Camuffo et al., 2024). Gans et al. (2019) underscore the importance of considering more alternative hypotheses to avoid prematurely committing to a suboptimal course of action.

There has also been substantial work on the final stage of choice selection. Choice selection can involve commitment to the idea or strategy, terminating the venture, or pivoting to new directions (Camuffo et al., 2020). One approach has been to see selection as an optimization process, identifying the option that yields the highest value relative to cost. However, Gans et al. (2019) and Levinthal (2017) emphasize the role of deliberate choice in this selection process.

Intriguingly, however, there has been less insight on the actual process of testing—the experiment itself (Gans (2023), Agrawal et al. (2021)).[1] This is surprising, given that this testing stage of experimentation is arguably its most distinctive aspect, with hypothesis generation and choice selection forming core components of most strategic decision-making approaches (Newell and Simon, 1976). Although the inputs (ideas and hypotheses) and outputs (choices and performance) of experimentation have received significant attention, there is less understanding of how the design of experiments results in different data and how this influences entrepreneurial learning.

In particular, there are two key gaps. First, the term "experiments" refers to a wide range of activities. They can be conceptual or real-world implementations (Agrawal et al., 2025). They can involve conversations with mentors (Cohen et al., 2019; Miller et al., 2024), upvotes on a forum (Cao et al., 2023), the release of a "minimimally viable product" (Ries, 2011b), or a formal A/B test (Thomke, 2020). In essence, experiments encompass any activity that might provide an opportunity to acquire information, which can be tightly or loosely tied to the hypotheses they are designed to test. Although these experiments generate different types of data, the literature often abstracts this data into simple signals. However, the experiments entrepreneurs undertake do not yield simple signals in practice. They must learn from the actual data generated, and we know less about this process.

Second, learning from experiments is often implicitly assumed to be coupled with the act of undertaking the experiment itself. Since experiments are designed to acquire information, they generally result in some information. Yet, how entrepreneurs process and interpret this data and the conditions under which it translates into learning are much less understood. Akin to the broader organizational learning literature where practices associated with success are reinforced and those associated with failure avoided (Contigiani and Levinthal, 2019; Levitt and March, 1988; Argote, 2012), experimentation is often viewed informally as inherently resulting in learning, irrespective of the outcome, which overlooks complexities in how entrepreneurs process and learn from data to inform their decisions.

### 9.2.2 Experimentation Through the Bayesian Lens

The Bayesian entrepreneurship lens offers a framework that centers on the interplay between an entrepreneur's prior beliefs and how evidence from experimentation can update those beliefs (Agrawal et al., 2025). It provides a formal approach to assess how entrepreneurs form beliefs about an opportunity and how these beliefs evolve through experimentation—and the resulting implications for entrepreneurial strategy. This framework thus defines learning as updating entrepreneurs' beliefs and emphasizes that beliefs are central to both the inputs and outputs of experimentation: Entrepreneurs' beliefs shape their choices on how to experiment, and their own or others' beliefs are also the outcomes of experimentation.

This provides three key implications for thinking about experimentation. First, experiments represent the acquisition of information that can meaningfully reduce uncertainty on entrepreneurial beliefs about their hypothesis, which can help refine our understanding of what represents informative experiments. Second, the design of experiments involves choice, and the informational value of experiments hinges on these explicit and implicit choices. Third, experiments may be informative, but whether this results in learning depends on the nature of belief updating.

## 9.3 Why Might Experiments Not Result in Learning?

In this section, I highlight three key frictions that hinder learning from experiments, which interact in interesting and important ways when considered through the lens of Bayesian entrepreneurship.

### 9.3.1 Uninformative Experiments

How to experiment is a fundamental choice that entrepreneurs must make in the experimentation process. Once they have generated their hypotheses, there are many ways to test them among the wide range of activities referred to as experiments, all of which involve vastly different actions and data that result.

Much of the literature on how to experiment has focused on the fidelity of experiments as it relates to noise and bias, emphasizing that most experiments are noisy and biased, at times reinforcing the idea of information scarcity in early entrepreneurship (Agrawal et al., 2021, 2025; Gans, 2023). Moreover, it highlights that, in many cases, lower-fidelity experiments, referring to those with smaller sample sizes or biased samples, may be more useful, suggesting that choosing fidelity often trades off the cost of experimentation (Azevedo et al., 2020).

However, entrepreneurs make many other choices on experiments that affect what data is generated, whether explicitly or implicitly, which may not necessarily come at the expense of higher costs, slower speed, or more difficult execution. These choices are often less salient but can have a profound impact on the informativeness of the experiment: (1) the type of activity to pursue, (2) how they operationalize their hypotheses (e.g., what features an MVP has or what the A versus B looks like in an A/B test), and (3) the outcomes they collect from the activity. Just as more data may not always improve predictions (Kim et al., 2024), if the experiment is uninformative, little learning can result, even with the most astute hypothesis (McGrath and MacMillan, 1995; Pfeffer and Sutton, 2006; Rousseau, 2006).

Putting this idea within the Bayesian entrepreneurship framework highlights that experiments must be designed to generate informative data about the hypothesis. If the experiment does not provide any new information about the hypothesis, it does not change the entrepreneur's posterior beliefs compared with their prior. Mathematically, this can be represented by the likelihood being constant for all hypotheses: $P(E|H) = P(E)$, where $P(E|H)$ is the probability of the evidence $E$, given the hypothesis $H$, and $P(E)$ is the probability of the evidence across all possible hypotheses. If $P(E|H)$ is constant and equal to $P(E)$, the evidence $E$ does not provide any additional information to differentiate between hypotheses. Similarly, when evidence $E$ is independent of the hypothesis $H$—that is, the occurrence of the evidence does not affect the probability of the hypothesis being true or false, as in $P(E|H) = P(E)$ and $P(H|E) = P(H)$—the evidence $E$ does not change the distribution over the hypothesis $H$, leading to no update in beliefs.

This enables us to increase precision on what constitutes informative experiments. It emphasizes that the defining feature of experiments from other forms of information acquisition is its focus on a specific hypothesis (Murray and Tripsas, 2004; Agrawal et al., 2021), and crafting informative experiments requires making deliberate choices that ensure that the resulting data informs the hypothesis under investigation. This illuminates two pivotal choices for designing informative experiments that must be made in the "testing" stage following hypothesis generation.

The first key choice involves operationalizing the hypothesis in the experiment such that the activity accurately reflects its core idea. Consider the case of Dropbox. In its early days, CEO Drew Houston and his team faced a significant challenge in growing their company, which sought to offer a cloud-based file-sharing and synchronization service. Their hypothesis was that providing a seamless customer experience would increase demand for file synchronization. However, many people struggled to grasp the concept of storing files online and accessing them seamlessly across multiple devices (Ries, 2011b). Determined to validate their assumption that a superior user experience would attract customers, the Dropbox team experimented with an MVP explainer video that has since become renowned. Narrated by Houston himself, the video displayed what the seamless user experience would look like, demonstrating a problem many were unaware they had. The explainer video was a resounding success, driving "hundreds of thousands of people to the website" and expanding their beta waiting list from "5,000 people to 75,000 people literally overnight" (Ries, 2011b).

This success helped to validate their hypothesis and provided valuable insights into their target audience.

This example highlights that the activity entrepreneurs choose to pursue in their experimentation and how they design the experiment to operationalize their hypothesis are crucial. The hypothesis Houston had was not an easy one to test through simply any experiment: Just as potential customers had difficulty understanding the concept, describing it to investors or mentors for advice made it challenging to gather informative data. Houston had many meetings with venture capital investors but was met with skepticism, as they could not imagine his vision. Although Houston believed that "if the software 'just worked like magic', customers would flock to it," investors pointed out that the market was crowded with existing products that had yet to generate substantial profits—suggesting that file synchronization was not an actual problem that customers had (Ries, 2011b).[2] Moreover, another experiment of even the same activity type may not have generated informative data. Had Dropbox chosen to experiment with a buggy MVP that was launched to early users rather than a video showcasing what the seamless experience could be like, it may have resulted in very different data because the experiment was independent from the hypothesis. By focusing on a video that could showcase the seamless user experience—the core idea behind their hypothesis—Dropbox was able to collect data that enabled them learn about their hypothesis.

The second key experiment choice involves identifying and selecting experiment metrics to generate data that can evaluate the hypothesis. When experiments collect metrics that are independent of the hypothesis, they cannot be informative. Consider Coca-Cola's well-known experiment with New Coke. In the early 1980s, its rival Pepsi-Cola gained market share through their "Pepsi Challenge" campaign, where consumers taking blind taste tests learned, to their surprise, that they preferred the flavor of Pepsi. Hypothesizing that the taste of Coke was driving declines in market share, Coca-Cola undertook an experiment to inform their choice, conducting 190,000 blind product tests between the old Coke and a new, sweeter formula (Klein, 2015). The experiment found that more than half of the participants preferred New Coke over both the original formula and Pepsi. Armed with the results from the experiment, Coca-Cola introduced New Coke to the market and withdrew the original formula. However, this resulted in decreased market performance and widespread customer outrage, with thousands of angry phone calls a day and protests from grassroots groups such as "Old Cola Drinkers of America" (Klein, 2015). It turned out that the old Coke had an emotional value for customers, evoking personal memories, and changing the formula did not result in higher sales or market share. What went wrong? Coca-Cola chose an activity that could inform their hypothesis and operationalized it by testing two formulas that directly mapped to its core idea. However, the metric they chose for the experiment was taste, not sales: They asked whether tasters preferred the taste of New Coke to the old Coke, not whether they would buy New Coke over the original. In this case, the metric they chose, taste, turned out to be a poor proxy of sales and market share—which was the key outcome critical to evaluating their hypothesis. Although this example centers on choosing the wrong metric, it is worth noting that once we have chosen the right metric for the hypothesis, considerations of noise and bias can also become important. If tasters were a biased sample of never-tried-Coke drinkers, one could imagine that, even with better metrics, the resulting data may have been less informative, especially if executives were not able to take into account the potential bias.[3]

Uninformative experiments that do not result in learning may seem like a rather benign problem. However, it is worth highlighting when this may have much more severe consequences. In the Bayesian framework, priors can influence experiment choices. For example, Gans (2023) shows that if an entrepreneur has a strong conviction about a particular hypothesis, they may be more likely to choose an experiment biased toward positive news rather than one that may be more likely to challenge their beliefs. This can lead to a self-reinforcing cycle where prior beliefs shape experimental choices, which, in turn, provide confirming evidence, further entrenching the initial assumptions. This underscores that focusing solely on generating hypotheses and theories—without thinking carefully about how experiment choices affect the data that are generated from it—may distort their learning.

Moreover, the fact that priors influence the choice of experiment is especially concerning when considering that individuals often do not follow precise Bayesian updating processes in practice. For example, when there is noninformative evidence, they should not update their beliefs. However, Kieren and Weber (2020) show that we tend to process noisy information signals in a reference-dependent manner based on our prior beliefs. They provide experimental evidence that, instead of recognizing that uninformative signals do not carry information, individuals update their beliefs based on the valence of the signal relative to their prior expectations, distorting learning.

Together, these insights highlight that poor choices on experiments can result in uninformative experiments that impede learning. In addition to considerations of noise and bias, entrepreneurs make choices on the type of experiment, how hypotheses are operationalized, and what metrics are collected—all of which are key determinants of whether experimentation results in learning about the chosen hypotheses.

### 9.3.2 (In)attention to Data from Experiments

Even when entrepreneurs have selected an informative experiment, they may fail to learn from it due to problems of attention. This attention problem can manifest in two key ways: inattention and selective attention to data.

First, entrepreneurs may struggle to identify and attend to key insights. For example, in an experiment with entrepreneurial firms, Kim (2025) found that managers knew competitor prices were important and believed they already knew this information, but 58 percent held outdated knowledge and could not state competitors' current prices. When they became aware of up-to-date prices, they changed their pricing and quality decisions, and their performance measures increased by 8–15 percent. Applied to the context of entrepreneurial experimentation, this suggests that entrepreneurs may examine data at a specific point in time but fail to continuously monitor it as the experiment progresses. For example, an entrepreneur might design an appropriate experiment, collecting customer and competitor reactions, but fail to track competitors after the initial analysis, assuming they remain static. McGrath and MacMillan (1995) illustrate that Polaroid fell into this exact pitfall, resulting in a $200 million loss from Polavision instant movies. In this case, Polaroid assumed that a $7 three-minute cassette would compete effectively against a $20 half-hour videotape, failing to attend to the rapidly declining costs of videotaping and playback technologies.

A related but distinct issue is that entrepreneurs may be selectively attentive. Hanna et al. (2014) highlight the critical role of noticing in learning, finding that 85 percent of seaweed farmers do not know the size of their seaweed pods, a key determinant of performance, despite being knowledgeable about many other aspects of their business. Even when provided with informative data, they fail to learn from it without being provided with a summary that explicitly highlights the relationship between pod size and performance. The key difference between this and inattention described in the first point is that inattention highlights the failure to continue attending to data, whereas selective attention emphasizes the failure to notice critical insights even when attending to the data.

As an example, consider the case of Kozmo. In the early 2000s, Kozmo was heralded as the epitome of the ambitious start-up, aiming to revolutionize the way people received goods by offering one-hour deliveries of various items, including food, basics, movies, and games. Kozmo ran many experiments, testing and iterating on its offerings to optimize its business model: They tested leasing out warehouse space to deliver orders by bike (Moser, 2021), equipping couriers with motorized scooters, and even expanded their offering to sell cigarettes (Bensinger, 2012). For these experiments, they collected data on delivery times, order volumes, customer satisfaction, and more. In recent years, the world has come to resemble Kozmo's vision: DoorDash was valued at $72 billion on its stock market debut (Hall, 2020), Uber acquired alcohol delivery service Drizly for 1.1 billion to help "consumers get almost anything—from food to groceries to alcohol" (Morris, 2024), and retail giants Amazon, Walmart, and Target are in a shipping war to deliver the fastest delivery speeds (Hadero et al., 2023). Yet, Kozmo is nowhere to be found. What prevented it from becoming a trailblazer of the delivery space?

Although there are likely many reasons for its failure, a key factor stands out: despite running experiments and collecting a wide range of data, Kozmo failed to pay attention to a crucial metric—average order size (Cassidy, 2002). Many of Kozmo's deliveries were for low-value items, with its average order size being $12, when labor and overheads alone were $10 (Cassidy, 2002). Although their data from experiments showed how low their average order size was, Kozmo paid selective attention to growth metrics on customer and order numbers, neglecting the fundamental unit economics of each order, which led them to continue investing in marketing and expansion (Cassidy, 2002). In the last few months before it shuttered, it finally instituted minimum order sizes and delivery fees—but it was too late. Ultimately, despite raising more than $250 million in funding and achieving significant brand recognition, Kozmo shut down in April 2001, laying off its 1,100 employees (Bensinger, 2012; Sandoval, 2001). Kozmo's story highlights the dangers of selective attention: Despite running experiments and collecting data, they failed to notice the crucial insight that their average order size was too low to support their business model.

In a Bayesian framework, these problems of attention can lead to a failure to learn, even when the evidence is new and informative. Inattention and selective attention can both be modeled by ignoring a portion of the evidence, leading to slower or incomplete belief updating. Alternatively, inattention can be modeled as adding noise to the likelihood function:

$$P(H|E) = \frac{(P(E|H) + \epsilon)P(H)}{P(E)},$$

reflecting the idea that not all relevant aspects of the evidence are considered or are perceived incorrectly. The noise term, $\epsilon$, dilutes the influence of actual evidence on the posterior belief. Similarly, selective attention may introduce a parameter that limits the amount of new evidence that can be considered at each update. New information may be weighted less heavily in the updating process, reflecting a conservative adjustment to prior beliefs if the amount of evidence exceeds the processing capacity.

Although the Bayesian framework allows us to highlight the implications of these attention problems, it is important to emphasize that the Bayesian process itself can exacerbate them. Priors are likely to guide attention (Felin and Zenger, 2017; Zellweger and Zenger, 2023), and if an entrepreneur believes that an important dimension is unlikely to matter, they may be especially likely not to notice it to be able to learn whether it actually does matter. For example, Kozmo's founder CEO, Joseph Park, seems to have believed that allowing customers to order anything they wanted was key to success, which may have reinforced their blind spot on order size (Bensinger, 2012). Additionally, overconfident entrepreneurs may be less receptive to new information that challenges their views and fail to pay attention to it, effectively leading them to discount the evidence. Over time, the cumulative effect of these factors can lead to a significant divergence between the learning an inattentive entrepreneur takes away and the true state of the world—even when they have designed informative experiments.

### 9.3.3 Model Bias in Interpreting Data from Experiments

Even when entrepreneurs have selected an informative experiment and attended to the key data, they may still make incorrect inferences, hindering learning.

Much evidence suggests that priors bias how individuals update beliefs (Benjamin, 2019). This is highlighted in experiments on belief polarization, where people with different priors who observe the same evidence draw opposite conclusions and move further apart in their beliefs. For example, Lord et al. (1979) show that proponents and opponents of capital punishment report further divergence in beliefs after reading the same account of a study on capital punishment—a finding that has been replicated in diverse contexts, including consumer beliefs of brand quality (Russo et al., 1998).

Although this evidence of updating toward one's priors is often interpreted as evidence of a bias relative to Bayesian updating, this may, in fact, be consistent with and even exacerbated by Bayesian reasoning (Dixit and Weibull, 2007; Andreoni and Mylovanov, 2012; Benoît and Dubra, 2018). For instance, Baliga et al. (2013) attribute this to ambiguity aversion, and Benoît and Dubra (2018) show how this may manifest when people have private information about an ancillary matter that does not directly bear on the issue of interest but matters for the interpretation of evidence. One proposed mechanism that may be especially relevant for entrepreneurial experimentation is the lack of consideration of alternative hypotheses. Fischhoff and Beyth-Marom (1983) propose that this updating behavior may be explained by the fact that people assess how consistent a signal is with their favored hypothesis, without considering its consistency with alternative hypotheses, highlighting the importance of considering multiple alternatives (Gans et al., 2019).

Furthermore, Schwartzstein and Sunderam (2021) show that a wrong model can be more compelling than the truth, especially when the data are quite random under the true model, which biases learning and leads to suboptimal decisions. They consider a framework where persuaders cannot change the observed data but propose models for interpreting this data to a decision-maker, who chooses the model that makes the observed data most likely, given their prior. A key assumption of their framework is that a proposed model is more compelling than an alternative if it makes the observed data most likely, given their prior. This assumption corresponds to various ideas in the strategy literature on what people might find persuasive, such as narratives that help with "sensemaking" (Weick, 1995; Chater and Loewenstein, 2016; Pillai et al., 2024), which entrepreneurs may also fall prey to.

Although Schwartzstein and Sunderam's (2021) model focuses on persuading others, this model can be applied to persuading oneself, as illustrated by the xkcd comic strip on "curve-fitting methods."[4] It highlights that scientists might fit the same data with different curves, which can result in them persuading themselves with a particular narrative, as well as persuading others. Similarly, research in strategy and management shows how managerial beliefs are resistant to change in the face of contradictory facts (Argyris and Schön, 1997; Levitt and March, 1988; Denrell and March, 2001; Pillai et al., 2024) and subject to systematic myopias that impede learning (Levinthal and March, 1993; Levitt and March, 1988; Tripsas and Gavetti, 2000; Menon and Yao, 2024).

Applied to entrepreneurial experimentation, this work suggests that even with informative experiments and attention to data, learning from experiments in ways that improve strategies and firm performance may be difficult. Such examples abound, particularly among start-ups and their beliefs about their business models. Jennifer Hyman, the CEO of Rent the Runway, a clothing rental subscription service, has discussed how the belief that their core hidden business is dry cleaning—rather than high technology or fashion, as one might assume—was key to unlocking their growth, as it dramatically influenced their learnings and decisions (Hoffman, 2020). Without it, her early experiments may have led the company to outsource their drycleaning operations or solve customer problems through technology. Instead, her belief led her to focus on their human operations to improve the customer experience, investing in quality control functions that could only be monitored by humans. This belief led her to learn from her experiments to invest in specialized labor in their warehouses—expert seamstresses—and experiment with providing corporate benefits to these hourly workers that could help retain them and improve the business.

A more cautionary tale is Framebridge, a venture-backed start-up that sold custom framing for photos online. Despite burning through $20 million in losses and struggling to achieve economies of scale, they saw themselves as a technology business and believed they would eventually achieve scale efficiency like Zappos, Facebook, and Uber, given their growth numbers (Koning and Dadlani, 2022). This belief led its founder, Susan Tynan, to invest in robots for their warehousing and pursue growth driven by automation, even as she saw her experiments result in losses. Only later did she realize they were misinterpreting their data due to beliefs about their business being similar to technology unicorns. They shifted their model to see themselves as an on-demand manufacturing business, which changed how they interpreted their data and led them to make decisions that allowed them to reach profitability.

## 9.4 Learning from Experimentation in the Age of AI

The previous section highlighted the frictions that impede learning from experiments, emphasizing the importance of designing informative experiments, attending to data, and drawing correct inferences when using experimentation to improve strategies.

Recent advancements in AI suggest that it may play an increasingly important role in various domains, including entrepreneurial experimentation. A growing body of research demonstrates AI's superior performance relative to human experts across diverse tasks, including making predictions (Agrawal et al., 2018; Cowgill, 2019; Kim et al., 2024), writing (Noy and Zhang, 2023; Doshi and Hauser, 2024), and business ideation and mentoring (Girotra et al., 2023; Dell'Acqua et al., 2023; Otis et al., 2023).

Most relevantly for entrepreneurial experimentation, emerging research suggests that these advancements may impact all three stages of the experimentation process: hypothesis generation, testing via experiments, and choice selection. In Csaszar et al. (2024), we highlight that AI may change the strategic decision-making process by enhancing the generation and evaluation of strategies. We provide evidence from a leading European start-up accelerator that large language models (LLMs) can generate entrepreneurial strategies at a level comparable to, or even surpassing, those of human entrepreneurs. This raises the intriguing possibility that AI could expand the search space for hypotheses and potentially generate superior ideas. Corroborating this potential, Ludwig and Mullainathan (2024) and Mullainathan and Rambachan (2024) demonstrate machine learning algorithms' capacity to generate scientific theories through data-driven pattern recognition and anomaly detection, often identifying those overlooked by humans. Although these findings raise the possibility that hypothesis generation may no longer be the exclusive domain of human cognition, questions remain about AI's ability to develop truly innovative theories, especially in the realms of strategy and entrepreneurship (Boussioux et al., 2023; Camuffo et al., 2023).

AI's potential may also shape the design and implementation of informative experiments. Horton (2023) and Brand et al. (2023) introduce the concept of using LLMs to simulate human samples in real-world experiments. Their findings indicate that these AI-generated samples produce responses consistent with economic theory predictions and real-world willingness-to-pay estimates. This suggests that AI could dramatically expand the scope and reduce the cost of experimentation, potentially providing more informative data and facilitating easier implementation of complex experimental designs.

Finally, the last phase of experimentation may also be affected by AI. Csaszar et al. (2024) show that LLMs can provide evaluations of entrepreneurial business plans that correlate with those of experienced angel and venture capital investors. These AI evaluations exhibit higher inter-rater reliability than that observed across human investors, suggesting their potential value in the idea selection and evaluation process. That said, whether these evaluations contain true informative signal and what the equilibrium effects might look like remain key open questions.

Together, these findings suggest that AI may significantly aid in the entire process of experimentation. If so, this raises important questions about the evolving role of human entrepreneurs and the capabilities that may become increasingly valuable complements to AI-assisted experimentation.

One possibility is that the ability to learn effectively from data may become an increasingly scarce and valuable entrepreneurial skill. Although AI can assist in expanding the range of possible experiments and improving their design, the ultimate responsibility for choosing which experiments to conduct and updating prior beliefs based on the resulting data remains with the entrepreneur. The entrepreneur's capacity to synthesize AI-generated insights with their own domain knowledge and strategic intuition may also become a key differentiator in successful entrepreneurial experimentation.

Moreover, the role of prior beliefs may gain increased importance in this AI-augmented landscape. Different entrepreneurs may derive vastly different insights from the same set of (AI-assisted) experiments, depending on their initial beliefs and their ability to integrate new information. This raises the possibility that the capacity to develop novel, contrarian beliefs with causal logic may remain a crucial entrepreneurial skill (Camuffo et al., 2023; Zellweger and Zenger, 2023).

In sum, although AI advancements have the potential to transform entrepreneurial experimentation, the ability to choose informative experiments, make meaningful inferences from the data, and integrate these insights into novel and coherent strategies may remain critical skills for entrepreneurs—at least in the short- to medium-term. In the long run, it may be the case that AI will evolve to help automate this entire process and overcome many of these challenges, by helping to explore a variety of priors, attend to a large set of data, and update more efficiently by eliminating cognitive biases. Future research should explore how AI might shape the process and landscape of entrepreneurial experimentation, and how entrepreneurs can best leverage these technologies to enhance their learning processes and strategic decision-making capabilities.

## 9.5 Discussion and Conclusion

This chapter explores the critical role of testing in entrepreneurial experimentation and highlights how its effectiveness depends on a range of factors relating to how entrepreneurs design and learn from information.

These insights raise many more questions than they answer, and highlight new avenues for research on entrepreneurial experimentation. First, if learning from data is a complement to theory generation and choice, then we need more nuance on how entrepreneurs learn from data. The insights in this chapter suggest that paying attention to more theory-free data in the world may be in and of itself more valuable. Might too much emphasis on theory lead us to possibly underplay this value? As Ludwig and Mullainathan (2024) suggest, machine learning algorithms can often outperform human experts in identifying data-driven patterns and anomalies without relying on explicit theories. This suggests that entrepreneurs may benefit from a more data-driven approach to experimentation, using AI tools to identify promising opportunities and refine their strategies.

Second, we may need to teach entrepreneurs to be scientific not just in theory generation (Camuffo et al., 2020; Felin and Zenger, 2017) but also in the steps of inference. We often assume that entrepreneurs have the tools to measure and learn what they need, but might this be a larger friction than we realize? As an analogy, although scientists need to carefully design the theories they are testing, if they use a broken microscope, they will not learn about the hypothesis. What, then, are the microscopes in entrepreneurial experimentation? It may be that entrepreneurs need to develop a range of experimentation capabilities, including the ability to design informative experiments, interpret data, and make correct inferences from the available evidence. Identifying and addressing the specific tools and skills that entrepreneurs need to conduct effective experiments—especially when they hold on to specific theories—may be an important area for future research and education.

Lastly, this capability to design informative experiments and learn from data may explain why we see successful businesses that arrived there without clear theories. Maybe they lucked into a great theory, or maybe they were great at learning from data. It raises the question: What is the binding constraint on better entrepreneurship? Is it the ability to generate good theories, to learn from experiments effectively, or to make the right choices based on the available evidence? The answer likely lies in a combination of factors, including the entrepreneur's cognitive abilities, the richness of the available data, and the complexity of the problem space. Understanding how these factors interact and shape entrepreneurial success is a key challenge for future research.

## Acknowledgments

I thank Natalie Carlson, Ashley Craig, Susan Cohen, Rembrand Koning, Nety Wu, Dennis Yao, and participants in the Bayesian Entrepreneurship Workshop for valuable comments and conversations.

## Notes

1. Two key exceptions are the studies by Agrawal et al. (2021), which provides important theoretical insights on the trade-offs between testing the value of an idea or strategy in isolation versus in combination, and Gans (2023), which formally highlights biases in experimental choice. Both have highlighted the importance of noise and bias in experiment design.

2. This example also highlights how the data collected might involve correlated signals, which relates to the argument made in section 9.3.3: In this case, the advice from different investors are likely to be correlated, as they all observe the same evidence (existing file synchronization products that were failing). Research has documented how individuals fail to properly account for the correlation between different pieces of information, which is referred to as "correlation neglect" (Enke and Zimmermann, 2019; Enke, 2020). This may lead entrepreneurs to learn incorrectly from informative evidence and highlights how experiment choices can also interact with biases in learning: Advice may not be as valuable an activity to pursue when advisers' signals are correlated.

3. For example, Cao et al. (2023) show that female-focused products may observe biased feedback from early-user forums such as Product Hunt, whose user base is composed of a high proportion of male users. When such products launch on days when more women are active on the platform, this bias is reduced, suggesting that choices on experiment bias can help them gather more informative data. Furthermore, research on the use of algorithmic predictions in decision-making shows that even experienced managers may not be able to discern the existence of bias, let alone its size and direction (Biermann et al., 2022; Kim et al., 2024), suggesting that, finding ways to develop this capability may be another approach to address this problem.

4. https://www.explainxkcd.com/wiki/index.php/2048.

## References

Adner, R., and Levinthal, D. A. (2004). What is not a real option: Considering boundaries for the application of real options to business strategy. *Academy of Management Review*, 29(1), pp. 74–85.

Agrawal, A., Camuffo, A., Gambardella, A., Gans, J. S., Scott, E. L., and Stern, S. (2025). The foundations of bayesian entrepreneurship. *Bayesian entrepreneurship.* MIT Press, 2025. Chapter 1, pp. TBA.

Agrawal, A., Gans, J., and Goldfarb, A. (2018). *Prediction machines: The simple economics of artificial intelligence.* Boston, MA: Harvard Business Review Press.

Agrawal, A., Gans, J. S., and Stern, S. (2021). Enabling entrepreneurial choice. *Management Science*, 67(9), pp. 5510–5524.

Andreoni, J., and Mylovanov, T. (2012). Diverging opinions. *American Economic Journal: Microeconomics*, 4(1), pp. 209–232.

Argote, L. (2012). *Organizational learning: Creating, retaining and transferring knowledge.* Springer Science & Business Media.

Argyris, C., and Schön, D. A. (1997). Organizational learning: A theory of action perspective. *Reis* (77–78), pp. 345–348.

Azevedo, E. M., Deng, A., Montiel Olea, J. L., Rao, J., and Weyl, E. G., , (2020). A/B testing with fat tails. *Journal of Political Economy*, 128(12), pp. 4614–4672.

Baliga, S., Hanany, E., and Klibanoff, P. (2013). Polarization and ambiguity. *American Economic Review*, 103(7), pp. 3071–3083.

Benjamin, D. J. (2019). Errors in probabilistic reasoning and judgment biases. In B. D. Bernheim, S. DellaVigna, and D. Laibson (Eds.), *Handbook of behavioral economics—Foundations and applications* (Vol. 2, pp. 69–186). North-Holland/Elsevier. https://doi.org/10.1016/bs.hesbe.2018.11.002.

Benoît, J.-P., and Dubra, J. (2018). When do populations polarize? An explanation. MPRA Paper 86173, Universidad de Montevideo. https://mpra.ub.uni-muenchen.de/86173/.

Bensinger, G. (2012). In Kozmo.com's failure, lessons for same-day delivery. Wall street Journal Digits blog, December 3, 2012. https://www.wsj.com/articles/BL-DGB-25584.

Biermann, J., Horton, J. J., and Walter, J. (2022). *Algorithmic advice as a credence good.* ZEW-Centre for European Economic Research Discussion Paper (22-071).

Blank, S. (2013). Why the lean start-up changes everything. *Harvard Business Review*, 91(5), pp. 63–72.

Boussioux, L., Lane, J. N., Zhang, M., Jacimovic, V., and Lakhani, K. R., (2023). The crowdless future? How generative AI is shaping the future of human crowdsourcing. HBS Tech & Ops Mgmt Working Paper 24-005. https://ssrn.com/abstract=4533642.

Brand, J., Israeli, A., and Ngwe, D. (2023). Using GPT for market research. Marketing Science Institute Working Paper 23-131. https://www.msi.org/working-paper/using-gpt-for-market-research/.

Brynjolfsson, E., and McElheran, K. (2016). The rapid adoption of data-driven decision-making. *American Economic Review*, 106(5), pp. 133–139.

Camuffo, A., Cordova, A., Gambardella, A., and Spina, C. (2020). A scientific approach to entrepreneurial decision-making: Evidence from a randomized control trial. *Management Science*, 66(2), pp. 564–586.

Camuffo, A., Gambardella, A., Pignataro, A., et al. (2023). Framing strategic decisions in the digital world. *Strategic Management Review*, 4(2), pp. 127–160.

Camuffo, A., Gambardella, A., and Pignataro, A. (2024). Theory-driven strategic management decisions. *Strategy Science*, 9(4), pp. 382–396.

Cao, R., Koning, R., and Nanda, R. (2023). Sampling bias in entrepreneurial experiments. *Management Science*, 70(10), pp. 7283–7307. https://doi.org/10.1287/mnsc.2021.01740.

Cassidy, J. (2002). *Dot.con: The greatest story ever sold.* New York, NY: HarperCollins.

Chater, N., and Loewenstein, G. (2016). The under-appreciated drive for sense-making. *Journal of Economic Behavior & Organization*, 126, pp. 137–154.

Cohen, S. L., Bingham, C. B., and Hallen, B. L. (2019). The role of accelerator designs in mitigating bounded rationality in new ventures. *Administrative Science Quarterly*, 64(4), pp. 810–854.

Contigiani, A., and Levinthal, D. A. (2019). Situating the construct of lean start-up: Adjacent conversations and possible future directions. *Industrial and Corporate Change*, 28(3), pp. 551–564.

Cowgill, B. (2019). Bias and productivity in humans and machines. Upjohn Institute Working Paper (19-309). W.E. Upjohn Institute for Employment Research. https://doi.org/10.17848/wp19-309.

Csaszar, F., Ketkar, H., and Kim, H. (2024). Artificial intelligence and strategic decision-making: Evidence from entrepreneurs and investors. Strategy Science, 9(4), 322–345. https://doi.org/10.1287/stsc.2024.0190.

Dell'Acqua, F., McFowland, E., Mollick, E., Lifshitz-Assaf, H., Kellogg, K. C., Rajendran, S., Lisa, K., Candelon, F., and Lakhani, K. R. (2023). Navigating the jagged technological frontier: Field experimental evidence of the effects of AI on knowledge worker productivity and quality. Harvard Business School Working Paper (24-013).

Denrell, J., and March, J. G. (2001). Adaptation as information restriction: The hot stove effect. *Organization Science*, 12(5), pp. 523–538.

Dixit, A. K., and Weibull, J. W. (2007). Political polarization. *Proceedings of the National Academy of Sciences*, 104(18), pp. 7351–7356.

Doshi, A. R., and Hauser, O. P. (2024). Generative artificial intelligence enhances creativity but reduces the collective diversity of novel content. Science Advances, 10(28), eadn5290. https://doi.org/10.1126/sciadv.adn5290.

Eisenmann, T. R., Ries, E., and Dillard, S. (2012). Hypothesis-driven entrepreneurship: The lean startup. Harvard Business School Entrepreneurial Management Case (812-095).

Enke, B. (2020). What you see is all there is. *The Quarterly Journal of Economics*, 135(3), pp. 1363–1398.

Enke, B., and Zimmermann, F. (2019). Correlation neglect in belief formation. *The Review of Economic Studies*, 86(1), pp. 313–332.

Felin, T., and Zenger, T. R. (2017). The theory-based view: Economic actors as theorists. *Strategy Science*, 2(4), pp. 258–271.

Fischhoff, B., and Beyth-Marom, R. (1983). Hypothesis evaluation from a bayesian perspective. *Psychological Review*, 90(3), pp. 239–260.

Gans, J. S. (2023). Experimental choice and disruptive technologies. *Management Science*, 69(11), pp. 7044–7058.

Gans, J. S., Stern, S., and Wu, J. (2019). Foundations of entrepreneurial strategy. *Strategic Management Journal*, 40(5), pp. 736–756.

Girotra, K., Meincke, L., Terwiesch, C., and Ulrich, K. T. (2023). Using large language models for idea generation in innovation. *SSRN Electronic Journal*. https://doi.org/10.2139/ssrn.4526071.

Gomez-Uribe, C. A., and Hunt, N. (2015). The Netflix recommender system: Algorithms, business value, and innovation. *ACM Transactions on Management Information Systems*, 6(4), Article 13, pp. 1–19.

Hadero, H., D'Innocenzio, A., (2023, December 19). Amazon, Walmart, and Target speed up deliveries as they fight challengers Shein and Temu. *Fortune*. https://fortune.com/2023/12/19/amazon-walmart-target-speeding-deliveries-challengers-shein-temu-holiday-orders/.

Hall, C. (2020). DoorDash closes first day of trading with $72B valuation. Crunchbase News. https://news.crunchbase.com/venture/doordash-stock-pops-on-first-day-of-trading/.

Hanna, R., Mullainathan, S., and Schwartzstein, J. (2014). Learning through noticing: Theory and evidence from a field experiment. *The Quarterly Journal of Economics*, 129(3), pp. 1311–1353.

Hoffman, R., (Host). (2020, January 9). What's the hidden business behind your business? (with Jenn Hyman) [Podcast episode]. Masters of Scale. https://mastersofscale.com/jenn-hyman-the-secret-business-behind-the-business/.

Horton, J. J. (2023). Large language models as simulated economic agents: What can we learn from *Homo Silicus*? NBER Working Paper (No. 31122). https://doi.org/10.3386/w31122.

Kerr, W. R., Nanda, R., and Rhodes-Kropf, M. (2014). Entrepreneurship as experimentation. *Journal of Economic Perspectives*, 28(3), pp. 25–48.

Kieren, P., and Weber, M. (2020). *Expectation formation under uninformative signals*. Available at SSRN 3971733.

Kim, H. (2025). The value of competitor information: Evidence from a field experiment. *Management Science*, 71(4), pp. 3600–3621. https://doi.org/10.1287/mnsc.2022.04062.

Kim, H., Glaeser, E. L., Hillis, A., Kominers, S. D., and Luca, M., (2024). Decision authority and the returns to algorithms. *Strategic Management Journal*, 45(4), pp. 619–648.

Klein, C. (2015). Why Coca-Cola's "New Coke" flopped. History.com. https://www.history.com/articles/why-coca-cola-new-coke-flopped.

Koning, R., and Dadlani, A. (2022). *Framebridge (A): Reimagining custom framing*. Harvard Business School Case (723-352).

Koning, R., Hasan, S., and Chatterji, A. (2022). Experimentation and startup performance: Evidence from A/B testing. *Management Science*, 68(9), pp. 6434–6453.

Levinthal, D. A. (2017). Mendel in the C-Suite: Design and the evolution of strategies. *Strategy Science*, 2(4), pp. 282–287.

Levinthal, D. A., and March, J. G. (1993). The myopia of learning. *Strategic Management Journal*, 14(S2), pp. 95–112.

Levitt, B., and March, J. G. (1988). Organizational learning. *Annual Review of Sociology*, 14(1), pp. 319–338.

Lord, C. G., Ross, L., and Lepper, M. R. (1979). Biased assimilation and attitude polarization: The effects of prior theories on subsequently considered evidence. *Journal of Personality and Social Psychology*, 37(11), pp. 2098–2109.

Ludwig, J., and Mullainathan, S. (2024). Machine learning as a tool for hypothesis generation. *The Quarterly Journal of Economics*, 139(2), pp. 751–827.

McGrath, R. G., and MacMillan, I. C. (1995). Discovery-driven planning. *Harvard Business Review*, 73(4), pp. 44–54.

Menon, A., and Yao, D. (2024). Rationalizing outcomes: Interdependent learning in competitive markets. *Strategy Science*, 9(2), pp. 97–117. https://doi.org/10.1287/stsc.2018.0083.

Miller, A., O'Mahony, S., and Cohen, S. L. (2024). Opening the aperture: Explaining the complementary roles of advice and testing when forming entrepreneurial strategy. *Organization Science*, 35(1), pp. 1–26.

Morris, C. (2024). Uber's $1 billion bet on Drizly is getting shuttered after less than 3 years. Fortune, January 16, 2024. https://fortune.com/2024/01/16/uber-shuts-down-drizly-alcohol-delivery-1-billion-acquisition/.

Moser, W. (2021). Why dot-com disaster Kozmo never became Instacart. Marker (Medium), February 23, 2021. https://marker.medium.com/why-dot-com-disaster-kozmo-never-became-instacart-5283563dc79b.

Mullainathan, S., and Rambachan, A. (2024). From predictive algorithms to automatic generation of anomalies. NBER Working Paper (No. 32422).

Murray, F., and Tripsas, M. (2004). The exploratory processes of entrepreneurial firms: The role of purposeful experimentation. *Business Strategy over the Industry Lifecycle*, Emerald, pp. 45–75.

Nagaraj, A. (2022). The private impact of public data: Landsat satellite maps increased gold discoveries and encouraged entry. *Management Science*, 68(1), pp. 564–582.

Newell, A., and Simon, H. A. (1976). Computer science as empirical inquiry: Symbols and search. *Communications of the ACM*, 19(3), pp. 113–126.

Noy, S., and Zhang, W. (2023). Experimental evidence on the productivity effects of generative artificial intelligence. *Science*, 381(6654), pp. 187–192.

Otis, N., Clarke, R. P., Delecourt, S., Holtz, D., and Koning, R. (2023). The uneven impact of generative AI on entrepreneurial performance. HBS Working Paper 24-042. https://www.hbs.edu/ris/Publication%20Files/24-042_9ebd2f26-e292-404c-b858-3e883f0e11c0.pdf.

Pfeffer, J., and Sutton, R. I. (2006). Evidence-based management. *Harvard Business Review*, 84(1), pp. 62–74.

Pillai, S. D., Goldfarb, B., and Kirsch, D. (2024). Lovely and likely: Using historical methods to improve inference to the best explanation in strategy. *Strategic Management Journal*, 45(8), pp. 1539–1566. https://doi.org/10.1002/smj.3593.

Ries, E. (2011a). *The lean startup: How today's entrepreneurs use continuous innovation to create radically successful businesses*. Crown Currency.

Ries, E. (2011b, October 19). How dropbox started as a minimal viable product. TechCrunch. Accessed July 21, 2025, from https://techcrunch.com/2011/10/19/dropbox-minimal-viable-product/.

Rousseau, D. M. (2006). Is there such a thing as "evidence-based management"? *Academy of Management Review*, 31(2), pp. 256–269.

Russo, J. E., Meloy, M. G., and Medvec, V. H. (1998). Predecisional distortion of product information. *Journal of Marketing Research*, 35(4), pp. 438–452.

Sandoval, G. (2001). *Kozmo to shut down, lay off 1,100*. CNET. Retrieved from https://www.cnet.com/tech/tech-industry/kozmo-to-shut-down-lay-off-1100/.

Schwartzstein, J., and Sunderam, A. (2021). Using models to persuade. *American Economic Review*, 111(1), pp. 276–323.

Thomke, S. H. (2020). *Experimentation works: The surprising power of business experiments*. Boston, MA: Harvard Business Review Press.

Tripsas, M., and Gavetti, G. (2000). Capabilities, cognition, and inertia: Evidence from digital imaging. *Strategic Management Journal*, 21(10–11), pp. 1147–1161.

Weick, K. E. (1995). *Sensemaking in organizations*, Vol. 3. Sage.

Zellweger, T., and Zenger, T. (2023). Entrepreneurs as scientists: A pragmatist approach to producing value out of uncertainty. *Academy of Management Review*, 48(3), pp. 379–408.

# 10

## Charting (and Updating) the Path: A Bayesian Perspective on Entrepreneurial Learning

Joshua L. Krieger

### 10.1 Introduction

Entrepreneurs have a bias for action because entrepreneurial learning requires generating new information. The entrepreneur creates new evidence through actions, such as building prototypes, constructing software simulations, interviewing suppliers, running scientific experiments, and surveying customers, and so forth—that update his or her theories around an opportunity and/or level of conviction. However, the type of information produced and its subsequent interpretation vary greatly, depending on the nature of the entrepreneur's *motivation to learn*. Although skilled entrepreneurs excel at systematically resolving uncertainty and persuading others to believe in them, what evidence they seek depends on their personal convictions as well as their beliefs about the audiences they aim to influence. The venture's early experimentation will differ based on whether the key uncertainties lie in the validity of the venture's underlying (contrarian) theories versus the venture's ability to outperform competitors within the boundaries of more commonly held beliefs. Like a navigator charting the path in through unseen terrain, the entrepreneur not only has to choose the initial path but also must determine what signals will warrant further commitments to continue, shift direction, or turn around entirely.

To examine how entrepreneurs update their beliefs about their venture, we first must consider the entrepreneur's motivation for gathering information. This chapter groups those motivations into two broad categories: de-risking the venture's *viability* and choosing from *alternative development paths*. As argued below, those motivations and the strength of conviction behind the entrepreneur's prior beliefs shape the type of activities on the venture's road map. Furthermore, the inevitable asymmetries between the venture team's beliefs and other stakeholders' beliefs alter the nature and sequencing of information-generating activities.

By comparing the distinct motivations of entrepreneurial experimentation, this chapter aims to provide a common language and framework for venture investment and sequencing choices. Too often, entrepreneurs and investors justify venture development paths based off simple heuristics that prioritize a single element of commercial potential (e.g., total addressable market, attractive "beachhead" markets, experimentation costs, etc.). Rather than replacing those heuristics with a new one, the framework described below uses the concepts of Bayesian entrepreneurship to integrate these elements, highlight the trade-offs between them, and quantify the relative value of different development paths.

## 10.2 Defining Opportunities and Venture Beliefs

In articulating their expectations around any new venture, entrepreneurs and investors will often bundle beliefs of varying specificity and criticality (to the venture) into broad categories of risk and rewards.[1] Further, entrepreneurs may conceive of "opportunities" quite differently. Some might describe broad applications of a new technology (hammer with plausible nails) or as a problem "pain point" to be solved with all available tools (nails looking for hammers), whereas others might define their venture with detailed descriptions of particular business models (including value proposition, technology and operations, go-to-market, and cash flow plans). However, the process of learning about the venture depends on underlying assumptions about which and whose beliefs the entrepreneur aims to challenge and update through information gathering.

To fix ideas, I discuss an *entrepreneurial opportunity* as specific solution–problem pairs (e.g., robotic hands for packaging cleaning supplies). Thus, opportunities must embody a set of both technological and commercial hypotheses—although the degree of uncertainty may vary independently. Entrepreneurs, investors, and potential customers might also vary in their baseline beliefs about the "fit" of any particular solution–problem pair.[2] "Adjacent opportunities" take the same or similar technologies and apply them to a distinct commercial application. The differences between two adjacent applications may be subtle or obvious, but what makes them distinct is some divergence in either their technological development needs or commercialization steps.[3]

Thus, in discussing learning venture viability, the chapter will focus on updating beliefs over a particular opportunity—probability of success ($p_i$), cost ($CostDev_i$), and profitability conditional on success ($V_i$)—as well as the venture's overall viability ($p_{\text{overall}}$, $CostDev_{\text{overall}}$, potential $V_{\text{overall}}$ across all target applications).[4] The venture's "overall" expected values reflect either the sum of its various opportunities, if all are viable, or the expected value of the subset of opportunities that the entrepreneur expects to pursue. Meanwhile, the choice of alternative development paths involves the subset and sequence of *i*s that maximize the value of the opportunity, including learning spillovers and option value.

## 10.3 Venture Viability

The venture's viability refers to beliefs about whether pursuing the venture is a bet with positive net present value. Those beliefs are distinct from the question of whether a given entrepreneur—with his or her own skills, opportunity costs, access to resources, and risk preferences—*should* choose to pursue that venture in the first place. Aside from that personal choice, individuals might vary in their forecasts about the venture's viability due to factors such as asymmetric information about a particular opportunity and overall venture potential or different understanding of how key resource providers might view an opportunity or overall venture. Thus, a venture's viability is in the eye of the beholder, but these beliefs can still be decomposed into distinct components, each of which is subject to Bayesian updating.

### 10.3.1 Components of Venture Viability

Viability encompasses multiple subsidiary beliefs, which can be grouped into three main groups: technological, commercial, and development cost.

**Technological feasibility** is the belief of whether the proposed problem–solution pair will "work." That feasibility threshold might depend on whether a novel technology will achieve its most basic goals (e.g., Does the flying car actually take off and land?) or whether the technology or service can achieve a minimum performance threshold (e.g., Will the customer service chatbot answer questions as quickly as a human employee?). The technological feasibility belief is summarized by $p_i$ as the likelihood that the venture can deliver on this minimum performance threshold for a particular opportunity, with the assumption that the entrepreneur commits to maximizing the venture's technological potential for that given opportunity. The "strength" of this belief is reflected in the distribution of $p_i$.

**Commercial viability** implies a similar set of questions about the economic potential of the venture, but assuming all technological uncertainty has been resolved favorably. Will the willingness to pay exceed marginal costs? What is the break-even scale of production? What is the ultimate market size of the best version of the opportunity ($\max_{i\in\{1,2,\ldots,n\}} V_i$) or cluster of adjacent opportunities ($V_1 + V_2 \ldots + V_N$)? Again, the entrepreneur may be more or less certain about the commercial potential as reflected in the spread of the distribution(s) of each $V_i$.[5] The entrepreneur (and investors) must consider the time until the venture can receive payoff $V$, such that the expected payoff today is NPV $= V/(1+r)^t$, where $t$ is the expected time until the payoff can be realized and $r$ is the discount rate.[6]

**Development costs** ($CostDev_i$) refer to the company-building efforts necessary to pursue the venture "in earnest" (i.e., with full commitment to at least one particular opportunity and to the best of the entrepreneur's abilities). Even if technological and commercial feasibility beliefs are strong, the entrepreneur must still evaluate the cost of capital and time needed to set up the fully operational venture. Holding the nomical* $CostDev_i$ constant, the timing and shape of a venture's development cost curve can vary (e.g., front-loaded capital expenditures vs. steady increase in development costs over time), such that the true cost of development is expected to be:

$$CostDev_i = \sum_{t=0}^{N} \frac{c_t}{(1+r)^t},$$

where $c_t$ is expected development cost in period $t = 0$ for a development path of $N$ total periods. Forecasting matters here, as back-loaded costs should not be discounted as far in the future if the decision to pursue that opportunity commits the venture to that future costs (and precludes shifting that expenditure to other uses). Thus, the initial period $CostDev_i$ should be the irrevocable cost commitments associated with pursuing that project (i.e., "dollars locked up" in that opportunity). Section 10.4 discusses how intermediate milestones might allow ventures to delay some development costs commitments to later points in time, in which case those costs can be discounted at the initial decision point. Long timelines and/or large opportunity cost of locking up significant capital might limit the set of potential investors, even if those investors share the entrepreneur's beliefs about the potential for the mature version of the venture. For example, a scientist who invents a compelling early diagnostic test for dementia might face a very long road to commercialization, with multiyear clinical trials and regulatory scrutiny. Even a venture capital investor convinced of the large market opportunity might pass on the investment due to capital requirements and fund timeline considerations.

To summarize, a venture opportunity's viability can be broken into three main components of viability that together give a rough sense of the opportunity's net present value:

$$VentureValue_i = NPV(p_i \times V_i - CostDev_i) = p_i \times \frac{V}{(1+r)^t} - \sum_{t=0}^{N} \frac{c_t}{(1+r)^t}$$

### 10.3.2 Testing Viability

Learning about the venture's viability may involve focused experiments to resolve uncertainty about a particular component of viability for one (or more) opportunities or gathering information that informs multiple elements of viability.

For example, the founders of start-up Rest Devices wanted to understand the commercial value of embedding their thin flexible sensors within baby sleepwear to monitor sleep and breathing patterns. The application was attractive but was not the team's natural passion or area of expertise. Led by three MIT graduate students, none of whom had children or experience in the baby-gear market, they had initially believed the entrepreneurial opportunity was in sensors for sleep apnea before they learned that that commercial path would require a long and expensive regulatory gauntlet (Ghosh and Payton, 2016). The feasibility of the baby-wear sensors venture would rely on a number of factors: Could they produce accurate sensors and accompanying monitoring software that works reliably (technical)? Could they convince parents of infants to purchase their product at a profitable margin (commercial)? And assuming it was technically feasible, could they build out inventory and customer service systems to properly serve retailers and mass market customers (cost of development)?

The Rest team's early testing involved different activities to address the various components of the venture's viability. They created prototype versions of the baby-wear onesie outfits, equipped with sensors and a connected smartphone app, and tested it with the permission of some early-adopter parents. On the commercial side, they conducted an online survey and follow-up phone interviews of new parents to gauge interest and willingness to pay, and they brought their prototype to trade shows to get feedback and sales opportunities with retailers. In other words, their early evidence-generating activities were multipronged, testing multiple components of their viability expectations.

In other cases, testing can be more concentrated in a particularly critical component of viability. For example, Vaxess Technologies had a design for a new type of shelf-stable microneedle patch for vaccine delivery (Quelch and Rodriguez, 2014). The idea was rooted in published research and intellectual property from Tufts University and MIT. Despite academic research to support their approach, the viability of the venture would primarily depend on two sub-elements of technical feasibility: producing the tiny silk protein-based tips that painlessly penetrated the skin and slowly released the vaccine, and showing their safety and efficacy in clinical trials. Early experimentation focused on engineering and optimizing the 3D printing of the microneedle tips and then testing their efficacy in murine and simian studies. The focus on technical feasibility did not mean that Vaxess had no commercial questions.

Indeed, even if the vaccines proved effective and manufacturable at scale, the company would have to navigate partnerships, distribution, pricing, marketing, and negotiations with insurers and government health agencies. But the greatest uncertainties (weakest or most dispersed priors) were associated with the technology, and the team had confidence that a vaccine market would still be there and provide sufficient demand for high-performing vaccine delivery technologies.

### 10.3.3 Viability Belief Differences Across Key Parties

In addition to forming one's own priors over $p$, $V$, and *CostDev*, the entrepreneur also tries to understand the comparable beliefs held by key stakeholders and resource providers. That group of other players includes potential investors, customers, employees, suppliers, and regulators. I follow chapter 1 and focus on investors as the representative external party.

Entrepreneurs and investors might have divergent beliefs about the key components of the venture's viability. Those may arise from differences in their previous experiences as well as different access to information or expertise. By virtue of *choosing* to pursue (and persuade others about) their venture, the entrepreneur is relatively more optimistic than others about the opportunity. Frontier technology start-ups that need to prove viability related to both their technology and commercial prospects face an even more daunting challenge in convincing investors to take on multiple types of uncertainty and evaluate technologies more likely to lie outside of their expertise (Lerner and Nanda, 2020; Arora et al., 2024; Nanda, 2020).

That gap in beliefs is not a problem if both parties still believe that the business is viable—that is, sufficiently high $pV - CostDev$ to warrant investment. However, two possible scenarios emerge if the investor's viability beliefs are worse than the entrepreneur's and lower than the investor's threshold for investment.

First, if the entrepreneur holds "strong" priors (e.g., narrow distributions around their expected values of $p$, $V$, and *CostDev*), then the entrepreneur's main motivation in generating updating/learning is *persuasion*. See figure 10.1 for a chart describing the distinct "modes" of entrepreneurial learning. The process of generating new information in service of that goal may or may not align with the development road-map the entrepreneur would have followed without any resource constraints. To the extent that persuasion activities diverge from that optimal road map, the strong prior entrepreneur will want to take on experiments that directly address the component of viability where the entrepreneur and external party have their greatest disagreement and do so with minimal cost (e.g., "lean start-up" methodologies).[7] However, persuasion activities need not

| | Experimenting for. . . viability | Experimenting for. . . Optimal path |
|---|---|---|
| *Strong priors* | Persuasion | Updating development path |
| *Weak priors* | Learning to kill or commit | Pure exploration / Messing around |

**Figure 10.1**
Entrepreneurial learning modes.

take the company off its efficient road map. If persuasive activities are aligned with the natural company-building roadmap, then the strong prior entrepreneur will want to maximize the size/scope of experiments (e.g., experimenting at scale, larger prototypes, expanding manufacturing earlier), subject to resource and investor patience constraints.

The second situation is that the entrepreneur holds "weak" priors (e.g., wide distributions around their expected values of *p*, *V*, and *CostDev*). Despite holding more optimistic viability beliefs than the investor, the two parties now share the common goal to efficiently retire uncertainty and maximize the option value of their investments. Practically, that means striving to run cheap experiments focused on increasing precision over the venture's overall viability, targeted at efficiently maximing learning for cost of that experiment/activity. Updated beliefs may result in either a quick termination of the venture or the entrepreneur strengthening his or her convictions and the investor updating positively.

## 10.4 Choosing from Alternative Development Paths

Learning that a venture is a worthwhile pursuit to initiate, continue, or finance is only one "mode" of entrepreneurial learning. Success in venture building also depends on the ability to navigate development toward the most promising directions possible for a given venture. The previous section highlighted distinct motivations for learning about a venture's *viability* and how those motivations map to different information gathering and persuasion needs. But many entrepreneurial activities aim instead to rule in/out *different development options for the given venture* (i.e., adjacent opportunity paths, *i*s). Charting a venture road map with multiple potential development options is central to entrepreneurial choice, especially when the entrepreneur has already established conviction around the venture's overall viability and committed to pursuing the venture in some shape or form.

Choosing paths is especially important for novel technologies with potential utility across multiple application domains. Demonstrating viability along initial paths might be a prerequisite for gaining access to adjacent opportunities and spurring complementary investments as "general purpose technologies" (Bresnahan, 2010). But initial path failures might also quell such opportunities for follow-on exploration. Learning about any one path has implications (spillovers) for learning about adjacent paths. In other words, the entrepreneur is choosing between multiple *initial development paths* as well as possible *sequences* of development paths for adjacent opportunities.

To select the sequence of development paths, I propose a framework that addresses the independent viability of distinct paths as well as the degree of development spillovers (adjacency) between paths. The framework requires estimating three project features:

1. **Independent project-path characteristics**: Each potential project path $i$ has a development cost $c_i$, a probability of success $p_i$, and a potential profits conditional on development success $V_i$.

2. **Learning spillovers between paths**: Any pair of two possible paths have a correlation in their probability of success (e.g., the probability of success on path B, conditional on success in path A, is represented as $S_{B|A}$), which is increasing in their R&D overlap and decreasing in their difference in relative cost/complexity ($c_B - c_A$).

3. **Development "distance" between paths**: The distance is the additional cost ($c_{A \to B}$) of developing the technology for the second application after developing it for the (given) first application. This distance mechanically follows #2 above. Knowing the relative direction and cost of both paths is sufficient to know their distance. Project pairs with tighter correlations ($S_{B|A}$ closer to 1) will have shorter development distances ($c_{A \to B}$, $c_{B \to A}$).

### 10.4.1 Visualizing Paths

These features can be visualized as a set of vectors, where each vector represents a distinct project path (see figure 10.2). The length of each vector represents the total cost ($c_i$) of developing the venture for that particular opportunity/application. Those costs include labor, capital expenditures, R&D,

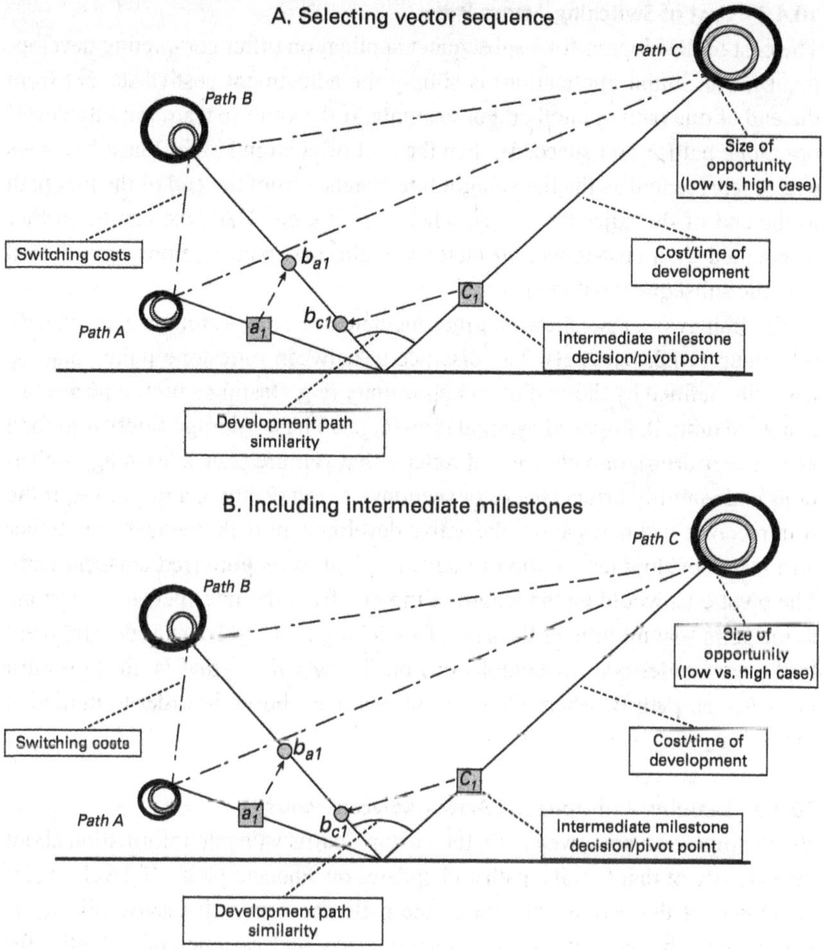


**Figure 10.2**
Development paths framework.

experimentation costs, legal, and commercial development (e.g., sales, marketing, demonstrations, etc.). At the end of each vector is a node, whose size represents the expected enterprise value ($V_i$) conditional on successfully developing the venture along that path.[8] Since initial estimates of $V_i$ might be highly uncertain, one can also visually represent the nodes as a range of potential outcomes with concentric circles for different percentiles in the expected payoff distribution. The degree of development overlap between any two paths is reflected by the angle between their two vectors.

### 10.4.2 Cost of Switching Across Paths

The cost of developing for a subsequent application (after completing development for an initial application) is simply the adjustment cost (distance) from the end of one path to another. For example, if the venture team initially develops along path A and succeeds, then the cost of pursuing path B next becomes $c_{A \to B}$, represented as the the straight line distance from the end of the first path to the end of the adjacent pathby, where $c_{A \to B} < c_B$. That cost can be further discounted by a cost-reduction factor $\gamma$ if the first path is more complicated than the subsequent path ($c_A > c_B$).

The framework also allows for intermediate milestones along any given path (see figure 10.2, panel B). The distance to/between milestone points may be naturally defined by the cost of an opportunity (e.g., testing a prototype aircraft in a wind tunnel), imposed by regulation (e.g., a Phase I dosage safety trial for a new cancer drug), or a choice that reflects that venture team's learning motivations and viability priors (see prior sections). Upon reaching a milestone, if the firm receives bad news about the active development path, the team can either terminate development (option to quit) or pivot to its preferred adjacent path. The pivot cost would be the length of the line from the intermediate milestone to the point that minimizes the sum of switching costs and cost of development to the next milestone (or completion) on the new line—that is, move toward point $b_{a1}$ on path B, where the location of $b_{a1}$ is chosen in order to minimize the term $c_{a_1 \to b_{a1}} + (c_{b_1 - b_{a1}})$.

### 10.4.3 Learning and Updating Across Adjacent Paths

By moving down any given path, the venture learns valuable information about the viability of that specific path and updates on adjacent paths. If development continues all the way to the end of the path, the firm will resolve all uncertainty about the technology, development costs, and commercial potential for that given path. The degree of adjacency between related venture opportunities captures the level of *spillover learning* across paths.

For both success and failure on the initial path (or milestone), the venture can update the probability of success for adjacent paths based on their development overlap. That overlap is the pair's path success correlation— denoted as the probability that path B is viable, given success on path B ($S_{B|A}$). That correlation is determined by two components: the angle between the two paths ($\theta_{A,B}$) and the relative lengths of the two paths (cost/complexity differences: $c_{A-B}$).[9]

To calculate the updated probability of success on adjacent lines, one simply applies Bayes' rule as follows:

$$P(B|A = success) = \frac{P(A|B) \cdot P(B)}{P(A)} = \frac{S_{(A|B)} \cdot P(B)}{P(A)}$$

$$P(B|A = failure) = \frac{(1 - P(A|B)) \cdot P(B)}{(1 - P(A))} = \frac{(1 - S_{(A|B)}) \cdot P(B)}{(1 - P(A))}.$$

Thus, the expected value of any given path includes the the risk-adjusted value of that given path, plus the probability weighted values of the expected *updated* value of adjacent paths following the initial path outcome. Conceptually, one can calculate the expected value of all plausible paths, including all intermediate milestones and switching pathways. In practice, the number of possible scenarios increases greatly as one adds paths and milestones, so the Bayesian entrepreneur may limit the set of possible paths to the most likely and limit intermediate decision points to the most significant milestones (rather than any opportunity to update).

Any venture that purports to develop a multi- or general purpose technology implicitly makes this optimization decision. But instead of articulating the sequencing trade-offs of this development paths framework, they often determine or justify their choice of initial paths by using heuristics such as total addressable/obtainable market size, targeting "beachheads" with high likelihood of success or establishing "basecamps" with great learning opportunities. However, different heuristics can sometimes point to different development sequences.

For example, RightHand Robotics spun out of Professor Robert Howe's Harvard engineering lab with a gripper technology that had beat competition in a DARPA competition for robotic hands picking up a variety of objects without preprogramming. The venture considered a wide range of potential commercial applications, including produce harvesting, bomb disposal, meal kit preparation, warehouse automation, and recycling (Eisenmann and Straaberg, 2013). RightHand could justify any one path by emphasizing a single selection heuristic or a simple weighted combination of factors.

Going straight for the largest addressable markets would point to produce harvesting or warehouse picking/packing However, both would require complex and capital intensive development to meet the high performance, computer vision, and mobility needs. Recycling might allow for a shorter path to market and more fault tolerance, but the profit potential was potentially much smaller, with only moderate learning applicable to other applications. Meal kit preparation was a growing market, with a clear need for stationary robotics to pick and scoop a small set of repeat items, but required application-specific engineering (e.g., food-safe grippers, operating in refrigerated rooms, identifying poor-quality ingredients) with little overlap with their other target applications. The development paths framework incorporates the risk and profit opportunities from each opportunity with the learning across applications to highlight the optimal *initial path* and allow the entrepreneur to analyze how sensitive that choice is to any given within- or across-path variables.

In addition to choosing initial paths, the development paths frameworks also applies to evaluating new opportunities for existing ventures. Young ventures often begin by pursuing a single market application, but then learn about adjacent market opportunities or even consider new technologies to serve their existing target markets. Properly evaluating these new options requires assessing their learning and real option value as well as their individual path expected values ($p_{newpath} \times V_{newpath} - c_{newpath}$). When new opportunities arise, the same approach as the initial path selection can be applied—visualizing the decision point as either a new origin or an unexpected intermediate milestone.

The team from weather analytics company Tomorrow.io faced multiple such moments of reassessing their path. One occurred early on when they were focusing their commercial efforts on serving the airlines and live events (e.g., professional sports and concerts). The carmaker Porsche offered them an opportunity to collaborate and use sensor data from their connected vehicles as new inputs into their weather models (Krieger et al., 2019). The decision to take on the pilot project required estimating implications that all feed into the path selection framework: (1) the time and cash costs of pursuing the pilot project ($c_i$), including opportunity costs to other current R&D efforts; (2) the probability that they win a bigger contract with Porsche based on the pilot project; (3) the value of the Porsche opportunity conditional on succeeding; (4) the potential value of adjacent automotive industry opportunities, conditional on a successful relationship with Porsche; and (5) the learning and spillover value of both success and failure on the Porsche pilot project on adjacent paths in automotive, shipping, airlines, and military applications. The company did pursue the collaboration, only to once again reevaluate its path choice with the emergence of another new (set of paths) around developing its own satellite constellation (Krieger et al., 2022).

## 10.5 Conclusion

This chapter highlights two primary "modes" of entrepreneurial learning: determining viability and optimizing path direction. However, the two modes are not mutually exclusive. Even if the entrepreneur has enough conviction to start down a particular development path, the necessary activities will still generate new information that leads to updating about venture viability (for the team and investors). Sophisticated entrepreneurs might select initial paths and experimental milestones that resolve questions of overall venture viability while also trying to adjudicate choices between adjacent market applications paths.

Whether committed to pursuing the venture and trying to navigate the best path or still trying to learn whether the venture is viable, the type of experimentation and information gathering depends on the venture's greatest uncertainties. The approach in this chapter emphasizes three key expected values for each opportunity: probability of success, cost of development, and profitability (conditional on success). However, the framework also shows that ranking projects based on their individual expected values is not sufficient for charting the venture's learning direction. The optimal sequence of paths also depends on the strength of the priors (i.e., spread of the distribution) for each of these variables, the ability to (cost-effectively) update on those priors, and the spillover learning value for adjacent opportunities.

This approach not only helps entrepreneurs navigate the uncertainties inherent in venture creation but also aligns their strategies with the expectations of key stakeholders. The visualization of development paths and the incorporation of learning spillovers offer a robust method for making informed decisions about initial and subsequent venture trajectories. Ultimately, by moving beyond simplistic heuristics, this framework empowers entrepreneurs to systematically evaluate and pivot their strategies, thereby enhancing their ability to build successful and innovative ventures.

## Notes

1. A central challenge of generalizing about entrepreneurial experimentation is that the core beliefs behind a venture vary greatly across industries and technologies, and entrepreneurs rarely articulate those beliefs as precise hypotheses to be tested.

2. For example, an individual may see merit in addressing a given problem and may even think the proposed solution would work but still not find the match of the two compelling (e.g., "willingness to pay for solving that warehouse problem is sky high, but I've seen robotic systems that look more effective than this design").

3. For simplicity in this chapter, I abstract away from (the many) business model choices that might exist within the given opportunity or across the set of related adjacent entrepreneurial opportunities—for example, degree of vertical integration, payment models, licensing, and sales channels—and assume that each opportunity is tied to a particular dominant business model. However, one could also represent "adjacent" opportunities as similar problem-solution pairs that differ mainly in their business model hypotheses if those business model distinctions require a different set of commercial development activities.

4. $V$ here is a payoff or profit measure, conditional on the venture achieving a reasonable threshold of "success" in terms of both its initial technical hypotheses and ability to gather sufficient resources to develop the venture.

5. The shape or skewness of the distribution of $V_i$ likely differs by industry. For example, some fields might tend to have "longer-tailed" payoff distributions if they tend to have a small number of "blockbuster" winners.

6. $V$ itself could also be considered a time-discounted expected value based on the forecasted stream of cashflows produced by the venture starting in time $t$.

7. Notably, persuading the investor is not merely about moving mean expectations of viability but may also be about increasing the investor's expected variance (e.g., probability of long-tail payoff outcomes) such that the investor is willing to embrace the venture's option value in its portfolio.

8. More specifically, $V_i$ is the net present value of cashflows of the venture after it has established its viability. The level of experimentation and company-building activities needed to establish that viability will vary across settings and opportunities. The venture's products or services may still evolve after establishing viability, but here I assume that the entrepreneur has a prior over the expected captured value (i.e., serviceable obtainable market) that the mature venture can achieve along that given application path.

9. In practice, an estimate of the percentage of project overlap direction can be converted into the angle of overlap, or vice versa. For example, 80 percent development overlap can be converted to a 36° vector angle because $(1-(36°/180°)=80\%)$. Even though $\theta_{A,B}=\theta_{B,A}$, $S_{A|B}$ and $S_{B|A}$ need not be equal, since the amount of learning can also be scaled by the relative complexity of the two paths—for example, if $c_B - c_A > 0$, then any updating on $p_B$ could be scaled by that difference, $S_{B|A} = \frac{c_A}{c_B} \times (1-(\theta^{\circ}_{A,B}/180°))$.

## References


Arora, A., Fosfuri, A., and Rønde, T. (2024). The missing middle: Value capture in the market for startups. *Research Policy*, 53(3), p. 104958.

Bresnahan, T. (2010). General purpose technologies. *Handbook of the Economics of Innovation*, Vol. 2. Elsevier, pp. 761–791.

Eisenmann, T., and Straaberg, S. (2013). *RightHand Robotics: Choosing the first market*. Harvard Business School Case Study, pp. 824-006.

Ghosh, S., and Payton, C. (2016). *Rest devices*. Harvard Business School Case Study, pp. 816-076.

Krieger, J. L., Nagaraj, A., and Barnett, J. (2022). *Tomorrow.io goes to space*. Harvard Business School Case Study, pp. 822-005.

Krieger, J. L., Stanton, C., and Barnett, J. (2019). *Forecasting climacell*. Harvard Business School Case Study, pp. 9-820-044.

Lerner, J., and Nanda, R. (2020). Venture capital's role in financing innovation: What we know and how much we still need to learn. *Journal of Economic Perspectives*, 34(3), pp. 237–261.

Nanda, R. (2020). Financing innovation "tough tech". In S. Dutta, B. Lanvin, and S. Wunch-Vincent (Eds.), *The Global Innovation Index* Cornell University, pp. 113–119.

Quelch, J. A., and Rodriguez, M. (2014). *Vaxess Technologies, Inc.* Harvard Business School Case Study.

# 11

## Advice and the Bayesian Entrepreneur

Susan Cohen and Rembrand Koning

### 11.1 Introduction

Bayesian entrepreneurship starts from the premise that entrepreneurs' beliefs guide their theorizing, experimentation, and choices (Agrawal et al., 2025). Since each entrepreneur has unique beliefs based on their own set of past experiences, cognitive ability, and disposition, individual entrepreneurs are likely to take different actions, even if they stumble upon the same entrepreneurial opportunity. Beyond initial beliefs, Bayesian entrepreneurship is rooted in the idea that experimentation can be a valuable tool to update beliefs and improve choices. Building on recent research (Kerr et al., 2014; Agrawal et al., 2025), we define a Bayesian experiment as any purposeful information-gathering activity that tests the promise of a new venture.

Although Bayesians can learn from multiple forms of experiments, the study and practice of experimentation in entrepreneurship has largely focused on tests involving customers (Ries, 2011; Eisenmann et al., 2012). This idea was seeded in the popular lean start-up methodology, which encourages entrepreneurs to develop and share minimum viable products to test customers' reactions to a product's core value propositions. Although, newer academic frameworks widen the lean start-up lens, the examples and cases used to motivate these approaches—such as consumers shopping for tattoo artists or a new pair of shoes—largely highlight how entrepreneurs benefit from experimenting with users. Yet, these broader frameworks—including the Bayesian approach—also allow entrepreneurs to experiment by testing ideas in other ways, including with advisers.

Advice can fundamentally alter the trajectory of a nascent venture. Yet, how and what the Bayesian entrepreneur learns from advice is unclear. In this paper, we highlight how the fundamentally *social* nature of advice yields unique challenges and opportunities in the context of Bayesian learning.[1] First, information from an adviser is shaped not only by the entrepreneurs' beliefs but also by the adviser's own unique beliefs. As a result of these differing beliefs, a Bayesian entrepreneur is less likely to seek out advice, and when they do, that advice is likely to be biased. Second, the presence of differing beliefs also has benefits. Because advisers draw on their own unique theories and experiences, advice is especially beneficial in broadening and extending an entrepreneur's theories of value creation and capture. Third, since advice is cheap but correlated, Bayesian entrepreneurs can iterate on their beliefs quickly but with greater potential for over confidence, herding, and bubbles. Fourth and finally, the iterative and heterogeneous nature of advice promotes more efficient experimentation and so in and of itself encourages Bayesianism.

The founding, growth, and pivots of the direct-to-consumer online framing start-up Framebridge illustrates the promise and perils of learning from advice (Koning and Dadlani, 2023). Framebridge's founder, Susan Tynan, gained conviction in her idea for an online framing business after multiple expensive and frustrating trips to local picture-framing stores. Her initial idea was to build a smartphone app where customers selected a frame, mailed their art, and then received back a framed piece ready to hang. Yet, early advice on her idea was often biased: Investors—most of them men—who had never thought or tried to frame anything themselves advised against building an online business in the framing category. More broadly, much of the early feedback and advice Tynan received was focused on "why this [idea] wouldn't work." As we explain below, negative advice on early ideas is common for Bayesian entrepreneurs determined to launch new businesses.

Despite early negative signals, Tynan persevered and raised a million-dollar seed round from a group of venture capitalists. Momentum in the round was sparked by her first investor, Dayna Grayson, whose belief in the idea was motivated by her own prior experience trying to search for a framing service. Tynan used the seed funding to launch a minimum viable framing business out of a small local warehouse, validating that there was a real market for online app-based framing. Based on these test results, the board invested $65 million over three years so Tynan could scale quickly. Although Tynan's original pitch deck hand-waved at the idea that scale would drive defensible margins, it was advice from her board that helped her create a new theory about how to capture value: invest in expensive Kiva-style robots, such as those used in Zappos and Amazon fulfillment centers, to position Framebridge as the "Zappos for Framing" and fuel growth and defensible margins from scaled automation.

Unfortunately, the "Zappos for Framing" theory was wrong. In just three years, annual revenues exceeded $20 million, but losses also exceeded $20 million a year. As it turns out, selling shoes online has fundamentally different economics than online custom framing. Shoeboxes are relatively uniform, requiring little variation in the packing, shipping, and handling processes, whereas pieces of framed art are essentially unique, each requiring different handling, framing techniques, and operational care. Why was Tynan's theory of value capture so off? As we argue below, when multiple advisers share similar backgrounds and networks—as Tynan's advisers of online tech executives did—advice can lead entrepreneurs to incorrectly apply theories from other firms and become overconfident in their beliefs, which reduces further learning.

Fortunately, one of Susan's closest advisers, Tim O'Shaughnessy, expressed concerns about Framebridge's long-term viability. Concerned over the mounting losses, O'Shaughnessy's skepticism injected enough doubt into Tynan to trigger a search for alternative strategies for how to capture value from the core Framebridge idea. Her search led to a manufacturing executive from the door fabrication industry, and thus with low network overlap with her board. He advised Tynan to experiment with a robot-free operational strategy with lower fixed costs and steeper learning curves. The operational experiment yielded promising improvements in metrics that previously were unappreciated. She doubled down on a new "custom manufacturing" strategy, inspired not by tech unicorns but rather by door fabrication companies. The pivot yielded both profits and continued growth. Although the "correlated" advice Tynan relied on to develop her Zappos-inspired robot-based strategy almost bankrupted the venture, the new strategy based on advice from a true outsider illustrates how, when properly structured and considered, advice can act as a valuable catalyst for better experimentation and learning.

## 11.2 The Bayesian Entrepreneur: A Primer

At the heart of the Bayesian approach to entrepreneurship lies the concept of priors—the beliefs an entrepreneur holds about their idea (Parker, 2006; Agrawal et al., 2025). An entrepreneur's priors are not single point estimates but rather probability distributions that represent the range of possible outcomes and their likelihood, reflecting the uncertainty inherent in entrepreneurial ventures and allowing for nuanced updating as new information is acquired. Priors, also referred to as beliefs, are subjective constructs shaped by an entrepreneur's background, experiences, circumstances, cognitive abilities, self-efficacy, and so on.

Importantly, the decision to pursue an entrepreneurial venture hinges on the strength of these priors. A Bayesian entrepreneur must have sufficiently strong beliefs in their idea to take action, as weak priors are unlikely to result in meaningful entrepreneurial activity. For instance, an individual contemplating leaving a stable job for an entrepreneurial pursuit must not only believe in their idea but also be convinced that this path will yield more favorable outcomes than remaining in wage employment.

In the Bayesian framework, these initial beliefs serve as the starting point for a dynamic updating process. Based on their priors, Bayesian entrepreneurs develop theories about how to create and capture value. For example, early scooter rental companies such as Bird and Lime may have theorized that if they offered an inexpensive and convenient way for students to get around campus, enough students would pay to ride them to drive economies of density and network effects. Bayesian entrepreneurs next articulate hypotheses based on these theories, conduct experiments to test their hypotheses, and finally update their beliefs through the theory-guided interpretation of experimental data. Theories often center around consumer behavior, but can also consider a wide array of factors related to entrepreneurial strategy, and serve as the "likelihood function" that helps entrepreneurs map the data they collected into inferences (Felin and Zenger, 2017; Agrawal et al., 2025). As entrepreneurs conduct experiments to gather new information through various means—including advice, as we will explore in this chapter—they update their priors, constantly refining their beliefs and theories, including their strength, and adjusting their strategies accordingly. This iterative process of belief and theory updating lies at the core of Bayesian entrepreneurship, driving decision-making and shaping the entrepreneurial journey.

## 11.3 Advice Aversion and Bias

Our first insight is that the Bayesian entrepreneur is *advice averse*. A decade of research has documented that entrepreneurs often fail to seek, receive, and listen to advice, despite the fact that advisers are willing to provide advice and the advice they provide is often valuable (Miller et al., 2024; Grimes, 2018; Parker, 2006; Dimitriadis and Koning, 2022). Nascent entrepreneurs often keep their precious ideas private, or in "stealth" mode, rather than seeking feedback (Bennett and Chatterji, 2019). One reason entrepreneurs keep their ideas private could be that entrepreneurs of all stripes—from founders of high-tech startups to small businesses in emerging markets—appear to be under-networked and so often lack access to advisers who could help them iterate on their ideas (Bennett and Chatterji, 2019; Dimitriadis and Koning, 2024). But even when given access to advisers, such as through entrepreneurial support programs, entrepreneurs struggle to generate and incorporate relevant advice (Miller et al., 2024; Dimitriadis and Koning, 2022).

Given the potential value of external information, why would the Bayesian entrepreneur be advice averse? Crucially, getting advice requires acting on beliefs. If an entrepreneur's beliefs are weak, entrepreneurs will fail to seek advice. Why might an entrepreneur have weak beliefs? If an entrepreneur was pushed into starting a business, perhaps because of a lack of outside options, they may not believe strongly in the business idea or even have the necessary business theories that advice can inform. And of course, entrepreneurs on the sideline still lack conviction in their idea, as otherwise they would have entered. With low conviction, an entrepreneur will see little reason to seek advice, even from a cheap source such as a peer or mentor—a fact that is confirmed in Bennett and Chatterji's (2019) survey of nascent US entrepreneurs and Dimitriadis and Koning's (2022) study of Togolese small and medium businesses.

Conversely, entrepreneurs with high conviction believe in the potential of their idea more than others and are likely to prematurely satisfice (Cohen et al., 2019) or filter advice to confirm their priors (Nickerson, 1998). Not seeking advice may be rational in the short term for these entrepreneurs because advisers bring their own set of biased experiences, priors, and mental models, which are likely not only different from the entrepreneurs (Bacon-Gerasymenko and Eggers, 2019) but also less enthusiastic. An entrepreneur with strong priors may find negative advice distracting rather than helpful. Moreover, given strong priors, even if solicited, negative advice is not likely to change the entrepreneur's theory (Grimes, 2018; Nickerson, 1998). As a result, left to their own devices, high-conviction entrepreneurs are likely to avoid seeking out what they view as unduly pessimistic and skeptical advisers prone to producing false negatives.

Entrepreneurs with exceptionally novel ideas also may not benefit from advice. The aim of the entrepreneur is to build a new business that can meet the needs of customers in the future. Yet the experience that advisers draw on is rooted in the past (Bacon-Gerasymenko and Eggers, 2019). As a result, if the theories and data of the past no longer apply—for example, because of a technological disruption fundamentally shifting how firms compete (Tripsas and Gavetti, 2000)—then the entrepreneur may only receive historically biased advice that undermines forward-looking performance. Similarly, because advice is often proffered by former entrepreneurs who were successful, advice is often riddled with selection biases that can favor especially risky strategies.

Drawing inferences from past successes has another unintended consequence, specifically for underrepresented founders. Highly successful and experienced entrepreneurs are disproportionately male and white (Guzman and Kacperczyk, 2019), presenting several related challenges. What works for a white male entrepreneur may not work for others. In some cases, the adviser's bias is rooted in taste-based discrimination by the adviser against outgroups (Brooks et al., 2014). Ironically, simple taste-based discrimination is easy for the Bayesian entrepreneur to debias. If advisers are more likely to give harsher advice to a female entrepreneur, she can adjust the signal upward, given she knows the bias exists. However, it is more difficult to correct bias affecting the content of advice. If, for example, advisers focus advice to a female entrepreneur on how not to fail rather than how to grow (Kanze et al., 2018), then debiasing the signal is potentially impossible. Or, as in the case of Susan Tynan at Framebridge, based on a flawed analogy. For the Bayesian entrepreneur, a merely downwardly biased assessment can be adjusted with the appropriate mental model; however, bias in the types of information cannot.

Similarly, a bias toward providing less informative advice cannot be "adjusted" by the Bayesian entrepreneur into a stronger signal. Given that entrepreneurial action rests on an entrepreneur having sufficiently strong beliefs, receiving less detailed and informative advice is more likely to result in inaction and failure (Cao et al., 2024). For example, an innovator working on a novel breast pump may seek advice on how to commercialize his venture. If his male advisers have little experience of breastfeeding themselves, and little understanding of the breastfeeding industry, even if they are generally supportive of his idea, their feedback can, at best, be positive but vague. With vague information, the inventor's priors may never strengthen enough to fully commit to the venture, leading him to prematurely abandon the idea. More broadly, given that the informativeness of an adviser depends on their experiences and backgrounds, gaps in who has relevant entrepreneurial experience are likely to lead to systematic biases in the value of advice for different types of founders and different types of ideas (Cao et al., 2024).

## 11.4 Advice Broadens Theory

Given the potential biases outlined above, why even try to learn from advice? One compelling reason is that advice can be especially effective at shaping theories, causal beliefs about how a venture will create and capture value. Such theories are crucial because entrepreneurs cannot test every aspect of their nascent businesses with customers due to resource constraints, time pressure, and the irreducible complexity inherent in building a new business (Agrawal et al., 2021). Instead, the Bayesian entrepreneur prioritizes where to experiment based on the theories they believe (Felin and Zenger, 2017).

Advice is especially valuable because the different priors and mental models of an adviser can help the entrepreneur evaluate aspects of their strategy that are difficult to assess otherwise (Bacon-Gerasymenko and Eggers, 2019). For example, when the founders of Rent the Runway approached Diane von Fürstenberg, a famous designer, with their theory about why designers should allow the firm to rent their high-end designer dresses for a fraction of the purchase price, she warned that designers would likely be concerned about cannibalizing their core businesses. Although a "user test" with von Fürstenberg might have resulted in a simple "no," her advice was more nuanced. She offered an alternative theory informed by her decades of experience in the fashion business: position dress rentals as a marketing channel to build relationships with younger, aspirational customers. The new framing worked. Similarly, one of the firms in Miller et al. (2024)'s study, Agri Lignin, was deciding between two potential markets for their Lignin product: "biodegradable mulch films" or "pulp and paper trays." Advisers cautioned against both markets due to long sales cycles and government regulations. Instead, they suggested a third, novel, market opportunity: garden nurseries. The firm updated their priors, ran additional customer-oriented tests, and pivoted to offer lignin pellets to garden pot manufacturers.

That even "simple" businesses such as Rent the Runway have complex entrepreneurial theories, with multiple moving parts and intricate interactions, compounds the value of advice. Unlike user experiments, which often narrowly focus on testing customer-facing elements of a strategy, experiments tested with advisers can evaluate multiple parts and alternatives of an entrepreneurial strategy simultaneously, albeit with lower fidelity. Advisers are particularly effective in informing comprehensive theories quickly and inexpensively, including those that extend beyond customer interactions to encompass strategic partnerships, distribution, supply, hiring, and organizational culture. These broader strategic elements, often encapsulated in little "t" theories, are critical for a venture's success and can be efficiently learned from advisers, including peers, who bring a diversity and breadth of experience (Hallen et al., 2020).

The complexity of entrepreneurial strategy partially stems from the fact that great business theories often follow "weakest link" or O-ring type behavior: A clever entrepreneurial strategy in a promising market, built around a disruptive technology, and with an amazing team will still fail if customer acquisition costs are simply too high to achieve profitability. Experienced advisers are particularly well suited to identify these fatal flaws, having observed hundreds of start-up ideas and the types of O-rings that doomed them (Bacon-Gerasymenko and Eggers, 2019). This expertise can be invaluable in helping entrepreneurs avoid common pitfalls and refine their strategies. For example, Cohen et al. (2019) observed a start-up in the solar industry that initially followed a distribution model, believing that combining quality products with industry knowledge would lead to premium prices. While the start-up gained traction, an experienced adviser identified a critical flaw: The distribution model's heavy reliance on inventory would require an amount of capital that would almost surely deter investors. The company pivoted and serviced the same customer base but with a cloud-based technology solution that improved their customers' operations, thus requiring no inventory. By adapting their approach based on this guidance, the start-up was able to test and validate a new theory with existing customers, averting a fatal flaw in their original strategy.

Although advisers can draw on a breadth and depth of experience to invalidate and broaden a Bayesian entrepreneur's theories, advice alone, even a lot of advice, cannot validate theories. In this regard, traction is king. Data showing users want the product, will pay for the product, and that the business model is profitable will always be better at persuasion, especially because hard evidence like customer purchases can strip away the biases and blind spots present in social learning. That said, as we discuss below, advisers can also help improve the experiments entrepreneurs run. Thus, advice is a complement to, not a substitute for, customer experiments.

## 11.5 Advice Enables (Correlated) Iteration

At the heart of the Bayesian view is the "iterative process of belief adjustment" (Agrawal et al., 2025). Despite the risk of receiving biased signals, an entrepreneur who learns from advice has the advantage that they can iterate on their beliefs much faster than relying on experiments with users. Advice, for better and worse, is "cheap" talk. Not only is the advice itself relatively cheap, but because advice often focuses on critiquing the design of the business, suggests shifts in strategic direction, and what customer experiments to run next, testing new ideas through advice often only requires changing a pitch, business model canvas, or strategy memo. Although writing is not easy, user experiments are harder and much more costly. As a result, advice can allow an entrepreneur to iterate over both big strategic pivots and small tactical changes, efficiently exploring a broader space of potential ideas.

In fact, recent research shows that a key benefit of accelerators is that they structure advice to facilitate this iterative updating process (Cohen et al., 2019). At the most basic level, accelerators push entrepreneurs to overcome advice aversion by normalizing and even requiring entrepreneurs to seek advice. For example, some accelerators arrange meetings with upwards of a hundred different mentors in the first month of their programs.

Why would accelerators require entrepreneurs to test their theories with so many mentors? Although the founders of the solar distribution company mentioned above heeded advice, this is not always the case. When an entrepreneur's priors are particularly strong, they likely will reject non-confirming advice. For example, in another case studied in Cohen et al. (2019), the entrepreneur's idea was to sell excursions to travelers—activities such as ziplining and tours. His theory rested on the idea that active travelers would purchase excursion tickets in advance of their trip. He ran initial tests with travelers, who responded favorably to the idea. But an adviser cautioned that customer acquisition cost was going to be prohibitively expensive. Based on the positive response from user tests, and his strong prior, he initially ignored the advice. However, after meeting with dozens of mentors in quick succession who offered nearly identical warnings, he conceded and updated his beliefs: His initial theory was not going to work. With the help of his advisers, he identified a new business-to-business distribution strategy, selling the same excursion tickets but through well-established travel partners who would add his offering as a valuable complement to their existing products—a theory that worked.

An additional benefit of learning from a multitude of mentors in rapid succession is that aggregating information across advisers can mitigate the impact of adviser bias. By averaging the signals from many advisers, the entrepreneur can recover a debiased signal and still potentially benefit from the diversity of ideas each adviser brings to the business. However, reducing bias through information aggregation is not without its downsides. As with investment committee decisions, relying on averages risks reducing the variance of outcomes. An entrepreneur who leans into learning from dozens of advisers and incorporating what appear to be individually valuable pieces of advice may end up with a good, but not exceptional, business. This tension suggests that accelerators, entrepreneurs, and advisers might benefit from experimenting with alternative ways of aggregating and iterating on advice that might better balance the value of learning "from the crowd" with the benefits from identifying outlier strategies.

Of equal concern, aggregating advice can lead to false confidence in the form of overly strong beliefs. Talking to two advisers is not necessarily akin to running two experiments. Precisely because advice is cheap and easy, advisers often learn from each other, study the same business cases, and share experiences from working in the same firms. As a result, what appear to be two independent sources of information are not. Within the context of advice networks, fully accounting for this nonindependence is computationally burdensome for network theorists, let alone entrepreneurs. As a result, models of Bayesian social learning—such as the canonical DeGroot model—often assume agents are naive Bayesians who treat each signal they receive as independent (Golub and Jackson, 2010). Indeed, behavioral research on "correlation neglect" largely confirms this assumption, even in the presence of market feedback and competition (Enke and Zimmermann, 2019). As a consequence of this correlation neglect, learning from multiple related advisers is likely to yield overconfident entrepreneurs whose beliefs are stronger than they ought to be. With stronger beliefs, entrepreneurs will be both more likely to shut down promising venture ideas prematurely and double-down on fatally flawed ideas. Within the context of accelerators, variation in how connected, and thus correlated, advisers are may well explain performance heterogeneity both across programs and within accelerators over time (Hallen et al., 2020).

Intriguingly, the network structure of the "market for judgment" may play a key role in determining when iterating with advice aggregates into accurate signals and when it is more likely to lead to bubbles; and the inverse, areas with promise but that remain underexplored (Golub and Jackson, 2010). When influence is distributed and entrepreneurs are naive Bayesians, the market for advice should yield accurate signals of an idea's potential. However, if certain advisers retain disproportionate influence, then iterating on advice can result in signals biased to the idiosyncratic views of these influential advisers.

More broadly, although advice enables cheap "experimentation" and so more iterations, the social nature of advice is not equivalent to merely lowering the cost of start-up or user experiments (Koning et al., 2022; Kerr et al., 2014). From the Bayesian perspective, the fact that advice is the result of an interdependent social process suggests that advice networks, markets for judgment, accelerators, and even the broader structure of start-up ecosystems may determine how advice impacts entrepreneurial learning and choice. Indeed, the quick and steep investment in robots at Framebridge illustrates how relying on advice from closed epistemic communities can result in adopting poor strategies, whereas advice from even a single outsider can radically improve a start-up's situation. In fact, the structured nature of programs such as the Creative Destruction Lab and Techstars suggests that institutions can play a pivotal role by designing the structure of advice networks to gain the benefits in terms of iterations and theory development while minimizing bias and problems of correlation neglect, which may escalate over time as mentors socialize (Cohen et al., 2019; Agrawal et al., 2021). Left unstructured, getting advice from multiple advisers is likely to yield more choice and action, although with more bad advice and a higher risk of bubbles and crashes (Goldfarb and Kirsch, 2019).

## 11.6 Advice Improves Experimentation

The fourth way advice informs the Bayesian entrepreneur is by improving the entire process of experimentation. We have argued that entrepreneurs are less likely to seek advice, largely because of concerns that the advice they receive will be biased. However, despite this risk, advice can both broaden an entrepreneur's theories and allow entrepreneurs to quickly test and iterate on their beliefs. With more developed priors and better theories, entrepreneurs can design and run more informative experiments, be they A/B tests with customers or interviews with suppliers. Further, advice can improve the efficiency of experimentation and the interpretation of test results.

First, advice broadens the scope of experimentation (Miller et al., 2024). Advice serves as a powerful antidote to confirmation bias and premature satisficing, particularly when structured to challenge the Bayesian entrepreneur's strong priors. As discussed above, programs such as the Creative Destruction Lab and Techstars flood the entrepreneur with advice not only to help them weaken priors on existing theories, but, more importantly, to generate the new theories which become the basis for new experiments. For example, when Sabeer Bhatia and Jack Smith launched the free, ad-supported email company, Hotmail, initially, it struggled to acquire new customers. Tim Draper, a venture capitalist and investor, offered a creative solution: adding a simple tagline to every outgoing email, "Get your free email at Hotmail." Each email sent became a marketing message. The company tested the idea, fueling a viral marketing loop that attracted thirty million customers in less than three years. Without this clever idea that spawned the term "viral marketing," Hotmail would have been unlikely to achieve the same level of success.

Second, advice improves experimental design and interpretation (Agrawal et al., 2021). As Bayesian entrepreneurs move from theorizing to testing, the fact that advisers hold different beliefs and mental models plays a crucial role in designing informative experiments. Experienced advisers, drawing from their diverse backgrounds and past experience, can guide entrepreneurs in designing more informative and resource-efficient tests (Bacon-Gerasymenko and Eggers, 2019; Cohen et al., 2019). For instance, in the case of Framebridge, an adviser from outside the tech start-up ecosystem suggested experimenting with a robot-free strategy focused on internal operational metrics that the founder Tynan would have likely discounted or simply overlooked. Turning to interpretation, Kim (2025) highlights how theories often blind entrepreneurs from seeing the signal in the data. Advisers, by expanding the entrepreneur's likelihood model(s), can help reduce this problem of inattention. For example, a startup in Cohen et al. (2019)'s study nearly discarded information generated from user tests indicating that increasing price increased demand. An advisor explained that buyers of luxury goods and services like his use price as a proxy for quality.

Third, advice helps prioritize critical and especially persuasive experiments. To thrive, a Bayesian entrepreneur not only needs to increase their own prior but also must persuade others of their venture's potential. Although advice can identify fatal flaws in a theory and offer alternatives, it offers, at best, weak signals of future potential. The Zappos case exemplifies this: Investors were initially skeptical about online shoe sales. Only experiments showing strong actual customer traction could convince investors of the viability of their theory. However, knowing why investors were skeptical presumably helped the Zappos team design an experiment with the potential to convince and persuade other stakeholders.

## 11.7 Conclusion: Advice and the Bayesian Entrepreneur

The topic of advice is a curious one. Entrepreneurial advice is valuable, but entrepreneurs are reluctant to get advice. Our exploration through a Bayesian lens illuminates this puzzle. The very same strong priors that motivate entrepreneurs to act on their ideas also make them resistant to advice. Bayesian entrepreneurs with high conviction in their ventures are likely to view advisers as unduly pessimistic, prone to producing false negatives. Conversely, entrepreneurs with weak priors may not seek advice at all. This Bayesian perspective explains why entrepreneurs often keep ideas private and struggle to incorporate advice even when it's readily available.

Yet, when Bayesian entrepreneurs do get advice, it offers significant value when properly structured. It can substantially improve both theorizing and experimentation. Advisers, drawing from diverse experiences and mental models, can help entrepreneurs evaluate aspects of their strategy that are difficult to assess otherwise. At a meta level, advice can enhance the entire experimentation process by broadening its scope, improving design and interpretation, and prioritizing critical tests. As seen in cases such as Framebridge and Rent the Runway, well-structured advice can lead to pivotal strategic shifts and more efficient learning cycles.

Yet, advice can backfire. The social nature of advice can lead to correlation neglect and overconfidence, particularly when advisers share similar backgrounds or when advisers are especially revered. Moreover, advice is fraught with biases—biases that are likely to disproportionately affect underrepresented founders. Although the Bayesian entrepreneur can adjust advice for this bias when the advice is simply harsher and biases are known, when advisers offer different amounts or types of information based on founder affiliations or ideas, these biases can impact what underrepresented entrepreneurs learn.

For entrepreneurs, these insights suggest three useful practices. First, Bayesian entrepreneurs should take stock and actively seek to challenge their own priors, even (and especially) when conviction is high. This might likely involve deliberately seeking advisers with differing backgrounds or expertise, although with the understanding that this advice will contain a mix of valuable signal and biased information. Second, entrepreneurs should structure advice seeking to maximize learning instead of confirmation—for instance, arranging intensive bursts of multiple adviser meetings instead of prematurely stopping after an adviser says an idea is promising. Within the conversation, asking advisers to help identify potential "fatal flaws" in the business model is more valuable than asking them to validate an investment pitch. Third, while valuing advice, entrepreneurs should remember that it complements, rather than replaces, direct market testing.

For researchers, we see three especially promising avenues for future work. First, there is a distinct lack of evidence on how "correlated" advice is and its broader role in entrepreneurial herding, bubbles, and crashes (Goldfarb and Kirsch, 2019). We need to understand whether talking to ten advisers is closer to running ten unique experiments or if the same information is merely being repeated across conversations. Additionally, it's important to identify the factors that shape when entrepreneurial advice is more or less correlated. For example, research focused on determining the optimal composition of an adviser pool could provide crucial guidelines for both entrepreneurs and entrepreneurial training programs, ensuring a diverse range of insights that mitigate biases.

Second, future research could explore the applicability of these principles to various organizational contexts and methodological approaches. For example, qualitative studies might explore how managers in established corporations seek and incorporate advice on their strategic theories, especially those around new innovations and technological change. Experiments could be designed to train advisers to focus their guidance effectively: Rather than offer positive feedback, advisers should challenge and expand them as well as offer insight on how to improve the effectiveness of experimentation.

Third, many of the arguments above rest on having access to a deep pool of diverse and experienced advisers. Yet, outside of top accelerators and thriving ecosystems such as Silicon Valley, advisers can be hard to access, limiting the ability of many entrepreneurs to iterate on their beliefs. How can we scale helpful advisers to all? One promising approach pioneered in Otis et al. (2024) is to leverage generative artificial intelligence tools to build artificial advisers for entrepreneurs. These advisers can be designed to have different priors, data, and mental models. Beyond the managerial and policy promise of artificial advisers, these tools also have the potential to help researchers more precisely manipulate and measure the mechanisms at the heart of both entrepreneurial advice and the Bayesian learning process.

## Acknowledgments

We thank participants at the Bayesian entrepreneurship conference and Marta Villamor Martin for their invaluable comments and conversations.

## Notes

1. Miller et al. (2024, p. 2) define advice as "information provided by any party outside the firm drawn from their knowledge or experience intended to inform firm decisions." Advice can come from formal advisers such as board members and investors as well as informal advisers such as seasoned entrepreneurs, entrepreneurs at similarly staged ventures, managers from incumbent firms, and professionals such as accountants and lawyers (Cohen et al., 2019; Garg and Eisenhardt, 2017).

## References

Agrawal, A., Camuffo, A., Gambardella, A., Gans, J. S., Scott, E. L., and Stern, S. (2025). The foundations of Bayesian entrepreneurship. *Bayesian entrepreneurship.* MIT Press.

Agrawal, A., Gans, J. S., and Stern, S. (2021). Enabling entrepreneurial choice. *Management Science*, 67(9), pp. 5510–5524.

Bacon-Gerasymenko, V., and Eggers, J. P. (2019). The dynamics of advice giving by venture capital firms: Antecedents of managerial cognitive effort. *Journal of Management*, 45(4), pp. 1660–1688.

Bennett, V. M., and Chatterji, A. K. (2019). The entrepreneurial process: Evidence from a nationally representative survey. *Strategic Management Journal*, 44(1), pp. 86–116.

Brooks, A. W., Huang, L., Kearney, S. W., and Murray, F. E. (2014). Investors prefer entrepreneurial ventures pitched by attractive men. *Proceedings of the National Academy of Sciences*, 111(12), pp. 4427–4431.

Cao, R., Koning, R., and Nanda, R. (2024). Sampling bias in entrepreneurial experiments. *Management Science*, 70(10), pp. 7283–7307.

Chatterji, A., Delecourt, S., Hasan, S., and Koning, R. (2019). When does advice impact startup performance? *Strategic Management Journal*, 40(3), pp. 331–356.

Cohen, S. L., Bingham, C. B., and Hallen, B. L. (2019). The role of accelerator designs in mitigating bounded rationality in new ventures. *Administrative Science Quarterly*, 64(4), pp. 810–854.

Dimitriadis, S., and Koning, R. (2022). Social skills improve business performance: Evidence from a randomized control trial with entrepreneurs in Togo. *Management Science*, 68(12), pp. 8635–8657.

Dimitriadis, S., and Koning, R. (2024). Networking frictions and entrepreneurial learning in developing economies. *Management Science*, 71(4), pp. 3127–3145.

Eisenmann, T., Ries, E., and Dillard, S. (2012). *Hypothesis-driven entrepreneurship: The lean startup.* Harvard Business School Entrepreneurial Management Case (812-095).

Enke, B., and Zimmermann, F. (2019). Correlation neglect in belief formation. *The Review of Economic Studies*, 86(1), pp. 313–332.

Felin, T., and Zenger, T. R. (2017). The theory-based view: Economic actors as theorists. *Strategy Science*, 2(4), pp. 258–271.

Garg, S., and Eisenhardt, K. M. (2017). Unpacking the CEO–board relationship: How strategy making happens in entrepreneurial firms. *Academy of Management Journal*, 60(5), pp. 1828–1858.

Goldfarb, B., and Kirsch, D. A. (2019). *Bubbles and crashes: The boom and bust of technological innovation*. Stanford University Press.

Golub, B., and Jackson, M. O. (2010). Naïve learning in social networks and the wisdom of crowds. *American Economic Journal: Microeconomics*, 2(1), pp. 112–149.

Grimes, M. G. (2018). The pivot: How founders respond to feedback through idea and identity work. *Academy of Management Journal*, 61(5), pp. 1692–1717.

Guzman, J., and Kacperczyk, A. (2019). Gender gap in entrepreneurship. *Research Policy*, 48(7), pp. 1666–1680.

Hallen, B. L., Cohen, S. L., and Bingham, C. B. (2020). Do accelerators work? if so, how? *Organization Science*, 31(2), pp. 378–414.

Kanze, D., Huang, L., Conley, M. A., and Higgins, E. T. (2018). We ask men to win and women not to lose: Closing the gender Gap in startup funding. *Academy of Management Journal*, 61(2), pp. 586–614.

Kerr, W. R., Nanda, R., and Rhodes-Kropf, M. (2014). Entrepreneurship as experimentation. *Journal of Economic Perspectives*, 28(3), pp. 25–48.

Kim, H. (2025). Learning from data in entrepreneurial experimentation. *Bayesian entrepreneurship*. MIT Press.

Koning, R., and Dadlani, A. (2023). *Framebridge (A): Reimagining custom framing*. Harvard Business School Case (723-352).

Koning, R., Hasan, S., and Chatterji, A. (2022). Experimentation and start-up performance: Evidence from A/B testing. *Management Science*, 68(9), pp. 6434–6453.

Lerner, J., and Malmendier, U. (2013). With a little help from my (Random) friends: Success and failure in post-business school entrepreneurship. *The Review of Financial Studies*, 26(10), pp. 2411–2452.

Miller, A., O'Mahony, S., and Cohen, S. L. (2024). Opening the aperture: Explaining the complementary roles of advice and testing when forming entrepreneurial strategy. *Organization Science*, 35(1), pp. 1–26.

Nickerson, R. S. (1998). Confirmation bias: A ubiquitous phenomenon in many guises. *Review of General Psychology*, 2(2), pp. 175–220.

Otis, N., Clarke, R., Delecourt, S., Holtz, D., and Koning, R. (2024). *The uneven impact of generative AI on entrepreneurial performance*. SSRN Scholarly Paper. Rochester, NY.

Parker, S. C. (2006). Learning about the unknown: How fast do entrepreneurs adjust their beliefs? *Journal of Business Venturing*, 21(1), pp. 1–26.

Ries, E. (2011). *The lean startup: How today's entrepreneurs use continuous innovation to create radically successful businesses*. Currency.

Tripsas, M., and Gavetti, G. (2000). Capabilities, cognition, and inertia: evidence from digital imaging. *Strategic Management Journal*, 21(10–11), pp. 1147–1161.

# IV

## Entrepreneurial Pedagogy and Practice

# 12

## Entrepreneurial Strategy Pedagogy and Practice: Learning, Commitment, and Choice

Erin L. Scott

### 12.1 Introduction

Over the past several decades, the entrepreneurship classroom has been transformed—from one that was often shaped around the hard-earned learnings of a seasoned entrepreneur to one that leverages wide-ranging, rigorous academic and practitioner frameworks and tools. Drawing on a wave of influential approaches to entrepreneurial learning, experimentation has become a central feature of entrepreneurship research, practice, and pedagogy. Eric Ries's successful entrepreneurship book *The Lean Startup* and tools centering design thinking into the entrepreneurship opportunity identification and development process have widely promoted the importance of iterative learning and testing in entrepreneurship (Ries, 2011). These toolkits seek to assist entrepreneurs in navigating the uncertainty associated with an early-stage idea by embracing practical steps for continuous learning and experimentation. Given the centrality of navigating uncertainty in entrepreneurship, it is no surprise that these tools have established themselves as the cornerstone of modern entrepreneurship education in both case and project-based entrepreneurship courses.

Yet, despite the myriad frameworks for effective experimentation and the meaningful reduction in the costs to do so (e.g., due to access to cloud computing or no-code solutions), entrepreneurs often find that their experiments provide limited clear, actionable insights to guide their venture forward. Frustrated with the experimentation process and lacking clarity on how to proceed with their idea, even in ecosystems where experimentation is supported through targeted programs, mentorship, and funding, as well as celebrated as a foundation of entrepreneurial culture through mantras such as "fail fast," there is a temptation for students and entrepreneurs alike to eschew experimentation altogether.

If entrepreneurs know that entrepreneurial experimentation is critical to startup success and possess the tools and resources to conduct these experiments, why do their experiments often fail to provide them with the guidance they seek? Why do the very real and proven benefits of hypothesis-driven learning and experimentation fail to translate into clear action for so many entrepreneurs? At a fundamental level, entrepreneurs operate in environments where theoretical principles of effective experimentation compete with the personal and practical exigencies of venture founding, survival, and growth.

At founding, the beliefs and priors that resulted in the initial entrepreneurial idea are inextricably linked to the entrepreneur, how they will approach the process of entrepreneurial learning and experimentation, and the evidence they will need moving forward. The intersection of an entrepreneur and their idea, or more broadly, entrepreneurial stakeholders (such as potential employees or investors) and their beliefs about said idea, cannot be separated from the design of effective entrepreneurial experiments. Entrepreneurial experimentation is not simply about the revelation of information, it is about persuasion. Different stakeholders, including the founder(s) themselves, may require distinct evidence and experimental designs that confront the differing dimensions and levels of uncertainty associated with a given idea in order to persuade them to move forward and in what direction.

Effective experimental design also requires balancing the desired learning with the required commitments to productively execute the experiment. Beyond the time and financial resources (which may be non-trivial), some experiments may require strategic commitments, including recruiting co-founders, staff, and investors, tailoring the technology or early product, or establishing a relationship, reputation, or brand with customers. While these commitments may allow the entrepreneur to effectively learn, they may also limit their freedom to act on the resulting information by having already implicitly cemented them to a given path (or otherwise closed alternative paths) through those very commitments.

Even the most effective experimental designs are unlikely to yield direct guidance as to the explicit path forward. Entrepreneurial experimentation is rarely a "path to the epiphany." It requires judgment and choice. Though substantively reducing uncertainty, effective entrepreneurial experimentation still leaves some degree and dimensions of uncertainty remaining regarding the idea itself as well as paths forward. Entrepreneurial experimentation does not absolve the entrepreneur from choice, instead, it informs that choice.

Effective entrepreneurial experimentation must center the relevant stakeholder, most significantly the founder(s), in the design and evaluation process, balance the trade-offs of learning and commitment, and highlight the role of judgment in choosing a path forward that reflects the information gained through the experimentation process while also grounded in the entrepreneur's objectives, skills, and constraints. By putting founder choices center stage, entrepreneurial strategy offers students and entrepreneurs a systematic, choice-oriented approach to undertake the process of exploring and evaluating the core strategic choices while balancing learning and commitment, enabling them to effectively choose a path forward to move their idea to the market.

## 12.2 Choice Matters

Far from struggling to identify a single path forward for their venture, entrepreneurs often surface multiple potential paths to create and capture value from their idea. Although lack of prior strategic commitments allows entrepreneurs the freedom to pursue any number of these paths in isolation, an entrepreneur cannot pursue all these paths at the same time. Resource constraints as well as strategic inconsistencies among these alternative pathways limit an entrepreneur's ability to pursue multiple paths (at least initially). As a result, entrepreneurs must make choices as to how to move forward in the face of uncertainty about the value of the idea itself as well as the value of the alternative paths.

Specifically, as an entrepreneur explores a given idea, they face many alternatives that cannot be pursued at once, and so they must adopt, implicitly or explicitly, a process for entrepreneurial learning and choice. Choosing between these alternatives requires knowledge that can only be gained through learning and experimentation. Yet, the process of learning and experimentation inevitably results in (at least some) commitment that forecloses strategic options. By synthesizing a choice-oriented approach to entrepreneurship, entrepreneurial strategy offers founders a pathway to move beyond this paradox of entrepreneurship by providing a sequential decision-making process to test specific value-creation and value-capture hypotheses, particularly in those contexts where experimentation necessitates partial commitment. Over the past decade, my colleagues and I have undertaken research, practice, and pedagogy whose objective is to provide rigorous yet actionable foundations for the field of "entrepreneurial strategy" (Gans, 2018; Gans et al., 2019, 2024). This work builds upon a long line of research-driven and practice-based frameworks emphasizing the role of choice in entrepreneurial action (Kirzner, 1976; Shane, 2003; Ries, 2011). And over the past several years, there has been a flurry of both academic research and practitioner insights identifying the specific roles that strategy and choice play in the founding of new ventures (e.g., Osterwalder and Pigneur, 2011; Camuffo et al., 2020, 2024; Felin et al., 2020, 2024; Coali et al., 2024). Ultimately, the central strategic challenge facing an entrepreneur is how to choose the path forward.

## 12.3 These Choices Matter

Although entrepreneurs face countless choices in moving their idea into the market, entrepreneurial strategy focuses on the core domains of choice for any start-up: the choice of a customer; the choice of technological trajectory; the choice of founding team, resources, and capabilities; and the choice of commercialization strategy. These reflect the key initial questions founding teams must address early on: Who is our beachhead customer? How will we leverage emerging innovations? How will we build out our team? How do we get into the marketplace? Within each of these core domains of choice—customer, technology, organization, and competition—lies layers of assumptions as to how the start-up will create and capture value. Taken together, they form an entrepreneurial strategy, or an alternative path forward, for a given idea and the earliest instantiation of entrepreneurs distilling their early learnings and explorations into hypotheses to be tested (see figure 12.1).

Synthesizing research and practice into an actionable, choice-based framework, entrepreneurial strategy guides entrepreneurs to formulate these hypotheses and understand the key trade-offs and short- and long-term consequences of alternatives within these core domains of choice. For instance, a choice orientation reconciles perceived inconsistencies in the tactical advice for choosing and leveraging a beachhead customer segment between the market adoption S-curve and the Crossing the Chasm literatures. The market adoption S-curve posits that entrepreneurs should identify innovators and early adopters for their beachhead customer and then ride the inevitable adoption process across adopter categories and into the mainstream market. By contrast, Crossing the Chasm implores entrepreneurs that they must actively manage the transition from the beachhead customer to the mainstream market else they find themselves consigned to their beachhead customer and never transitioning to the mainstream market. A choice orientation sharpens the connection between these two sets of empirical evidence and literatures, clarifying the strategic guidance for entrepreneurs. Simply put, for a given idea, there are multiple, equally viable market adoption S-curves, and the choice of beachhead customer reflects (implicitly or explicitly) a choice as to the mainstream market they will seek.

A choice of beachhead customer, therefore, reflects not only a hypothesis about the relative value creation for this initial target customer but also a hypothesis as to the value creation for their target mainstream customer segment and a hypothesis as to how the choice of initial beachhead customer will enable the entrepreneur to reach this mainstream customer. For instance, the entrepreneur may choose an accessible, yet poorly served beachhead customer in order to receive quick, valuable feedback and engage in iterative productive

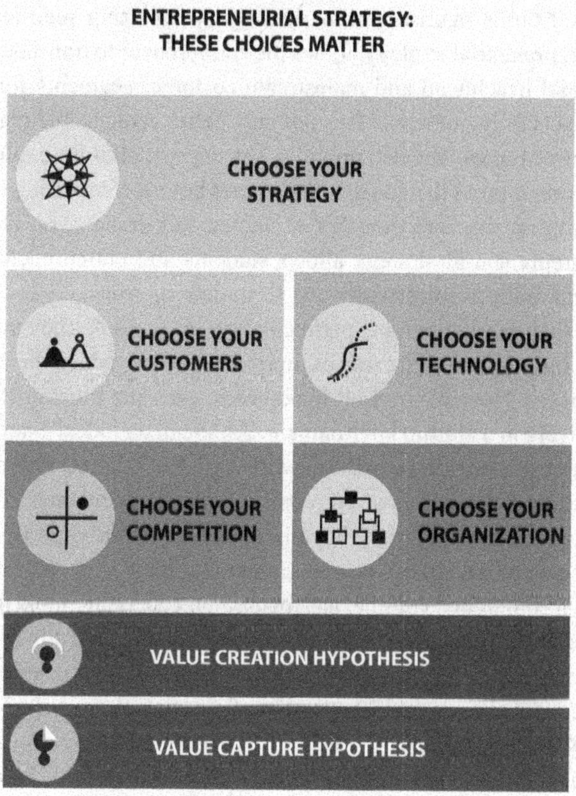


**Figure 12.1**
Choice domains in entrepreneurial strategy.

development to rapidly improve an emerging technology trajectory along key value drivers for the mainstream customer segment as the startup expands to serve those mainstream customers. Alternatively, the entrepreneur may seek beachhead customers that serve as strong, credible references for the mainstream customer segment and have similar market needs as those mainstream customers, even if the hurdles to initially engage and convert such a beachhead customer is considerably higher. Entrepreneurs engaging a beachhead customer without choosing a mainstream target market or a hypotheses-driven approach to facilitate that transition—for instance, through strategic learning or referencing—may limit the value of the beachhead customer, increase their likelihood of unintentionally making irreversible strategic commitments that limit the ability of the venture to serve other potential customer segments, including their target mainstream market. By synthesizing such research and practice, entrepreneurial strategy allows the entrepreneur to translate their alternative potential beachhead and mainstream customer segments for their idea into defined sets of hypotheses. This not only better cements the playbooks for these alternative routes, but also provides a stronger toolkit for exploration and experimentation around the choice of customer before committing to any one.

By focusing on the core domains of choice and drawing on research and practice, entrepreneurial strategy allows students and entrepreneurs to better isolate critical areas of uncertainty and formulate rigorous, concrete hypotheses to guide their learning and experimentation. As a result, they avoid wasted time, effort, and resources on areas of limited value and reduce their frustration with the process. However, not all hypotheses, even the most critical, can be tested effectively in a startup environment. Although an "ideal" start-up experiment would test a critical hypothesis with high-fidelity and low opportunity (including strategic) cost, most entrepreneurship experimentation involves a trade-off between criticality of the hypothesis, fidelity of the experiment, and the opportunity cost of executing the experiment. An increasing field of research, including this conference volume, has focused on generating tools for conducting high-fidelity experiments in the entrepreneurial context, given the unique dynamics relative to other business experiments (see, among others, Murray and Tripsas, 2004; Nanda and Rhodes-Kropf, 2016; Cohen et al., 2019a; Koning et al., 2022; Cao et al., 2024). Yet, even as experimental tools and techniques improve, it is important for entrepreneurs to acknowledge that not all assumptions can be tested and that good ideas have multiple paths forward to create and capture value. As a result, in most cases, even well-designed and executed entrepreneurial experiments will rarely yield a singular path forward. Entrepreneurs should seek to design experiments not to yield an unlikely deterministic solution, but rather to reduce uncertainty, sharpen their hypotheses, and inform their choice as to the path forward.

Proactively managing this process of experimentation involves prioritizing and sequencing experiments to focus first on testing the most critical hypotheses with the highest fidelity and lowest opportunity cost experiments and then escalating the strategic commitment as the entrepreneur learns more about their idea and alternative strategic paths forward. The founding of Vera Wang Bridal provides a clarifying example of an entrepreneur sequencing their experiments to maximize learning with low strategic commitment. As a designer at Ralph Lauren, Vera Wang was unable to find wedding dresses for her upcoming wedding that matched her more mature, streamlined style. Given that an increasing percentage of consumers, like herself, were getting married at an older age,

Wang was convinced that there was a market opportunity for "modern" high-end wedding dresses (Beard, 2019). However, her colleagues at Ralph Lauren, although well positioned to meet this opportunity, were not convinced this market need extended beyond Wang herself. Nonetheless, like many user innovators, Wang was convinced this opportunity was promising. She quit her job and opened a high-end wedding dress store on 5th Avenue in New York City. Although a significant personal commitment, the choices she made in opening the store highlighted a clear decision both to delay core strategic commitments until she learned more about her innovation and target customer and to utilize high-fidelity experiments with a high threshold for success to enable that learning. Specifically, Wang filled the store with the traditional dresses of high-end designers and included only one of her own modern design (Beard, 2019). In the months that followed, Wang was able to observe how consumers interacted with her design, the comments they made, and their ultimate purchasing decisions. As she integrated this feedback into future designs, she filled the store with more and more of her own designs. By the time she launched Vera Wang Bridal, her experiments had allowed her to understand the changing consumer preferences and demand in the wedding dress industry, tailor her vision of the modern wedding dress to that market, and receive credible high-fidelity signals as to the market viability of her dresses against the established category leaders. Even today, the aesthetic she crystallized in those early experiments remains the hallmark of her iconic dresses and global design house, which generates more than $700 million in sales annually (Milnes, 2024).

By contrast, consider the countless entrepreneurs whose first substantive experiment is launching a product into the market with considerable commitments around customer, technology, team, and competition. These entrepreneurs often focus on *The Lean Startup*'s agile product development embracing Ries's build-measure-learn feedback loop, which emphasizes the important of developing a minimum viable product (MVP) to test critical hypotheses with real customers, while de-emphasizing Ries's discussion of the pre-MVP learning required to effectively understand the features and characteristics of a viable product and yield the benefits of the high-fidelity MVP experiment (Ries, 2011). It is not uncommon for these product launches to miss the key overarching lesson of Ries's framework—namely, that an MVP is not simply a product, it is a set of hypotheses. As a result, even when the MVP surpasses the high threshold required to provide a positive signal of market interest, without a clear hypothesis to benchmark against, the information from the experiment may prove difficult for the entrepreneurs to interpret to drive future strategic choices for the venture, including the next iteration of the product. By not articulating concrete hypotheses before choosing their early product features and product releases, flawed MVP experiments, even when resulting in successful sales, may result in limited and haphazard learnings. Given this lack of actionable learning and clarity regarding the path forward combined with the often significant strategic commitments involved, it is not surprising that many an entrepreneur leaves this process frustrated.

Even when their early products are hypothesis-driven, entrepreneurs should approach with caution the commitments involved in developing and releasing these MVP experiments. For instance, a team may commit to building specific capabilities as they recruit early team members to develop the MVP or may commit to certain technological trajectories in prioritization of key performance metrics over other metrics for their MVP. A product release may similarly commit the venture to a certain value proposition or customer segment, limiting the venture's ability to entirely rebrand or dramatically pivot to alternative value propositions. Even failed experiments may be detrimental to the reputation of the founding team and the venture. Ultimately, an MVP allows both the entrepreneurs and their customers to learn about each other, and the customer may learn the start-up is unlikely to effectively solve their market need or is simply not to be trusted, regardless the viability of subsequent iterations of the product. There are many tools an entrepreneur may use to balance these commitments with the benefits of the MVP experiment however, doing so requires entrepreneurs to be concrete as to the information the experiment will yield, how it will drive strategic choice moving forward, and the trade-offs relative to low-commitment experiments to gain similar information. By choosing to experiment small to choose big, Wang's hypothesis driven, sequential process of experimentation with increasing degrees of strategic commitment allowed her to effectively navigate these trade-offs and demonstrate compelling value for customers before choosing to launch her products and venture at scale.

Entrepreneurial strategy's emphasis on prioritizing early no- and low-opportunity cost learning and experimentation should not be read as an excuse to avoid commitment. Rather, entrepreneurial strategy argues that the value of experimentation is far higher when guided by strategic priorities, as this allows for more effective entrepreneurial learning, staging of irreversible commitments, and, ultimately, choice. In evaluating the commitments, it is important for entrepreneurs not to conflate the personal commitments involved for learning with the strategic commitments required. Although personal commitments are an important consideration, in some cases, personal commitments, such as opening a wedding dress store, enable more effective entrepreneurial experimentation, which, in turn, yields more efficient and fruitful information to drive the entrepreneur's choices, including more efficient staging of their personal commitments to the venture. While opening a retail store on 5th Avenue is not in the reach of every entrepreneur and certainly not every class project, the value of hypotheses-driven, sequential experimental design empowers students and entrepreneurs of all types as they balance their personal and strategic commitments and move their ideas into action.

In designing experiments, entrepreneurs should also be cognizant that the criticality and fidelity of a given experiment will be influenced by the stakeholder interpreting the results. Founder(s), key partners, early hires, and investors bring their own priors and beliefs to the idea and may require different evidence to meaningfully reduce their uncertainty, sharpen their hypotheses, and inform their choice about the idea and venture. As well, their distinct personal opportunity costs and commitments may influence their timeline allowed for early entrepreneurial learning and experimentation. As a result, the design, sequencing, and timeline of these experiments will be influenced by the beliefs and constraints of the key stakeholders the entrepreneur needs to influence, most significantly the entrepreneur themselves. For instance, students considering translating their entrepreneurship class projects into a venture must design early experimentation processes that weigh their personal opportunity costs, for instance bypassing internships to work on the idea or committing to the venture full-time after graduation instead of traditional employment, as well as the strategic commitment inherent in those experiments. Entrepreneurial strategy equips these students with greater clarity as to the critical hypotheses and associated learnings required for them to commit to their idea and venture full-time as well as how to evaluate the relative fidelity and commitments involved across potential experiments. This enables them to design experiments and sequence those experiments in a manner that balances the learnings, commitments, and choices of the venture as well as the entrepreneur's pressing personal student commitments and deadlines.

The case of student start-up Taxie, launched from a project-based entrepreneurship course, is illuminating in demonstrating how much information students can gather through effective entrepreneurial experimentation and sequencing and how that shapes both their strategic choices and personal career decisions (Castillo and Scott, 2022). Although initially exploring an electric vehicle (EV) taxi company, the Taxie team ultimately came together to explore leasing EVs to rideshare drivers to lower emissions while increase wages for rideshare drivers (from premium vehicles earning more per ride from rideshare companies). Although the team had considerable experience in the electric vehicle space, their experience in the rideshare industry was limited. As a result, their early experimentation focused on learning enough about the rideshare industry to be able to articulate clear hypotheses. Following traditional customer discovery processes, their early experimentation process prioritized no to low strategic commitment learning, involving countless interviews with rideshare drivers and eventually becoming rideshare drivers themselves. At the end of this learning process, the student team was able to articulate a concrete business model, choose a beachhead customer, choose their technology, and state the value-creation and value-capture hypotheses under which the business model rested. At the same time, key dimensions of uncertainty remained regarding the perceived (vs. actual) value proposition by their target beachhead rideshare drivers, the limits of the EV technology in transitioning to the rideshare context (including driving range and degree of charging infrastructure), fleet operations and operating margins, as well as the saliency of potential competitive threats by large incumbents (including the rideshare companies themselves).

In evaluating whether to move their project out of the classroom and into the field, the team considered the uncertainty associated with their identified hypotheses and their ability to meaningfully reduce that uncertainty before making significant strategic commitments (including building the team and fundraising) and personal career commitments (including declining to participate in recruiting for full-time post graduation jobs).

Ultimately, this focus resulted in the team purchasing two used Tesla sedans and launching a limited pilot of their proposed service in order to make a decision to commit before graduation. Confident in their ability to sell these cars at roughly the purchase price in a few months, these cars allowed the team to test their hypotheses by recruiting potential drivers, operating the vehicles, and supporting two full-time rideshare drivers. The team rapidly updated their hypotheses, enhanced their understanding of the rideshare customer segments, formed concrete assumptions as to the likely evolution of EVs in rideshare capacities and the associated required charging infrastructure, and generated considerable market and consumer interest with a lengthy waitlist of interested rideshare drivers.

Buoyed by positive signals, the team evaluated whether to continue the process of incremental fleet growth or to commit to fundraising to rapidly scale their fleet. In debating this early inflection point for the team, the founders returned to their hypotheses, noting what information had been gained to date and what information incremental fleet growth would provide them. It immediately became clear that, for their key hypotheses, continued incremental growth would provide only limited new information to guide their choices relative to what they had already learned. For instance, although their hypotheses as to fleet operations and operating margins at scale were sharpened by operating their two-car fleet, it was clear to the team that growing to even a dozen cars and drivers would do little to clarify their remaining uncertainty as to how those margins would adjust at scale. Instead of accepting this as simply an uncertainty that would only be resolved by launching and scaling the venture, the founders again focused on identifying low-commitment experiments to gain the needed information. In this case, they surfaced a creative path forward: recruiting an experienced rideshare executive, drawing on their own experience, to evaluate their early Taxie data and better extrapolate the likely operation adjustments and operating costs at scale. In so doing, the team was able to see that even under the best of credible scenarios, the margins were likely to be very low, even withstanding likely competitive entry. Impressively, before committing to the venture full time, before committing to fundraising, before committing to building a large fleet and operations staff, and before committing to foreclose post graduation career opportunities, and in only a handful of months, the student team was able to demonstrate a key weakness with their idea that otherwise would have taken years to identify through either continued incremental fleet growth or rapid scale. Having chosen not to go forward with the startup, they sold their Tesla vehicles back with final costs for the experiment (outside of their time) less than that typically required for the popular start-up Facebook ad experiments. Additionally, they effectively leveraged their industry learnings in their full-time recruiting efforts.

By focusing on the core choices, entrepreneurial strategy provides entrepreneurs with rigorous tools and an actionable framework to formulate and test their hypotheses, understand the commitments involved with those experiments, and empower them to choose when and how to make those trade-offs. Importantly, although the process of learning and experimentation may yield information regarding key hypotheses that may chart a path ahead (as in Vera Wang Bridal) or shutter the venture (as with Taxie), the entrepreneur must interpret that information and actively make a choice. In most cases, entrepreneurial experimentation will not validate or invalidate a single "true path." Rather, the process of learning and experimentation will offer insights into the value of the idea and surface multiple potential viable strategic paths to value.

## 12.4 The Choices Matter Together

Although entrepreneurs may be able to test some core domains of choice in isolation, ultimately the choices that underlie an entrepreneurial strategy are not made in isolation. Entrepreneurial strategy addresses the interdependencies between these key domains of choice and yields a synthetic framework for choosing an overall entrepreneurial strategy using the Entrepreneurial Strategy Compass (Gans, 2018; Gans et al., 2024).

The Entrepreneurial Strategy Compass moves beyond simply highlighting clear examples of proven start-ups with demonstrated strategic coherence and instead offers entrepreneurs with a framework to formulate a coherent path forward for their idea and their venture. A key component is providing clarity and rigor as to these alternative paths by synthesizing research and practice within each distinct strategy, connecting those to concrete core choices and hypotheses, and specifying a clear path to action for each strategic route forward.

The Entrepreneurial Strategy Compass places the entrepreneur and their idea centrally and uses the choice of competition to distinguish four strategic playbooks for start-ups (see figure 12.2). For a given idea, an entrepreneur chooses with whom to compete and how to compete. The choice of who to compete

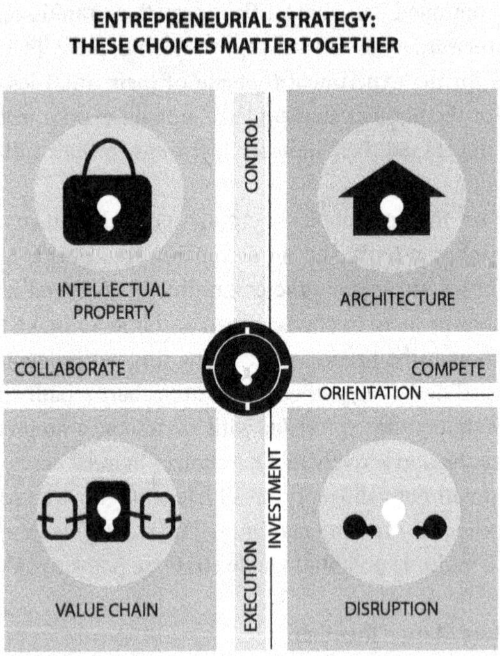


**Figure 12.2**
The entrepreneurial strategy compass.

with reflects a choice to orient toward collaboration (i.e., competing to partner with established firms) or toward competition (i.e., competing with established firms). The choice of how to compete reflects a choice to prioritize investing in control (e.g., focusing on establishing formal intellectual property rights) versus investing in execution (e.g., focusing on rapid, iterative innovation). These two dimensions of choice of competition—who to compete with and how to compete—result in four entrepreneurial strategies: intellectual property strategy, disruption strategy, value chain strategy, and architectural strategy. These strategies represent distinctive sets of choices regarding customer, technology, organization, and competition that reflect different value-creation and value-capture hypotheses. For a given idea, entrepreneurs are typically able to translate their idea to at least three of these alternative strategic playbooks.

Each strategy involves unique strategic commitments at the earliest stages and, therefore, requires the entrepreneur to take a different approach to prioritizing early learning and experimentation. Entrepreneurial strategies premised on execution and rapid iteration (i.e., disruption and value chain strategies) allow for more early flexibility and adjustment to market feedback than those strategies emphasizing control (i.e., intellectual property and architectural strategies). These execution-oriented strategies equip the entrepreneur to readily build iterative capabilities as well as products tailored to the evolving needs of customers and, as a result, offers the entrepreneur the ability to enhance value creation, although this requires sharp hypotheses as to potential novel customer segments and needs. By contrast, strategies premised on control are far more commitment oriented and require substantial early exploration and learning before designing and launching product experiments. To exert and maintain control, the entrepreneur must make significant early commitments, including investing in layers of intellectual property rights, product designs that limit imitation and encourage lock in, and acquisition of rare and valuable critical resources. This involves choices as to both what to protect and what degree to invest in those respective protections. These choices and commitments anchor the identity of the venture and provide a foundation for sustainable competitive advantage, if those early commitments prove to ultimately be a foundation to create value for customers.

Similar trade-offs exist in the commitments across strategies oriented toward collaboration (i.e., intellectual property and value chain strategies) versus those focused on competition (i.e., architectural and disruption strategies). Although collaborative strategies enhance value creation by leveraging the existing capabilities of the partner organization, including their relationship with their end user, they can involve early commitments, such as technology development choices to align with the partner's constraints, which may affect the start-up's bargaining position within that collaboration. Although this is a nontrivial commitment, it also provides considerable clarity as to the end user and the value proposition the partner provides to that end user. This allows the entrepreneur to focus early experimentation and learning on other dimensions of uncertainty. By contrast, although competitive strategies allow flexibility to create new value for new customers, there tends to be greater uncertainty as to customer segmentation and the relative value propositions across customers. Competitive strategies require the entrepreneur to make early commitments as to the core of the venture and the structure of novel value chains to deliver these new goods, even while they are still testing their value-creation hypotheses with early, novel sets of customers.

Taken together, these distinct commitments, as well as those commitments involved with the specific core choices and value-creation and value-capture hypotheses within each strategy, result in differing freedom to pivot should those choices and hypotheses be proven incorrect. In particular, an architectural strategy, with an investment in control and orientation toward competition, is the most ambitious entrepreneurial strategy and involves a significant degree of early commitments that will constrain the venture's options and subsequent performance. This strategy will require far more information to guide early choices and before an entrepreneur may be willing to commit to the strategy relative to other strategies. As a result, distinct entrepreneurial experimentation toolkits will be needed in evaluating these strategies and, once a strategy has been chosen, to execute and refine along the chosen path. More specifically, entrepreneurs should be cautious in applying advice and toolkits for entrepreneurial experimentation without understanding the context of that advice and the consequences for the strategic pathways they are considering. For instance, a successful serial entrepreneur who has repeatedly embraced a disruption strategy playbook may provide experimentation tactics that, while effective in the disruption strategy context, does not fully account for the significant early commitments involved in intellectual property and architectural strategies and may unintentionally limit these ventures. The Entrepreneurial Strategy Compass provides students and entrepreneurs with the judgment to sort such advice.

But how do entrepreneurs choose their entrepreneurial strategy?

As the Entrepreneurial Strategy Compass encapsulates, good ideas often provide multiple paths to create and capture value. As a result, the process of early-stage entrepreneurial exploration, learning, and experimentation is unlikely to provide a clear "path to the epiphany" highlighting the most viable route forward. Rather, the objective of the learning and experimentation process

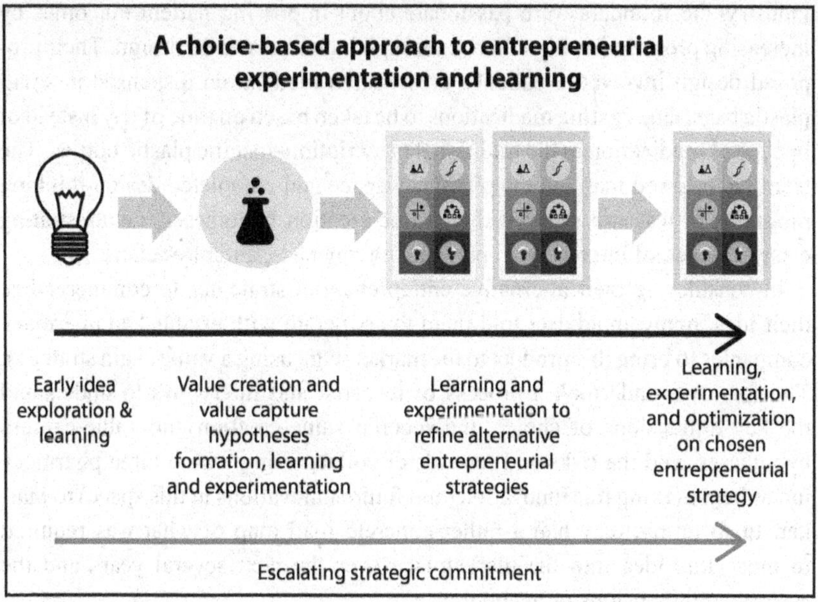


**Figure 12.3**
"Test two, choose one" decision-making rule.

should be to synthesize the distinctive choices, hypotheses and trade-offs of the alternative entrepreneurial strategies without committing the venture to any one path. Contrary to being a source of inertia, the existence of multiple viable alternative entrepreneurial strategies should serve as a catalyst for strategic action—compelling the entrepreneurs to have greater confidence in their underlying idea and commit to a particular path while forgoing other potentially promising options.

The "test two, choose one" decision-making rule guides entrepreneurs in translating their ideas into alternative entrepreneurial strategies, testing those hypotheses and ultimately making a choice to commit to one strategy, leaving other potentially viable paths behind (at least temporarily). As an entrepreneur moves through the "test two, choose one" process they move through distinct phases of entrepreneurial learning and experimentation, prioritizing no and low commitment learning and experimentation as they explore the early idea exploration, formulate alternative value creation and value capture hypotheses, and generate viable entrepreneurial strategies (see figure 12.3). The experience of Elliot Cohen and T. J. Parker as they considered an idea generated at a campus hackathon highlights the dynamics of this process both in the classroom and in practice (Stern and Fehder, 2019). Drawing on their knowledge of the pharmacy industry, the founders were passionate about improving patient outcomes by increasing prescription compliance through human-centered design. Their proposed design involved a patient's prescription medication dispensed in small plastic bags, aggregating medications to be taken based on time of day instead of by type of medication in the traditional prescription-specific plastic bottles. The team interviewed many in the pharmacy space and completed design-thinking programs to test and sharpen their value-creation hypotheses, demonstrating clear evidence of interest from patients, pharmacists, and physicians.

In considering their alternative entrepreneurial strategies to commercialize their idea, many an adviser told them to cooperate with established pharmacy companies to bring this product to the market (e.g., using a value chain strategy). The founders undertook a process of research and interviews to understand the key dimensions of choice, the specific value-creation and value-capture hypotheses, and the risks and rewards of collaborating with a large pharmacy incumbent to bring this innovation (and future innovations in this space) to market. In so doing, they had a rather concrete road map of what was required to move this idea into the marketplace over the next several years and the associated dimensions of uncertainty.

Leveraging the Entrepreneurial Strategy Compass, the entrepreneurs then decided to consider what a more competitive orientation might involve for their idea. Drawing on their industry experience and the knowledge they already gained in exploring the idea and articulating a preliminary, although well-defined, cooperative strategy, the entrepreneurs were able to quickly articulate a strategy to launch their own online pharmacy company, using their design innovation to create new value for customers within a new value chain. They understood this strategy would require significant commitments in intellectual property, acquisition of state-by-state pharmacy licenses, and development of software and robotic capabilities to sort, fill, and ship their novel prescription products. Seeking out advice from those in relevant industries, the founders sharpened their understanding of the key domains of choice, the specific value-creation and value-capture hypotheses, and the key risks to this strategy.

As they considered pursuing a cooperative strategy, reflecting the advice of their many mentors, as well as a more ambitious competitive strategy, reflecting their own entrepreneurial objectives, the entrepreneurs conducted further tests, including low-commitment test to sharpen their understanding of differences between these two strategies as well as remaining areas of uncertainty associated with each strategy. These experiments allowed the entrepreneurs to better understand the choices and commitments required to make these potentially viable entrepreneurial strategies a reality. Rather than illuminating the clear path ahead, these experiments armed the entrepreneurs with a clearer picture of the potential paths ahead. Only then did the entrepreneurs step back to consider which strategy reflected their interest in choosing entrepreneurship, in choosing this opportunity, and their skills and objectives in realizing that opportunity. Ultimately, they chose a competitive strategy, although it involved a rockier and more ambitious path forward architecting a new value chain to provide consumers with their novel prescription service, PillPack. Equipped with a clear road map, they began to make substantial commitments to give their chosen strategy the best chance of success, including diligently acquiring state-by-state pharmacy licenses and seeking patient investors with experience in the pharmaceutical industry. Having "tested two, chosen one," the entrepreneurs continued to experiment as they refined their chosen strategy and reconciled new information from the market with their hypotheses. In 2018, PillPack was acquired by Amazon for $753 million (Farr, 2019).

As Pillpack demonstrates, "test two, choose one" decision rule allows entrepreneurs to balance learning with commitment as they design and sequence their experiments so they are able to leverage the information gained to guide their choices moving forward. Although entrepreneurs must ultimately create and capture value, and the Entrepreneurial Strategy Compass is a proven playbook for articulating and testing alternative strategies to do so, it is ultimately the entrepreneur who must choose and execute a given strategy. By confronting the distinct opportunities and challenges, including dimensions of uncertainty that remain unresolved after no- and low-commitment experimentation, entrepreneurs choose a strategic path that reflects their personal ambitions for entrepreneurship. Indeed, it may not be overconfidence or optimism that leads entrepreneurs to choose a more uncertain strategic path or one fraught with considerable risk. Similarly, it might not be the result of ineffective experimentation that results in an entrepreneur prioritizing one customer segment over another. Rather, they may make these with full awareness of the tradeoffs of their chosen path because it aligns with their objectives and skills. The choice to be an entrepreneur, the choice of which opportunity to pursue, and the choice of which type of success to seek are far more personal than career objectives in traditional employment. The objective of entrepreneurial strategy is not to identify a singular, correct strategic path for an idea. Instead, it is to equip entrepreneurs to make an informed choice as to how they may translate their idea to reality, realizing their objectives, leveraging their skills, and balancing their personal constraints.

Entrepreneurship is personal. Clear hypotheses, frameworks, and experimentation provides entrepreneurs the judgment to choose. Too often, entrepreneurs and students find their process of entrepreneurial learning and experimentation does not move their idea forward. By highlighting the role of choice, the critical choices to test, rigorous and actionable frameworks to structure their hypotheses and learning, and a process for sequencing early experimentation balanced with commitment, entrepreneurial strategy enables choice and action.

Entrepreneurial strategy is about deliberate learning, intentional commitment, and informed choice. Although modern entrepreneurship education and practice has embraced iterative learning, entrepreneurs and students must recognize that experimentation alone does not guarantee progress. Without clearly formulated hypotheses, a structured approach to testing, and an understanding of how commitments shape future strategic flexibility, even the best-designed experiments can lead to misguided confidence, misaligned choices, or misdirected action. The "test two, choose one" decision rule alongside the Entrepreneurial Strategy Compass framework offers entrepreneurs and students a rigorous and actionable process to explore alternative strategic pathways, extract meaningful insights, and make deliberate commitments while balancing uncertainty and flexibility. By synthesizing rigorous research, practitioner insights, and real-world case studies, entrepreneurial strategy equips entrepreneurs and students with the tools to navigate uncertainty, translate learning into action, and ultimately chart a coherent path forward. In the end, successful entrepreneurial experimentation is not just about testing ideas—it is about the judgement to make strategic choices that transform those ideas into reality.

## References

Beard, A. (2019, July–August). Life's work: An interview with Vera Wang. *Harvard Business Review*, 97(4), 164–166.

Camuffo, A., Cordova, A., Gambardella, A., and Spina, C. (2020). A scientific approach to entrepreneurial decision making: Evidence from a randomized control trial. *Management Science*, 66(2), pp. 564–586.

Camuffo, A., Gambardella, A., Messinese, D., Novelli, E., Paolucci, E., and Spina, C. (2024). A scientific approach to entrepreneurial decision-making: Large-scale replication and extension. *Strategic Management Journal*, 45(6), pp. 1209–1237.

Cao, R., Koning, R., and Nanda, R. (2024). Sampling bias in entrepreneurial experiments. *Management Science*, 70(10), pp. 7283–7307.

Castillo, G. and Scott, E. L. (2022). Taxie (Entrepreneurial Strategy Case). MIT Sloan School of Management.

Coali, A., Novelli, E., Sirigiri, A., and Spina, C. (2024). A scientific approach to decision-making: Key tools and design principles. *Journal of Business Venturing Design*, 3, p. 100023.

Cohen, S. L., Bingham, C. B., and Hallen, B. L. (2019a). The role of accelerator designs in mitigating bounded rationality in new ventures. *Administrative Science Quarterly*, 64(4), pp. 810–854.

Cohen, S. L., Bingham, C. B., and Hallen, B. L. (2019b). The role of accelerator designs in mitigating bounded rationality in new ventures. *Administrative Science Quarterly*, 64(4), pp. 810–854.

Farr, C. (2019). The inside story of why Amazon bought PillPack in its effort to crack the $500 billion prescription market. *CNBC*. https://www.cnbc.com/2019/05/10/why-amazon-bought-pillpack-for-753-million-and-what-happens-next.html

Felin, T., Gambardella, A., Novelli, E., and Zenger, T. (2024). A scientific method for startups. *Journal of Management*, 50(8), pp. 3080–3104.

Felin, T., Gambardella, A., Stern, S., and Zenger, T. (2020). Lean startup and the business model: Experimentation revisited. *Long Range Planning*, 53(4), p. 101889.

Gans, J. (2018, February 28). A better way to bring science to market. *MIT Sloan Management Review*.

Gans, J. S., Scott, E. L., and Stern, S. (2024). *Entrepreneurship, choice and strategy*. W. W. Norton & Company.

Gans, J. S., Stern, S., and Wu, J. (2019). Foundations of entrepreneurial strategy. *Strategic Management Journal*, 40(5), pp. 736–756.

Garg, S., and Eisenhardt, K. M. (2017). Unpacking the CEO–board relationship: How strategy making happens in entrepreneurial firms. *Academy of Management Journal*, 60(5), pp. 1828–1858.

Kirzner, I. (1976). *The economic point of view: An essay in the history of economic thought*. Sheed Andrews McMeel.

Koning, R., Hasan, S., and Chatterji, A. (2022). Experimentation and start-up performance: Evidence from A/B testing. *Management Science*, 68(9), pp. 6434–6453.

Miller, A., O'Mahony, S., and Cohen, S. L. (2024). Opening the aperture: Explaining the complementary roles of advice and testing when forming entrepreneurial strategy. *Organization Science*, 35(1), pp. 1–26.

Milnes, H. (2024). Vera Wang on why she's selling her brand after 35 years. *Vogue Business*. https://www.voguebusiness.com/story/fashion/vera-wang-sells-brand-to-management-group-whp-global

Murray, F., and Tripsas, M. (2004). The exploratory processes of entrepreneurial firms: The role of purposeful experimentation. *Business strategy over the industry lifecycle*. Emerald, pp. 45–75.

Nanda, R., and Rhodes-Kropf, M. (2016). Financing entrepreneurial experimentation. *Innovation Policy and the Economy*, 16(1), pp. 1–23.

Osterwalder, A., and Pigneur, Y. (2011). *Business model generation: A handbook for visionaries, game changers, and challengers*. John Wiley & Sons.

Ries, E. (2011). *The lean startup: How today's entrepreneurs use continuous innovation to create radically successful businesses*. Crown.

Shane, S. (2003). *A general theory of entrepreneurship: The individual-opportunity nexus*. Edward Elgar.

Stern, S., and Fehder, D. (2019). PillPack (Entrepreneurial Strategy Case). MIT Sloan School of Management.

# 13

## What Helps Entrepreneurs Learn: Reflections, New Directions, and Open Questions

Chiara Spina

### 13.1 Introduction

What helps entrepreneurs learn what is necessary for success? As the economic and social benefits of entrepreneurship become increasingly clear (Baumol and Strom, 2007; Lerner, 2013; Fairlie and Miranda, 2017), governments, universities, and support institutions worldwide have rolled out numerous programs intended to provide aspiring founders with the tools they need to succeed (Shepherd and Gruber, 2021). However, these numerous resources and approaches have also highlighted a critical gap in our approach to entrepreneurial learning: the lack of a unifying framework that can guide entrepreneurs in making informed decisions across the diverse and often unpredictable scenarios they face.

It is against this backdrop that the concept of Bayesian entrepreneurship is emerging as a promising paradigm. At its core, Bayesian entrepreneurship provides a structured yet flexible approach to decision-making under uncertainty, acknowledging the unique challenges of the entrepreneurial context. Studies informing the Bayesian entrepreneurship framework offer a powerful lens through which to view the entrepreneurial journey—not as a linear path from idea to execution but rather as an iterative process of learning and adaptation (Gans et al., 2019; Agrawal et al., 2021; Camuffo et al., 2020). However, effectively teaching this iterative approach presents significant challenges. It is far from obvious how to teach an approach that creates a structured process to continuously refine the understanding of both the problem space and potential solutions. This challenge is complicated by the need to guide entrepreneurs in forming initial priors from which to build their iterative learning process.

This chapter examines effective approaches to entrepreneurial learning from a Bayesian entrepreneurship perspective, integrating cutting-edge research with practical insights from the field. I begin by examining the current state of knowledge on what works in entrepreneurial learning, with a particular focus on evidence from randomized controlled trials (RCTs). RCTs are research designs where participants are randomly allocated to various treatment or intervention groups in their real-world settings. This method enables researchers to examine the impact of interventions on individuals or situations (Harrison and List, 2004). Studies using RCTs offer the most rigorous causal evidence available (Duflo and Banerjee, 2017) on the impact of various knowledge-related interventions. The literature review of existing RCTs highlights three important facts. First, although knowledge itself remains valuable, studies show that "know-what" interventions alone do not dramatically improve outcomes, except in cases of very intensive programs. The shift toward "know-how" interventions reflects a growing understanding that entrepreneurial success relies more on the ability to update beliefs and adapt strategies, aligning with the Bayesian emphasis on iterative learning. Second, various interventions, including teaching, mentorship, and consulting, consistently show that providing structured guidance on gathering and interpreting new information is effective in the entrepreneurial context. This aligns with the Bayesian learning focus on systematic evidence collection and analysis to refine business hypotheses. Lastly, across different studies, interventions that help entrepreneurs form initial assumptions (priors) and then systematically update these beliefs based on new evidence tend to be most successful. This directly mirrors the Bayesian approach of starting with a prior belief and updating it as new information becomes available. Taken together, these findings position Bayesian entrepreneurship as a promising unifying framework for entrepreneurial learning, integrating the most effective elements identified in recent experimental research.

Building on this review, I zoom in on a set of interventions that have provided growing evidence of the effectiveness of a scientific approach to decision-making (Camuffo et al., 2020, 2024), an entrepreneurial method that embodies Bayesian learning. I share key considerations that informed the design of effective pedagogical tools in research studies that compared groups of entrepreneurs who were taught the scientific approach with groups of entrepreneurs who received comparable content but without the scientific approach through RCTs. Although the structure of these studies is well-documented in published research and working papers, this chapter provides a unique behind-the-scenes perspective, focusing on critical considerations often omitted from manuscripts, including the description of some tools developed to teach a scientific approach. This insider view offers valuable context to understand the nuances of implementing a Bayesian-inspired approach to entrepreneurship education.

Drawing from patterns that emerged during extensive fieldwork involving thousands of entrepreneurs, this chapter presents a series of carefully selected case studies and observations. These examples not only support the theoretical foundations of the scientific approach but also reveal intriguing patterns encountered during the implementation of the studies. These cases uncover emerging patterns and unresolved questions, pointing toward promising new directions for future research on entrepreneurial learning. Overall, the chapter combines theoretical foundations, empirical evidence, and practical applications to offer a comprehensive view of the scientific approach's potential to significantly impact entrepreneurial education and practice. The chapter concludes by discussing open questions, aiming to encourage further research to advance our understanding of effective entrepreneurial decision-making and learning processes.

## 13.2 What Helps Entrepreneurs Learn: Insights from the Experimental Literature

The importance of entrepreneurial learning has long been recognized in both the academic literature (Minniti and Bygrave, 2001; Chen et al., 2018) and practical guides for founders (Ries, 2011; Eisenmann, 2021). A rich body of work, spanning theoretical models (Bingham and Davis, 2012), qualitative case studies (McDonald and Eisenhardt, 2020), and empirical analyses (Parker, 2006), has sought to unpack the processes through which entrepreneurs acquire the knowledge and skills necessary for new business success. In recent years, there has been a growing emphasis on the rigorous evaluation of knowledge-related interventions (Spina and Hasan, 2024), with RCTs emerging as a gold standard for assessing the causal impact and precisely estimating the effectiveness of various approaches to enhancing entrepreneurial knowledge and capabilities.

RCTs offer unique value in the context of entrepreneurial learning for several reasons. First, they allow to isolate the specific effect of a learning intervention to be isolated, balancing at baseline the several other factors that might influence entrepreneurial outcomes (Duflo and Banerjee, 2017). Second, by randomly assigning participants to treatment and control groups, RCTs help mitigate selection bias (Di Stefano and Gutierrez, 2019)—a primary concern in studying entrepreneurship, where self-selection into programs or activities can often skew results (Yang and Aldrich, 2012). Finally, RCTs provide a clear benchmark against which to measure the relative effectiveness of different approaches, enabling more informed decisions about resource allocation in entrepreneurship support programs (McKenzie and Woodruff, 2014).

Multiple methods were used to find relevant published studies and recent working papers in entrepreneurship. These included a search in the EBSCO database, focusing on journals in management, economics, and sociology. Additionally, Google Scholar was used to find works citing these studies and to discover relevant working papers. The results of this search are summarized in table 13.1, providing an overview of RCTs in entrepreneurship focused on learning interventions, categorized by the type of knowledge they examine. Following a classification system adapted from Middleton and Donnellon (2014), the studies are classified into three main categories: know-what, know-how, and know-why. The know-what category refers to factual knowledge about business and entrepreneurship. It includes basic business concepts, financial literacy, and general business practices, often the focus of traditional business education. The know-how category encompasses practical skills and techniques for running a business. It involves the application of knowledge to real-world situations and includes skills such as decision-making, problem-solving, and strategic thinking. The know-why category relates to the mindset, motivation, and psychological aspects of entrepreneurship. It includes personal initiative, resilience, and the ability to understand and adapt to the entrepreneurial process. This classification is helpful in understanding the evolution of entrepreneurship education and support, as well as the relative effectiveness of different interventions and approaches.

The table reveals three key interesting facts about RCTs on entrepreneurial learning. First, studies initially focused on know-what interventions, such as traditional business training programs. However, the results from these studies were often limited or mixed. A key early study in this area was conducted by Karlan and Valdivia (2011), who evaluated the effectiveness of entrepreneurial training for microfinance clients in Peru. Their study involved a sample of more than 4,500 borrowers, randomly assigned to either a control group or an intervention group. The intervention group received up to twenty-two weekly training sessions, each lasting thirty to sixty minutes. Although the study found significant improvements in certain business practices and knowledge, it did not show a meaningful increase in business revenue. Similarly, Bruhn and Zia (2013) conducted an RCT in Bosnia Herzegovina with 445 young entrepreneurs. Treated entrepreneurs attended a program where they covered six modules learning business planning as well as marketing and finance concepts. The authors did not find the training had an impact on either survival or performance. More broadly, the impact of know-what interventions on entrepreneurial outcomes remains modest across a number of RCTs, as highlighted by McKenzie et al. (2023). More positive effects on adoption and performance are, however, observed for interventions that are more intensive (as in Anderson et al., 2018,

**Table 13.1**
Overview of RCT-based studies in entrepreneurship focused on learning interventions

| Knowledge examined | Study | Intervention type | Takeaway |
|---|---|---|---|
| Know-what | Acimovic et al. (2020) | SMS-based inventory management | No impact of inventory recommendations |
| Know-what | Anderson et al. (2018) | Finance vs. marketing training | Marketing training more effective than finance training |
| Know-what | Berge et al. (2015) | Business training | Positive effects for male entrepreneurs, limited for females |
| Know-what | Bruhn and Zia (2013) | Business and financial literacy training | Improved business practices, no significant impact on performance |
| Know-what | Calderon et al. (2020) | Business literacy training | Positive impacts on profits and business expansion |
| Know-how | Cole et al. (2024) | Rule-of-thumb training via voice messages | Limited impacts on business performance |
| Know-what | Custódio et al. (2020) | Executive education in finance | Improved financial practices for medium and large firms |
| Know-what | De Mel et al. (2014) | Business training for female entrepreneurs | Improved business practices, mixed results on performance |
| Know-what | Drexler et al. (2014) | Rule-of-thumb training vs. standard accounting training | Rule-of-thumb training more effective for less educated |
| Know-what | Giné and Mansuri (2021) | Business training | Improved business practices, limited impact on profits |
| Know-what | Karlan and Valdivia (2011) | Basic business training | Improved business knowledge, no significant impact on revenue |
| Know-what | Mano et al. (2012) | International Labour Organization business training | Increased business survival, imprecise effects on profits and sales |
| Know-what | McKenzie and Puerto (2021) | GET Ahead business training for women | Improved business practices and profits after three years |
| Know-what | Valdivia (2015) | Business training plus technical assistance | Positive impacts on sales and profits |
| Know-how | Agarwal et al. (2025) | Theory-based approach | Positive impact on pivoting |
| Know-how | Anderson and McKenzie (2022) | Consulting, insourcing, outsourcing | Insourcing and outsourcing more effective than traditional training |
| Know-how | Åstebro and Hoos (2021) | Business model design, lean start-up method | Analytical skills more important for entrepreneurial outcomes |
| Know-how | Bloom et al. (2013) | Management consulting | Significant improvements in productivity and profitability |
| Know-how | Brooks et al. (2018) | Mentoring vs. classroom training | Mentoring more effective short term |

(continued)

**Table 13.1**
(continued)

| Knowledge examined | Study | Intervention type | Takeaway |
|---|---|---|---|
| Know-how | Bruhn and Zia (2013) | Consulting for small and medium-sized enterprises | Positive impacts on employment and firm growth |
| Know-how | Cai and Szeidl (2018) | Peer interactions | Significant improvements in firm performance |
| Know-how | Camuffo et al. (2020) | Scientific approach to decision-making | Improved business performance and increased pivoting |
| Know-how | Chatterji et al. (2019) | Advice on employee management | Increased employee growth and survival rates |
| Know-how | Chatterji et al. (2023) | Structured schema for idea evaluation | Enhanced ability to assess and generate high-quality ideas |
| Know-how | Dimitriadis and Koning (2022) | Social skills training for networking | Enhanced networking led to improved financial performance |
| Know-how | Eesley and Wang (2017) | Entrepreneur mentoring | Increased likelihood of entrepreneurial careers |
| Know-how | Fafchamps and Quinn (2016) | Networking | Positive effects on business practices and performance |
| Know-how | Hasan and Koning (2019) | Information-driven discussions | Richer discussions led to better idea generation |
| Know-how | Higuchi et al. (2015, 2019) | Kaizen training | Improved management practices and business survival |
| Know-how | Iacovone et al. (2022) | Individual and group-based consulting | Group-based approach as effective as individual consulting |
| Know-how | Lafortune et al. (2018) | Role models or individual consulting | Personalized consulting more effective than role model training |
| Know-how | Novelli and Spina (2024) | Scientific approach | Increase employment and revenue for entrepreneurs with developed business models |
| Know-how | Santamaria et al. (2022) | Demand-side vs. resource-side activities | Prioritizing customer needs more effective for entry and revenue |
| Know-why | Alibhai et al. (2019) | Mindset training vs. personal initiative training | Mixed results, some positive effects for personal initiative |
| Know-why | Bjorvatn et al. (2020) | Edutainment | Increased interest in entrepreneurship, no impact on knowledge or actions |
| Know-why | Campos et al. (2017) | Personal initiative training | Significant improvements in business performance |
| Know-why | Gielnik et al. (2017) | Action-oriented approach | Increased entrepreneurial entry through enhanced self-efficacy |
| Know-why | Glaub et al. (2014) | Personal initiative training | Increased entrepreneurial performance |
| Know-why | Ubfal et al. (2019) | Personal initiative vs. traditional training | Mixed results, some positive effects for personal initiative |

for example). Although these results do not rule out that basic business knowledge can still be valuable for entrepreneurs, they suggest that this type of knowledge alone is not sufficient to improve entrepreneurial outcomes.

Second, there has recently been a shift toward studying know-why and know-how interventions. Know-why interventions, which target motivation, attitudes, mindset, and aspirations, have drawn heavily from psychological principles. Personal initiative training programs aim to develop a proactive entrepreneurial mindset, encouraging owners to search continuously for new opportunities, learn from errors, and think of ways to differentiate their business from others. Campos et al. (2017), for instance, conducted an RCT with 1,500 microenterprise owners in Togo, testing the impact of personal initiative training compared with both a control group receiving no training and a group receiving a traditional small business training approach. Personal initiative training was found to significantly outperform traditional business training and produced a significant impact on outcomes, including profits. At the same time, an increasing number of studies are exploring know-how interventions, focusing on decision-making and problem-solving skills. A common thread emerging from these studies is that interventions providing structured guidance on gathering and interpreting new information tend to be most effective, regardless of whether they are delivered through teaching (Santamaria et al., 2022), mentorship (Bennett and Chatterji, 2019), or consulting (Anderson and McKenzie, 2022).

Third, researchers have also examined other approaches to enhance entrepreneurial capabilities. A particularly promising line of research has emerged around teaching entrepreneurs to apply scientific thinking to their business decisions. This approach aims to equip entrepreneurs with a systematic framework to test assumptions and make evidence-based decisions, embodying the principle of structured information gathering and interpretation. A series of RCTs, which tested a structured way to teach entrepreneurs to think and act like scientists when approaching their business, showed positive effects on a range of outcomes. Camuffo et al. (2020) provided initial evidence that entrepreneurs trained in a scientific approach to decision-making achieved improved decision quality and venture outcomes, findings replicated in Camuffo et al. (2024). Novelli and Spina (2024) found that the benefits of this approach were particularly pronounced for companies at more advanced stages of business model development, suggesting a compounding effect, as the approach is applied over time. Spina and Battaglia (2023) highlighted how a focus on hypothesis falsification and clear theory of value articulation led to increased profitability among treated firms. Agarwal et al. (2025) further reinforced these findings, showing how a rigorous, theory-driven approach to opportunity evaluation led to more focused and successful pivots when necessary. These studies support the idea that interventions that help entrepreneurs form initial assumptions (priors) and then systematically update these beliefs based on new evidence tend to be most successful. Building on the evidence supporting the scientific approach and its coherence with effective entrepreneurial learning patterns, we now turn to examine the practical challenges and strategies involved in implementing this approach. The next section provides an insider's perspective on designing and delivering interventions that teach scientific thinking to entrepreneurs.

## 13.3 What Helps Entrepreneurs Learn: Insights from the Field

The RCTs related to the scientific approach were designed to teach entrepreneurs a systematic process of hypothesis formulation, experimentation, and evidence-based decision-making inspired by the scientific method, which is summarized in figure 13.1.

Designing the training interventions part of the RCTs related to the scientific approach required a deep understanding of the realities faced by entrepreneurs. To inform the design of the training interventions, field research was conducted and included interviews with entrepreneurs across various industries and stages of development, consultations with organizations supporting entrepreneurs (such as the Doers and the Lean Startup Machine), and reviews of curricula from leading entrepreneurship programs. This investigation allowed key priors about what works in entrepreneurship education to be refined. Three key patterns emerged that significantly informed the subsequent intervention design:

1. **The time-pressure paradox**: Entrepreneurs consistently communicate that they are under time pressure. This often leads them to rush important decisions. Paradoxically, this sense of urgency frequently results in wasted time and energy, as quickly made choices lead to costly mistakes or pursuit of unvalidated opportunities. This realization highlighted the need for

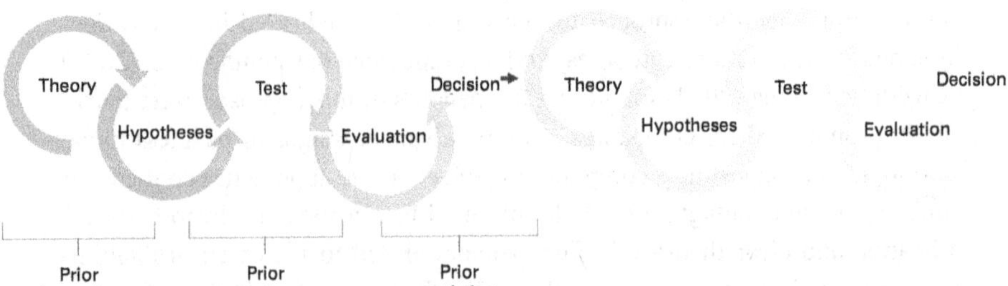


**Figure 13.1**
The scientific approach for decision-making.

tools that could help entrepreneurs make better decisions quickly, without sacrificing rigor.

2. **Cognitive barriers to scientific thinking**: Although most entrepreneurs intuitively understand the value of gathering data, many struggle with more abstract concepts such as theory development and hypothesis testing. The language of science often seems hard to understand, particularly for those without formal research training, creating a barrier to the adoption of more systematic approaches. This pointed to the need for frameworks that could make these concepts more accessible and immediately applicable.
3. **The challenge of habit formation**: Even when entrepreneurs understand and appreciate the value of a scientific and more systematic approach, many find it difficult to consistently apply these methods in their day-to-day decision-making. Entrepreneurs often rely on gut feelings and quick decisions due to time pressures, which can take precedence over more careful, long-term thinking. This suggested that any effective intervention would need to be intensive and sustained, with mechanisms for reinforcing new habits over time.

Building on these insights, it seemed important to develop tools and approaches designed to address the three patterns presented above. Consequently, three guiding principles were used to develop the teaching material used for the scientific approach RCTs:

1. **Blended learning**: Recognizing the need to counteract entrepreneurs' impatience with theoretical concepts, the content taught integrates teaching tools with hands-on application. Each session combines theoretical concepts followed right away by practical applications, effectively leveraging a "learning by doing" approach. In ensuring that every concept taught was immediately applied to the entrepreneur's own business, the goal was to maintain engagement and demonstrate tangible value in learning and applying theoretical concepts.
2. **Cognitive scaffolding**: To make the sometimes-abstract concepts of the scientific approach more understandable, a series of visual frameworks and structured tools was developed. These served as cognitive scaffolds, breaking down complex processes into more manageable steps and providing clear guidance for implementation. Coali et al. (2024) discuss some of these tools in depth and elaborate on principles that support teaching the scientific approach.
3. **Extended learning journeys**: Recognizing that changing how entrepreneurs approach their business and related decisions requires

sustained effort, the teaching material was designed to extend well beyond the classroom setting. Each program typically involved more than twenty hours of direct instruction, spread across several weeks. Between sessions, participants were given assignments directly relevant to their own businesses, conducting customer interviews, developing prototypes, running small-scale experiments, and so forth. This ensured a continuous application of the concepts and provided opportunities for instructor feedback on real-world implementation.

These three design choices proved crucial in helping entrepreneurs internalize and apply the scientific approach. The extended, applied nature of the learning journey allowed for the formation of new habits and the gradual integration of scientific thinking into day-to-day entrepreneurial practice. But the most effective choice was, in hindsight, the creation of tools that served as concrete anchors, making abstract concepts tangible and providing clear structures for decision-making processes. I describe some of these tools in the following section.

### 13.3.1 An Illustration of Key Tools of the Scientific Approach

This section introduces a set of original tools and templates designed to support entrepreneurs in implementing the scientific approach. Although not exhaustive (for a more comprehensive discussion, see Coali et al., 2024), these tools illustrate how to facilitate the process of prior formation, testing, and updating that forms the core of the scientific approach. Figure 13.2 provides a visual overview of these tools. This set of tools include:

**The theory map**: A visual tool for articulating and connecting the key elements of a business model, encouraging systematic thinking about value creation and capture. This tool helps entrepreneurs move beyond vague intuitions to clearly articulated theories about how their venture would succeed. It does this by presenting a set of questions directly related to the elements in other canvases (e.g., Why would potential users face the problem described in the business model canvas and how is this related to the proposed solution?). These questions guide entrepreneurs to identify key attributes and their causal connections. By encouraging simultaneous consideration of different attributes, the theory map promotes more thorough and holistic thinking about the business problem. It achieves this by causally connecting elements that are typically separated in standard canvases. This process helps entrepreneurs reflect on why specific attributes have been included, their importance to the business proposition, and how different elements are causally connected.

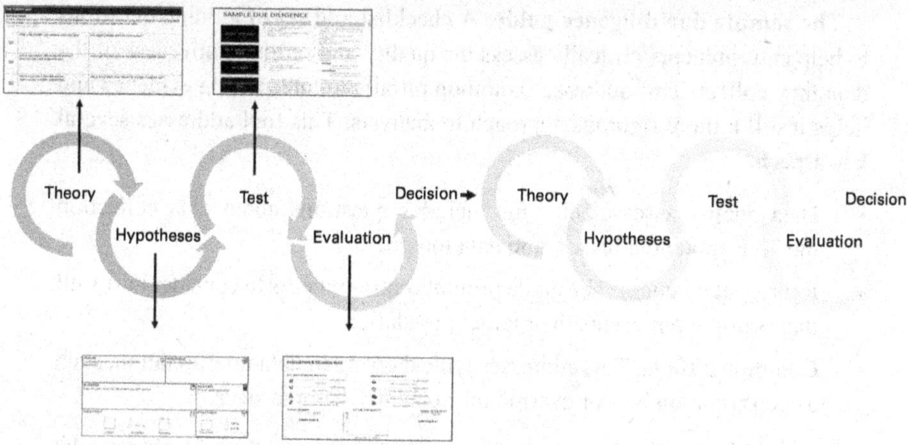


**Figure 13.2**
Tools supporting teaching the scientific approach.

**The hypotheses canvas**: A structured format for developing testable predictions, explicitly linking elements of the theory to specific evidence-gathering activities. The hypothesis canvas can be completed even before data collection begins, adhering to the principles of defining measures clearly linked to hypotheses and declaring outcomes ex ante. This structure bridges the gap between high-level theorizing and concrete action, helping entrepreneurs move from abstract ideas to specific, testable predictions about their business model. The canvas is divided into several key blocks:

- Theory and hypotheses: The upper two blocks prompt entrepreneurs to state the aspect of their theory relevant for testing and the related hypotheses under empirical investigation. This aligns with the design principle of theory integration.
- Data gathering and measures: The central blocks encourage entrepreneurs to list the chosen technique for data gathering and the specific measures used to test the hypotheses. This addresses the design principles of clear operationalization, appropriate technique selection, and pre-defined measures.
- Initial beliefs: The bottom-left block requires entrepreneurs to make explicit their initial beliefs, setting a threshold as if the entire population of targeted customers was available. This adheres to the principle of predefined priors.

**The sample due diligence guide**: A checklist and set of guiding questions to help entrepreneurs critically assess the quality and representativeness of the data they collect. This addresses common pitfalls in interpreting evidence and helps instill a more rigorous approach to analysis. This tool addresses several key aspects:

- Data quality assessment: This includes questions about data collection methods, potential biases, and data integrity.
- Representativeness: The guide prompts entrepreneurs to consider how well their sample represents their target population.
- Common pitfalls: This addresses typical errors in data interpretation, such as confirmation bias or overreliance on small sample sizes.

**The evaluation framework**: A framework for objectively analyzing results against predefined thresholds, reducing the influence of confirmation bias and encouraging more honest assessment of evidence. The framework also addresses unassessed facts and beliefs, taking into account factors that may not be easily measurable. This includes future opportunity costs, resources to commit, and estimate variability. These elements are rated on a green-to-red scale and summarized in an overall rating. A key feature of the framework is the threshold-setting matrix, labeled as "Set the Threshold V*." This matrix guides entrepreneurs in determining the appropriate decision-making threshold based on the accuracy of their data model and the level of unassessed facts and beliefs. The thresholds range from "very low" to "very high."

### 13.3.2 Obstacles to Scientific Thinking in Entrepreneurship

Although empirical studies are increasingly showing the effectiveness of the scientific approach and related tools, on average, in helping entrepreneurs absorb and use the scientific approach, it is helpful to point out several of the most common challenges observed among entrepreneurs with regards to forming, testing, and updating priors. These challenges have been observed in the RCTs both among entrepreneurs who were in the control group (and therefore were not exposed to a scientific approach) and among treated entrepreneurs who did not absorb the content of the training. A visual representation of these challenges and how they map to the scientific approach and Bayesian learning is provided in figure 13.3. These challenges emerging from fieldwork are consistent with research that shows that entrepreneurs might draw wrong conclusions due to sampling bias as they run initial tests to validate their business (Cao et al., 2024). They also align with research indicating that entrepreneurs are boundedly rational (Cohen et al., 2019; Bennett and Chatterji, 2019), emphasizing

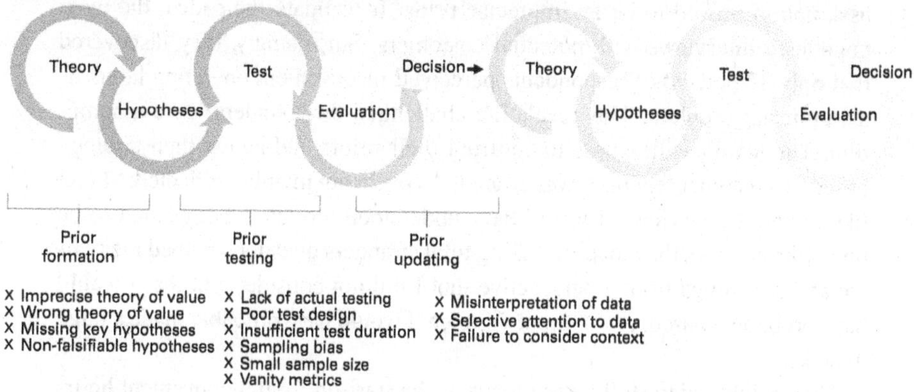


**Figure 13.3**
Common challenges in forming, testing, and updating priors in entrepreneurial decision-making.

the need for structured approaches to bridge this divide. Nevertheless, adopting structured approaches also presents some challenges, as noted by Leatherbee and Katila (2020).

Instead of detailing all challenges, I present some selected and representative cases from the fieldwork conducted as part of the scientific approach RCTs that describe common challenges and point to interesting open questions. The cases presented in this chapter are based on real events and experiences observed during the implementation of various RCTs related to the scientific approach. However, the names of the companies and individuals have been changed to protect their privacy and maintain confidentiality.

With regards to *prior formation*, entrepreneurs frequently have an imprecise theory of value. As a case in point, consider Washly, a start-up offering laundry pickup and delivery services. The idea for Washly was born from the personal experiences of its founders, two banking professionals frustrated by their inability to access laundry services due to late work hours. This led them to envision a digital platform connecting users with laundry services, similar to food delivery apps. The founders' initial priors were heavily influenced by their own experiences and perspectives as busy professionals. One of the founders, explained: "We had this idea because my friends and I typically get out of office pretty late and as such it is quite difficult for us to find an open laundry after 7:00 p.m., 7:30 p.m. The only laundries open after a certain time are the ones in the malls that typically close at 10:00 p.m."

This personal inconvenience led them to believe that the primary problem to solve was the inconvenient operating hours of laundries. However, this assumption proved to be an imprecise prior. To validate their idea, the team conducted interviews with potential customers. Surprisingly, they discovered that only 37 percent of respondents perceived inconvenient operating hours as a significant problem. This revelation challenged the founders' core assumption. The team's willingness to confront their priors and adjust their strategy based on customer feedback was crucial. As one team member reflected: "I feel like, without interviews, I would have understood too late what needed to be done [the changes they made]. Asking total strangers questions helped me realize and see things from a perspective that I did not consider and that I would have probably missed. . . . Without this step, I would have probably made a huge mistake."

This insight led to shifting the focus of the start-up from inconvenient hours to the lack of pickup and delivery services bookable online. The target market expanded beyond working professionals to include anyone who found it challenging to transport laundry, such as the elderly. The team's imprecise priors, however, were not limited to customer needs. They also made assumptions about laundry businesses, hypothesizing that these establishments struggled to expand beyond their local neighborhoods and lacked digital presence. These assumptions were largely confirmed through interviews, revealing another challenge for Washly's platform. "We realized retailers are kind of digital illiterates. . . . We recognized many laundries do not have tablets or devices to manage this kind of service," noted a team member. This realization led to another pivot, expanding the platform to consider partnerships with technology providers to equip laundries with necessary devices.

The example of Washly illustrates the danger of relying too heavily on personal experiences when developing a business idea. This appears to be a common challenge among founders, also described in the MiMoto case part of Spina et al. (2020). These instances raise an important open question related to prior formation: How does personal experience affect entrepreneurs' initial priors, and how can it best be leveraged to form accurate priors?

Although Washly's case highlights the importance of challenging assumptions in prior formation, entrepreneurs face different challenges when it comes to prior testing. A contrasting example is the case of Beerbox, which highlights how entrepreneurs often struggle to implement structured approaches to test their hypotheses. Beerbox offered an online craft beer subscription service, aiming to address perceived gaps in the market. The founder launched its service quickly and started making revenue early on. However, the founder's approach to market validation was largely reactive and based on sales performance rather than proactive testing of specific hypotheses. As the founder admitted: "I had a hard time in validating my idea with real data, I must be honest. Unconsciously,

I preferred setting up the operations before analyzing the market feedback. . . . My validation consisted in the feedback I would get through the sales. If sales were going good, then it meant I was doing good; if they were not, then it meant I was doing mistakes."

This approach reflects a lack of structured prior testing, which could have provided valuable insights before full-scale launch. The founder justified this decision by explaining the challenges of validating a business-to-consumer (B2C) business model: "The fact of not validating the model is normal in my opinion for a B2C business. In my case, considering the end result, maybe it was not a good idea, but how can you validate Beerbox? . . . People would say 'Yeah, nice idea!' and then they would not buy."

However, this explanation overlooks the possibility of conducting small-scale experiments or prototype testing to gauge potential customer interest and behavior. The lack of rigorous prior testing led to unforeseen challenges in the business model. The founder later realized that the service faced significant challenges with customer retention and perceived value: "The problem was that I underestimated the importance of ongoing value creation. Initially, customers were excited about the novelty of curated craft beer deliveries. But after a few months, many felt they weren't discovering enough new favorites to justify the ongoing cost. Our churn rate was much higher than anticipated."

This insight could have been gained through early prototype testing or customer interviews, potentially allowing for adjustments to the business model before full launch. As he relied solely on post-launch sales data for validation, the founder missed opportunities to refine the business model, understand customer needs, and test key assumptions before committing significant resources to their business. This example naturally highlights some unanswered questions, such as: What are the primary barriers that prevent entrepreneurs from conducting systematic experiments, even when they understand the potential benefits? How does an entrepreneur's risk perception influence their willingness to engage in early-stage experimentation? How do time pressure and resource constraints affect an entrepreneur's willingness to conduct thorough market tests?

After examining challenges in prior formation with Washly and prior testing with Beerbox, we now turn to a final case study that illustrates the difficulty entrepreneurs face in updating their priors, even when presented with contradictory evidence. Arya was conceived by a founder looking to create a high-end online retail store specializing in curated, quality fashion pieces with a focus on sustainability and ethical production. The founder's core theory was based on her observation of two key trends: the growing popularity of online shopping for luxury goods and an increasing demand for sustainable, ethically produced fashion items. She hypothesized that there existed an underserved market segment of fashion-conscious consumers aged thirty to forty, with high disposable income, who valued both luxury and sustainability in their clothing choices. Along with clear and testable hypotheses about her target demographic, the founder also prespecified a threshold for validation. Initially, the founder gathered evidence to validate her business model. She conducted extensive market research, including an online survey distributed to a thousand individuals in her target demographic, with more than four hundred responses, and a large number of in-depth interviews with potential customers alongside analysis of existing online luxury fashion retailers.

She found that most of the respondents did not prefer online shopping for luxury fashion, as the in-person experience of shopping is an integral part of the overall satisfaction they derive from purchasing high-end fashion items. In addition, although several reported valuing sustainability and ethical production, very few were already paying a premium for these products. When these results did not fully support her hypotheses, the founder struggled to update her prior beliefs. Instead of pivoting or adjusting her business, she persisted with her original concept. She reported: "I was so convinced of my idea that when the survey results came back negative, I couldn't believe it. I thought, 'These can't be the right people I'm asking. My target market is out there somewhere as this is not a one-of-a-kind concept.'"

Rather than interpreting the negative feedback as evidence against her initial theory, the founder attributed the lack of positive response to interviewing the wrong demographic. She continued to search for a target market that would confirm her preexisting beliefs about the viability of her business model.

The case of Arya underscores a critical challenge in entrepreneurial decision-making: the ability to balance passion and conviction with flexibility and openness to market feedback. In this specific case, the founder's unwillingness to update her priors based on extensive research data led to a prolonged "analysis phase" without moving toward actual implementation or pivoting to a more viable model. This example raises an interesting question for future research: What psychological factors contribute to an entrepreneur's resistance to updating their business model despite contrary market feedback? This question is vital because it addresses a common pitfall in entrepreneurship: the inability to pivot when necessary. Relatedly, it would be valuable to understand if the scientific approach should be tailored to resonate with entrepreneurs who have strong emotional attachments to their initial business concepts. Finally, this last case suggests that there is room to support entrepreneurs in generating alternative theories when initial hypotheses are not supported. New tools and techniques could be developed and tested as part of future studies.

Overall, these mini cases illuminate several interesting aspects of entrepreneurial decision-making that require further investigation. The next section will discuss emerging challenges and new directions for research in this field, taking into account the broader theoretical landscape of entrepreneurial decision-making.

## 13.4 New Directions and Open Questions in Entrepreneurial Learning

The scientific approach to entrepreneurship has shown promise in improving decision-making and outcomes for entrepreneurs across geographies and sectors. However, there are still many open questions regarding its implementation, efficacy, and scalability.

One area with significant potential for exploration is how best to teach this approach to entrepreneurs. Consider the question: How does the framing of the scientific approach impact its adoption by entrepreneurs? This is important because the way information is presented can significantly influence how it is received and implemented. A study might compare two groups of entrepreneurs: one taught the approach through cautionary tales of what not to do, and another through inspirational success stories that clearly show what to do. Researchers could track the adoption rates and effectiveness of the approach over time, potentially revealing whether positive or negative framing is more impactful in encouraging scientific thinking among entrepreneurs.

In addition, although studies show that the scientific approach leads to positive outcomes for entrepreneurs, we lack an understanding of synergies and complementarities that can further enhance its effect. The scientific approach primarily falls under the category of know-how knowledge transfer, providing entrepreneurs with practical skills for decision-making and hypothesis testing. However, its impact could potentially be amplified by combining it with other types of interventions. It is possible that integrating the scientific approach with interventions focused on entrepreneurial motivation and mindset (know-why) could yield powerful results. For instance, combining scientific approach training with personal initiative or resilience programs might not only equip entrepreneurs with analytical tools but also provide the perseverance needed to apply these tools consistently in the face of challenges. There might also be relevant complementarities with know-what interventions. Although the scientific approach provides a framework for decision-making, it could be enhanced by pairing it with targeted industry or market-specific knowledge (know-what). This combination might enable entrepreneurs to form more informed initial hypotheses and interpret data more effectively within their specific business contexts. Another intriguing possibility is the combination of the scientific approach with creativity-boosting interventions. Although the scientific approach provides rigor, creativity is crucial for generating novel hypotheses and innovative solutions. A dual-focus program might foster both analytical thinking and creative problem-solving, potentially leading to more innovative and well-tested business models.

If we focus on the challenge of habit formation presented in section 13.3 of this chapter, the timing and format of education also require additional investigation. We might ask: What is the impact of introducing the scientific approach to entrepreneurship earlier in the life of learners? Would that facilitate the adoption of this approach? Early exposure to these concepts could fundamentally shape how future entrepreneurs approach problem-solving and decision-making. A longitudinal study could follow students from secondary school through their early career years, comparing those who received early education in the scientific approach with those who did not. This could provide insights into how early exposure influences entrepreneurial tendencies, decision-making processes, and, ultimately, business outcomes. On the other hand, existing habits might hinder the adoption of the scientific approach. Experienced entrepreneurs might benefit greatly from structured scientific thinking. They may also be more resistant to changing their established way of doing things. An important question to study then is how the scientific approach can be effectively introduced to highly experienced entrepreneurs.

Another important avenue for future research relates to scaling teaching the scientific approach. All the RCTs describe intensive, in-person interventions conducted by instructors with small groups of about twenty to twenty-five start-ups. Traditional online and asynchronous learning methods might offer cost-effective scaling solutions, but they risk compromising the depth of learning and engagement that in-person courses offer. However, emerging technologies, particularly in the realm of artificial intelligence (AI), present a promising middle ground. AI-powered personalized learning systems could potentially revolutionize how we teach the scientific approach to entrepreneurship at scale. These systems could adapt to individual learners' needs, pacing, and learning styles, potentially replicating some of the benefits of personalized, small-group instruction in a more scalable format. This leads to a crucial research question: How effective is AI-powered personalized learning in teaching the scientific approach to entrepreneurship compared to traditional in-person and online methods?

The dynamics of team learning and knowledge diffusion within start-up teams is another rich area for exploration. Many start-ups are founded by teams, and the influence of individual learning on group dynamics can significantly impact a business's trajectory. A relevant question is: How does the adoption of the scientific approach by one team member influence the decision-making processes of the entire founding team? Relatedly, dynamics across start-up teams require further investigation. Peer networks could reinforce scientific thinking among entrepreneurs because peer influence can be a powerful force in shaping behaviors and maintaining new habits, as shown by Chatterji et al. (2019) and Dimitriadis and Koning (2022).

Lastly, the impact of the scientific approach on an entrepreneur's ability to persuade and secure resources is a critical area of study. We might ask: How does the use of scientific approach in business planning affect an entrepreneur's success in raising external funding and support? This question is vital because securing funding is often crucial for a start-up's success, and the way entrepreneurs present their ideas can significantly influence investors' decisions.

## 13.5 Conclusion: Toward a More Robust Entrepreneurial Decision-Making Framework

As we return to our opening question of this chapter—What helps entrepreneurs learn what is necessary for success?—we find that the answer is both more complex and more nuanced than we might have initially assumed.

The literature review in this chapter has highlighted a clear shift from traditional know-what interventions to more nuanced know-how and know-why approaches. The scientific approach, grounded in Bayesian learning principles, offers entrepreneurs a structured yet flexible path to form, test, and update their beliefs systematically. This approach, when properly implemented, leads to more informed decisions and faster adaptation to market realities. However, teaching this approach effectively presents significant challenges. Observations from field research have revealed that entrepreneurs often struggle with time pressures, abstract concepts, and the formation of new habits. To address these issues, specific tools such as the theory map and hypothesis canvas have been developed, with a view to making scientific thinking more accessible and immediately applicable to real business situations.

Despite these advances, obstacles remain. Many entrepreneurs still rely on imprecise theories of value and inadequate testing methods, and resist updating their beliefs when faced with contradictory evidence. The cases of Washly, Beerbox, and Arya clearly illustrate these challenges, emphasizing the complex interplay between cognitive biases, emotional investment, and rational decision-making in entrepreneurship.

Looking forward, several key questions demand attention: How can we effectively teach the scientific approach to experienced entrepreneurs? What role can AI-powered personalized learning play in scaling up this education? How does the adoption of scientific thinking by one team member influence an entire founding team's decision-making? These questions are crucial as we seek to refine and scale the teaching of scientific entrepreneurship.

Perhaps the most intriguing aspect of this evolving landscape is the potential for the scientific approach to bridge the gap between the art and science of entrepreneurship. By providing a structured framework for decision-making while encouraging creativity and adaptation, this approach may help reconcile the often-competing demands of intuition and analysis in entrepreneurial practice. Ultimately, our goal should be to equip entrepreneurs not just with a single decision-making framework but also with a rich toolkit of mental models and practical techniques that they can select and apply based on the specific challenges they face.

## References

Acimovic, J., Parker, C., Drake, D., and Balasubramanian, K. (2020). *Show or tell? Improving inventory support for agent-based businesses at the base of the pyramid*. SSRN.

Agarwal, R., Bacco, F., Camuffo, A., Coali, A., Gambardella, A., Msangi, H., Sonka, S., Temu, A., Waized, B., and Wormald, A. (2025). Does a theory-of-value add value? Evidence from a randomized control trial with Tanzanian entrepreneurs. *Organization Science*, 36(2), 601–625. https://doi.org/10.1287/orsc.2023.17590.

Agrawal, A., Gans, J. S., and Stern, S. (2021). Enabling entrepreneurial choice. *Management Science*, 67(9), pp. 5510–5524.

Alibhai, S., Buehren, N., Frese, M., Goldstein, M., Papineni, S., and Wolf, K. (2019). *Full esteem ahead? Mindset-oriented business training in Ethiopia*. World Bank Policy Research Working Paper (No. 8892).

Anderson, S. J., Chandy, R., and Zia, B. (2018). Pathways to profits: The impact of marketing vs. finance skills on business performance. *Management Science*, 64(12), pp. 5559–5583.

Anderson, S. J., and McKenzie, D. (2022). Improving business practices and the boundary of the entrepreneur: A randomized experiment comparing training, consulting, insourcing and outsourcing. *Journal of Political Economy*, 130(1), pp. 157–209.

Åstebro, T., and Hoos, F. (2021). Impact measurement based on repeated randomized control trials: The case of a training program to encourage social entrepreneurship. *Strategic Entrepreneurship Journal*, 15(2), pp. 254–278.

Baumol, W. J., and Strom, R. J. (2007). Entrepreneurship and economic growth. *Strategic Entrepreneurship Journal*, 1(3–4), pp. 233–237.

Bennett, V. M., and Chatterji, A. K. (2019). The entrepreneurial process: Evidence from a nationally representative survey. *Strategic Management Journal*, 44(1), pp. 86–116.

Berge, L. I. O., Bjorvatn, K., and Tungodden, B. (2015). Human and financial capital for microenterprise development: Evidence from a field and lab experiment. *Management Science*, 61(4), pp. 707–722.

Bingham, C. B., and Davis, J. P. (2012). Learning sequences: Their existence, effect, and evolution. *Academy of Management Journal*, 55(3), pp. 611–641.

Bjorvatn, K., Cappelen, A. W., Sekei, L. H., Sørensen, E. Ø., and Tungodden, B. (2020). Teaching through television: Experimental evidence on entrepreneurship education in tanzania. *Management Science*, 66(6), pp. 2308–2325.

Bloom, N., Eifert, B., Mahajan, A., McKenzie, D., and Roberts, J. (2013). Does management matter? evidence from india. *The Quarterly Journal of Economics*, 128(1), pp. 1–51.

Brooks, W., Donovan, K., and Johnson, T. R. (2018). Mentors or teachers? microenterprise training in kenya. *American Economic Journal: Applied Economics*, 10(4), pp. 196–221.

Bruhn, M., and Zia, B. (2013). Stimulating managerial capital in emerging markets: The impact of business training for young entrepreneurs. *Journal of Development Effectiveness*, 5(2), pp. 232–266.

Cai, J., and Szeidl, A. (2018). Interfirm relationships and business performance. *The Quarterly Journal of Economics*, 133(3), pp. 1229–1282.

Calderon, G., Cunha, J. M., and De Giorgi, G. (2020). Business literacy and development: Evidence from a randomized controlled trial in rural mexico. *Economic Development and Cultural Change*, 68(2), pp. 507–540.

Campos, F., Frese, M., Goldstein, M., Iacovone, L., Johnson, H. C., McKenzie, D., and Mensmann, M. (2017). Teaching personal initiative beats traditional training in boosting small business in West Africa. *Science*, 357(6357), pp. 1287–1290.

Camuffo, A., Cordova, A., Gambardella, A., and Spina, C. (2020). A scientific approach to entrepreneurial decision making: Evidence from a randomized control trial. *Management Science*, 66(2), pp. 564–586.

Camuffo, A., Gambardella, A., Messinese, D., Novelli, E., Paolucci, E., and Spina, C. (2024). A scientific approach to entrepreneurial decision-making: Large-scale replication and extension. *Strategic Management Journal*, 45(6), pp. 1209–1237.

Cao, R., Koning, R., and Nanda, R. (2024). Sampling bias in entrepreneurial experiments. *Management Science*, 70(10), pp. 7283–7307.

Chatterji, A., Delecourt, S., Hasan, S., and Koning, R. (2019). When does advice impact startup performance? *Strategic Management Journal*, 40(3), pp. 331–356.

Chatterji, A., Hasan, S., Larrick, R. P., and Masclans, R. (2023). *Taste before production: The role of judgment in entrepreneurial idea generation*. SSRN.

Chen, J. S., Croson, D. C., Elfenbein, D. W., and Posen, H. E. (2018). The impact of learning and overconfidence on entrepreneurial entry and exit. *Organization Science*, 29(6), pp. 989–1009.

Coali, A., Novelli, E., Spina, C., and Sirigiri, A. (2024). A scientific approach to decision making: Key tools and design principles. *Journal of Business Venturing Design*, 3, 100023.

Cohen, S. L., Bingham, C. B., and Hallen, B. L. (2019). The role of accelerator designs in mitigating bounded rationality in new ventures. *Administrative Science Quarterly*, 64(4), pp. 810–854.

Cole, S., Joshi, M., and Schoar, A. (2024). Heuristics on call: The impact of mobile-phone-based business-management advice. *The World Bank Economic Review*, 38(3), pp. 580–597.

Custódio, C., Mendes, D., and Metzger, D. (2020). The impact of financial education of executives on financial practices of medium and large enterprises. Swedish House of Finance Research Paper, working paper 70 pp.

De Mel, S., McKenzie, D., and Woodruff, C. (2014). Business training and female enterprise start-up, growth, and dynamics: Experimental evidence from Sri Lanka. *Journal of Development Economics*, 106, pp. 199–210.

Di Stefano, G., and Gutierrez, C. (2019). Under a magnifying glass: On the use of experiments in strategy research. *Strategic Organization*, 17(4), pp. 497–507.

Dimitriadis, S., and Koning, R. (2022). Social skills improve business performance: Evidence from a randomized control trial with entrepreneurs in Togo. *Management Science*, 68(12), pp. 8635–8657.

Drexler, A., Fischer, G., and Schoar, A. (2014). Keeping it simple: Financial literacy and rules of thumb. *American Economic Journal: Applied Economics*, 6(2), pp. 1–31.

Duflo, E., and Banerjee, A. (2017). *Handbook of field experiments*. Elsevier.

Eesley, C., and Wang, Y. (2017). Social influence in career choice: Evidence from a randomized field experiment on entrepreneurial mentorship. *Research Policy*, 46(3), pp. 636–650.

Eisenmann, T. (2021). *Why startups fail*. Penguin Random House.

Fafchamps, M., and Quinn, S. (2016). Networks and manufacturing firms in Africa: Results from a randomized field experiment. *The World Bank Economic Review*, 32(3), pp. 656–675.

Fairlie, R. W., and Miranda, J. (2017). Taking the leap: The determinants of entrepreneurs hiring their first employee. *Journal of Economics & Management Strategy*, 26(1), pp. 3–34.

Gans, J. S., Stern, S., and Wu, J. (2019). Foundations of entrepreneurial strategy. *Strategic Management Journal*, 40(5), pp. 736–756.

Gielnik, M. M., Uy, M. A., Funken, R., and Bischoff, K. M. (2017). Boosting and sustaining passion: A long-term perspective on the effects of entrepreneurship training. *Journal of Business Venturing*, 32(3), pp. 334–353.

Giné, X., and Mansuri, G. (2021). Money or management? A field experiment on constraints to entrepreneurship in Rural Pakistan. *Economic Development and Cultural Change*, 70(1), pp. 41–86.

Glaub, M., Frese, M., Fischer, S., and Hoppe, M. (2014). Increasing personal initiative in small business managers or owners leads to entrepreneurial success: A theory-based controlled randomized field intervention for evidence-based management. *Academy of Management Learning & Education*, 13(3), pp. 354–379.

Harrison, G. W., and List, J. A. (2004). Field experiments. *Journal of Economic literature*, 42(4), pp. 1009–1055.

Hasan, S., and Koning, R. (2019). Conversations and idea generation: Evidence from a field experiment. *Research Policy*, 48(9), pp. 103811.

Higuchi, Y., Mhede, E. P., and Sonobe, T. (2019). Short-and medium-run impacts of management training: An experiment in Tanzania. *World Development*, 114, pp. 220–236.

Higuchi, Y., Nam, V. H., and Sonobe, T. (2015). Sustained impacts of Kaizen training. *Journal of Economic Behavior & Organization*, 120, pp. 189–206.

Iacovone, L., Maloney, W., and McKenzie, D. (2022). Improving management with individual and group-based consulting: Results from a randomized experiment in Colombia. *The Review of Economic Studies*, 89(1), 346–371. https://doi.org/10.1093/restud/rdab005.

Karlan, D., and Valdivia, M. (2011). Teaching entrepreneurship: Impact of business training on microfinance clients and institutions. *Review of Economics and Statistics*, 93(2), pp. 510–527.

Lafortune, J., Riutort, J., and Tessada, J. (2018). Role models or individual consulting: The impact of personalizing micro-entrepreneurship training. *American Economic Journal: Applied Economics*, 10(4), pp. 222–245.

Leatherbee, M., and Katila, R. (2020). The lean startup method: Early-stage teams and hypothesis-based probing of business ideas. *Strategic Entrepreneurship Journal*, 14(4), pp. 570–593.

Lerner, J. (2013). The boulevard of broken dreams: Innovation policy and entrepreneurship. *Innovation Policy and the Economy*, 13(1), pp. 61–82.

Mano, Y., Iddrisu, A., Yoshino, Y., and Sonobe, T. (2012). How can micro and small enterprises in Sub-Saharan Africa become more productive? The impacts of experimental basic managerial training. *World Development*, 40(3), pp. 458–468.

McDonald, R. M., and Eisenhardt, K. M. (2020). Parallel play: Startups, nascent markets, and effective business-model design. *Administrative Science Quarterly*, 65(2), pp. 483–523.

McKenzie, D., and Puerto, S. (2021). Growing markets through business training for female entrepreneurs: A market-level randomized experiment in kenya. *American Economic Journal: Applied Economics*, 13(2), pp. 297–332.

McKenzie, D., and Woodruff, C. (2014). What are we learning from business training and entrepreneurship evaluations around the developing world? *The World Bank Research Observer*, 29(1), pp. 48–82.

McKenzie, D., Woodruff, C., Bjorvatn, K., Bruhn, M., Cai, J., Gonzalez-Uribe, J., Quinn, S., Sonobe, T., and Valdivia, M. (2023). Training entrepreneurs. *VoxDevLit*, 1(3).

Middleton, K., and Donnellon, A. (2014). Personalizing entrepreneurial learning: A pedagogy for facilitating the know why. *Entrepreneurship Research Journal*, 4(2), pp. 167–204.

Minniti, M., and Bygrave, W. (2001). A dynamic model of entrepreneurial learning. *Entrepreneurship Theory and Practice*, 25(3), pp. 5–16.

Novelli, E., and Spina, C. (2024). Making business model decisions like scientists: Strategic commitment, uncertainty, and economic performance. *Strategic Management Journal*, 45(13), pp. 2642–2695.

Parker, S. C. (2006). Learning about the unknown: How fast do entrepreneurs adjust their beliefs? *Journal of Business Venturing*, 21(1), pp. 1–26.

Ries, E. (2011). *The lean startup: How today's entrepreneurs use continuous innovation to create radically successful businesses*. Currency.

Santamaria, S., Abolfathi, N., and Mahmood, I. P. (2022). *Demand pull versus resource push approaches to entrepreneurship: A field experiment*. SSRN.

Shepherd, D., and Gruber, M. (2021). The lean startup framework: closing the academic–practitioner divide. *Entrepreneurship Theory and Practice*, 45(5), pp. 967–998.

Spina, C., and Battaglia, D. (2023). Do scientific entrepreneurs invest financial resources differently? Evidence from two RCTs. *Academy of Management Proceedings*, 2023(1), 17720. https://doi.org/10.5465/AMPROC.2023.17720abstract.

Spina, C., Camuffo, A., and Gambardella, A. (2020, November 18). Founders, apply the scientific method to your start-up. *Harvard Business Review*.

Spina, C., and Hasan, S. (2024, April 15). Field experiments in entrepreneurship and innovation (INSEAD Working Paper No. 2024/24/EFE). SSRN. https://doi.org/10.2139/ssrn.4794420.

Ubfal, D., Arráiz, I., Beuermann, D. W., Frese, M., Maffioli, A., and Verch, D. (2019). *The impact of soft-skills training for entrepreneurs in jamaica*. Discussion Paper 12325, IZA.

Valdivia, M. (2015). Business training plus for female entrepreneurship? Short and medium-term experimental evidence from Peru. *Journal of Development Economics*, 113, pp. 33–51.

Yang, T., and Aldrich, H. E. (2012). Out of sight but not out of mind: Why failure to account for left truncation biases research on failure rates. *Journal of Business Venturing*, 27(4), pp. 477–449.

# 14

## Introducing Causal Reasoning into Entrepreneurial Pedagogy and Practice

Andrea Coali, Claudia Frosi, Diego Jannace, Saeid Kazemi, and Abhinav Pandey

### 14.1 Introduction

The novel theorization around the concept of Bayesian entrepreneurship advanced in this book entails great normative potential. Agrawal et al. (2025) not only present a formal *description* of the thought process through which entrepreneurs make decisions but also advocate for the application of Bayesian reasoning models when making "low frequency-high impact" decisions (Camuffo, Gambardella, and Pignataro, 2024), thus including a strong *prescriptive* component. In the rapidly evolving landscape of entrepreneurship, this novel framework offers a promising solution for practitioners' growing need for a robust framework to guide decision-making under uncertainty. Indeed, a large body of research shows that entrepreneurs—and, more generally, managers and decision-makers—do not typically follow systematic routines when making decisions, ultimately leading to suboptimal choices (e.g., Bonabeau, 2003; Kahneman et al., 2019). This calls for the development and adoption of more structured approaches to decision-making (Davenport and Bean, 2018; European Business Review, 2021).

Bayesian entrepreneurship is embodied in several entrepreneurial methods (Agrawal et al., 2025). One of these is the scientific approach, pioneered by Camuffo et al. (2020) and described by Zellweger and Zenger (2023). As summarized by Novelli (2025), this approach has been empirically tested in several experimental studies (e.g., Agarwal et al., 2025; Camuffo, Gambardella, Messinese, et al., 2024; Coali, Gambardella, et al., 2024; Novelli and Spina, 2024). These studies translate a conceptually complex theory into practices that are directly actionable by entrepreneurs, addressing the key challenge of practically implementing a structured approach and delivering its principles pragmatically (Coali, Novelli, et al., 2024; van Aken, 2004).

The Bayesian entrepreneurship framework—particularly in its "deep Bayesian" formulation (Agrawal et al., 2025)—encourages entrepreneurs to think about theories of value supporting the development of their business ideas, which are based on causal logic and probabilistic reasoning. According to this approach, entrepreneurs should also translate their thought processes into hypotheses to be tested by means of well-structured experiments, which serve the purpose of updating their prior probabilities to make more informed decisions (Agrawal et al., 2025). Although these concepts might be familiar to academics, they may not be intuitive or practically implementable by entrepreneurs, who are typically oriented toward action rather than toward the extensive thinking processes required by a deep Bayesian approach (Shepherd and Gruber, 2021). Moreover, such an approach, which heavily relies on System 2 reasoning (Kahneman, 2013), must be reconciled with the dynamic and fast-paced environments of current businesses, where intuition- and heuristic-based decision-making has been found to be one of the most commonly adopted approaches (Miller and Ireland, 2005; Gareth Shepherd et al., 2024).

It follows that, although Bayesian and "scientific" approaches offer a powerful framework for decision-making under uncertainty, they also present significant challenges. How do we practically introduce causal and probabilistic reasoning into the broad entrepreneurial practice? How can this model be used to rationalize and formalize intuitive or experience-based decision-making? What tools or methodologies can effectively foster the application of the Bayesian entrepreneurship framework into entrepreneurial practice?

Recent academic literature has shown that these methods can be effectively taught to aspiring entrepreneurs and small ventures through extensive ad hoc trainings (e.g., Agarwal et al., 2025; Camuffo et al., 2020; Camuffo, Gambardella, Messinese, et al., 2024; Coali, Gambardella, et al., 2024). Such trainings, however, require significant resources—including, among others, classroom facilities, trained mentors, and teaching assistants—by the academic institution delivering it as well as an important time investment for participants. Given that entrepreneurs and managers are often unable or unwilling to invest dozens of hours into such training programs, the primary goal of this chapter is to explore the potential of simulation-based tools as a means to help decision makers in learning how to develop their own "theories of value." The development of theories of value represents, in fact, a fundamental part of the deep Bayesian learning process embodied in the application of the scientific approach.

The chapter begins with a brief discussion of the practical role of causal and probabilistic reasoning in decision-making, drawing from the decision-making and applied psychology literatures. Subsequently, the chapter presents a simulation-based tool managers and entrepreneurs can use to learn how to develop theories and to reason probabilistically. The chapter concludes with a discussion on the role of simulation-based tools in decision-making pedagogy and practice.

## 14.2 The Role of Causal and Probabilistic Reasoning in Practical Decision-Making

The backbone and fundamental aspect of the Bayesian entrepreneurship framework is the development of a theory of value (Agarwal et al., 2025; Felin and Zenger, 2009). As explained in the first chapter of this book (Agrawal et al., 2025), in the deep Bayesian learning formulation, entrepreneurs formulate a theory by explicitly identifying the so-called *attributes* (the key elements of the future state space set, with uncertain realizations; Camuffo, Gambardella, and Pignataro, 2024) and the related *causal relationships* among them. This allows entrepreneurs to develop a comprehensive framework that identifies a restriction on the parameters conceptualizing the different *probability distributions* of both attributes' realizations and causal linkages. The development of a theory in such a form necessarily requires entrepreneurs to think deeply about their ideas and the underlying assumptions that drive them, shifting from rapid action to deliberate contemplation.

However, this type of thinking and reflection is cognitively demanding. According to Kahneman (2013), it requires the activation of our System 2 mode of reasoning. This mode is engaged when uncertainty is high and decisions are complex, overriding the automatic responses associated with heuristic- and intuition-based choices made by the System 1 mode. Activating System 2 requires significant cognitive effort and resources, which can be limited by factors such as stress, cognitive load, and time pressure. Recent neuroscience evidence indicates that the regions of our brain involved in complex cognitive functions have high energetic costs (Castrillon et al., 2023). Research in decision-making has shown that individuals often revert to System 1 heuristics unless sufficiently motivated or capable of engaging System 2 (e.g., Stanovich and West (2000)). Furthermore, research has shown that entrepreneurs are typically oriented toward action rather than extensive thinking (Shepherd and Gruber, 2021).[1] Given these cognitive and temporal costs, it is necessary to determine whether the benefits of applying this framework outweigh them. If so, the challenge becomes understanding how to encourage its adoption among practitioners.

First, it is worth exploring whether the reasoning methodology prescribed by the deep Bayesian framework offers significant benefits. In the psychology of reasoning literature, it was 2009 when Over called for a shift from binary logics to Bayesian probabilistic theories as the standard of good inference. This shift implied focusing not only on structural reasoning based on logical forms but also on knowledge-rich inference based on content and background knowledge (the so-called priors). Adopting probabilistic reasoning means that decision-makers are never entirely certain of their logical premises, making reasoning an activity concerned with updating uncertain beliefs (Oaksford and Chater, 2020). This causal and probabilistic-based reasoning can simplify how individuals learn about decision-making cues. By using causal reasoning, decision-makers can streamline complex chains of relationships that appear in the environment, ultimately focusing on a small and manageable subset of cues, thereby improving decision quality (Garcia-Retamero and Hoffrage, 2006). Decision-makers who base their reasoning on relevant causal knowledge have been found to choose actions that increase the likelihood of achieving their desired outcome (Hagmayer and Witteman, 2017). In the entrepreneurship literature, research on the scientific approach discussed earlier in this book suggests that theory-based entrepreneurs experience improved performance compared with those following alternative approaches (Novelli, 2025). The benefits of applying a more deliberate and reflection-based approach on the development of new ventures and innovative ideas have been empirically tested (e.g., Santamaria et al., 2024), and research has shown how causal reasoning can complement and enhance resource-based approaches such as effectual reasoning (Smolka et al., 2018).[2]

The deep Bayesian framework attempts to rationalize and standardize the process of decision-making under uncertainty by utilizing one of the activities humans perform as experts, often without being fully aware of it: causal and probabilistic reasoning. By offering a standardized and clear methodology for decision-making, the approach also pushes theory and practice toward establishing a common language and a set of best practices that practitioners should adopt when making such decisions. The research cited above suggests that the benefits of such methodology can be substantial and can potentially outweigh the cognitive costs associated with its implementation. We believe that this is particularly true if an intuitive and time-efficient tool can be adopted to facilitate the absorption and application of the framework. Indeed, the crucial questions to be answered are *how* to explicitly introduce such reasoning in practical entrepreneurial decision-making activities and *how* to educate entrepreneurs in developing causal and probabilistic models. Given the cognitive costs of causal and probabilistic reasoning outlined above, and given entrepreneurs' preference toward action, it becomes paramount to identify ways to foster the application of such theory-based approach in practical decision-making activities. We believe that simulation-based tools offer an effective solution to this challenge.

## 14.3 Simulation Platforms as a Practitioner-Oriented Tool

Simulation-based tools have been used in classrooms for decades (Keys and Wolfe, 1990), for both teaching and research purposes. Simulations are argued to enhance the process of "generative learning" (Zantow et al., 2005)—namely, the process of generating relationships between one's knowledge and the information that one is trying to comprehend (Wittrock, 1985). Simulations, particularly computer-based ones, provide a risk-free environment where entrepreneurs can put in to practice theoretical knowledge and develop behavioral skills that can be replicated in real-world situations (Salas et al., 2009).

Helping entrepreneurs to absorb and apply the Bayesian framework primarily involves guiding them in formulating their problems causally and in reasoning probabilistically. Simulations offer two main advantages in this context. First, they help establish a taxonomy for Bayesian entrepreneurship, providing clear and applied definitions and constructs, as well as a common language for discussing entrepreneurial decision-making. By creating a standardized set of terms and definitions, simulations help to clarify the Bayesian approach, making it more accessible and easier to understand. Second, simulations offer a practical toolkit for practitioners. This toolkit includes templates, routines, and step-by-step guides that entrepreneurs can apply in real-world settings. For example, entrepreneurs can use simulation-based templates to map out causal relationships, identify key variables, and assign probabilities to different outcomes. These tools not only facilitate the application of Bayesian principles but also help entrepreneurs to systematically test and refine their business strategies. The practical nature of these tools ensures that theoretical concepts are translated into actionable steps that can be readily implemented in daily business routines.

Additionally, from a research and teaching perspective, computer-based simulations provide a common platform that can serve the academic purpose of bringing a more practical experience to classrooms and of allowing scholars to study entrepreneurial decision-making in a controlled setting. Researchers can design experiments within simulations to observe how entrepreneurs make decisions under varying conditions of uncertainty and complexity. This integration of theoretical and practical knowledge through simulations enables a more comprehensive understanding of entrepreneurial behavior and of the efficacy of the Bayesian framework in real-world decisions.

To illustrate the potential of simulation platforms as practitioner-oriented tools for implementing the deep Bayesian framework, the following paragraphs present the computer-based simulation platform developed by researchers at the ION Management Science Lab (IMSL) of SDA Bocconi School of

Management.[3] This platform articulates and applies the theoretical model and micro-foundations of theory-driven strategic management decisions developed in Camuffo, Gambardella, and Pignataro (2024).

Guided and inspired by the unraveling of world-famous cases of decisions under uncertainty, participants step into the roles of founders or top executives, simulating the strategists' process of reasoning, belief formation, testing, and updating. Each case immerses players in a strategic decision-making context characterized by uncertainty. They assume the role of the founder or CEO and learn a structured method to build a theory of value for the company. Importantly, and coherently with the deep Bayesian framework, participants do not simulate making decisions about specific actions. Rather, they focus on the earlier and more general phase of theory development, learning how to develop causal and probabilistic structures, how to use experimental data to update prior beliefs, and how to use such information to choose the preferred theory on which to base subsequent strategic actions.

Figure 14.1 shows an excerpt of the simulation page related to a strategic decision made by Red Bull's founder Dietrich Mateschitz back in 1982.[4] In this specific case, the player assumes the role of the founder, who, at the time, was evaluating whether to revamp his professional career by creating a new company producing energy drinks.

Each simulation begins by providing the player with background information about the company, the context, and the decision-maker whose role the player takes up. In the Red Bull case, the player is presented with the dilemma faced by Mateschitz. Mateschitz was a successful professional marketer, on track for an executive position at his previous company. However, after a trip to Thailand, he discovered a local energy drink that helped him combat jet lag, and he wondered whether the Western market would be receptive to such a product. This discovery prompted Mateschitz to explore the idea of founding his own brand, but he also saw significant opportunities in maintaining his current career trajectory. Consequently, he began developing a theory on why an energy drink brand in the Western world could be successful.

In most of the available case studies, the decision-maker envisions two different future scenarios and, thus, two different theories. For each theory, the player is provided with the attributes that the decision-maker considered relevant at the time of the decision. In the Red Bull case, the first theory relates to Mateschitz's idea of founding Red Bull. This theory relies on various attributes, such as product distinction and product adaptation, each of which can have dichotomous realizations (e.g., "yes" vs. "no"; "high" vs. "low"). Provided with this information, it is the player's duty to select the relevant attributes, causally link

**Figure 14.1**
The IMSL simulation platform: the Red Bull case-study.

them, and ultimately define their priors on both the attributes' realizations and causal links. This is accomplished using an interactive "theory building ground" that allows players to create their own causal structures in the form of Bayesian networks, as represented in figure 14.2.

In the "theory building ground," the player learns how to develop Bayesian networks in the form of causal graphs, which are visual representations that are commonly used to illustrate and analyze causal relationships between variables (Pearl, 1985, 1995; for a deep dive into the concept, please check the

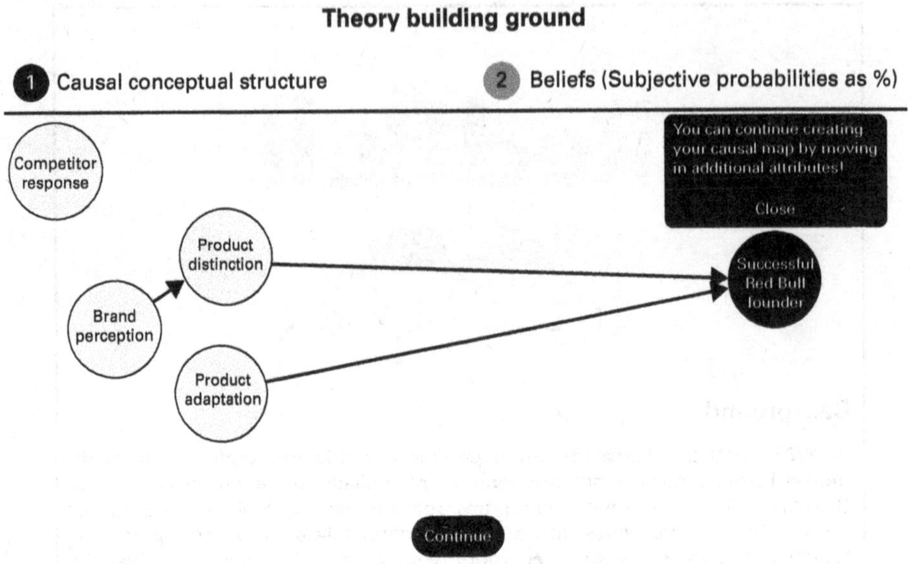


**Figure 14.2**
The theory building ground: creating causal connections.

appendix). In fact, the player can drag and drop attributes and choose how to causally connect them together. In figure 14.2, for instance, the player discarded the attribute "competitor response" and decided to causally link Red Bull success to two main antecedents, one of which also depends on the realization of the "brand perception" attribute. This approach allows the player to learn how to create causal graphs in an intuitive fashion. Once the player settles on a causal network, as illustrated in figure 14.3, the interface prompts them to formulate the probabilities both of attributes' realizations (depicted by the black and red ring around each attribute's circle) and of the causal links (depicted by the black circles juxtaposed on each link's arrow). Specifically, the player defines both the probability of the individual attributes' realization—namely, the prior subjective probability that an attribute actually happens—as well as the increase in probability associated with the causal link between two attributes.

The player should thus also determine their own beliefs about the degree to which causal links impact the probability of the realization of end attributes. Once this activity is completed, the backend simulation algorithm computes the expected probability of the theory formulated by the player and asks them to state their degree of confidence about the overall theoretical framework and the probability of the null hypothesis (figure 14.4), thus defining their "prior on

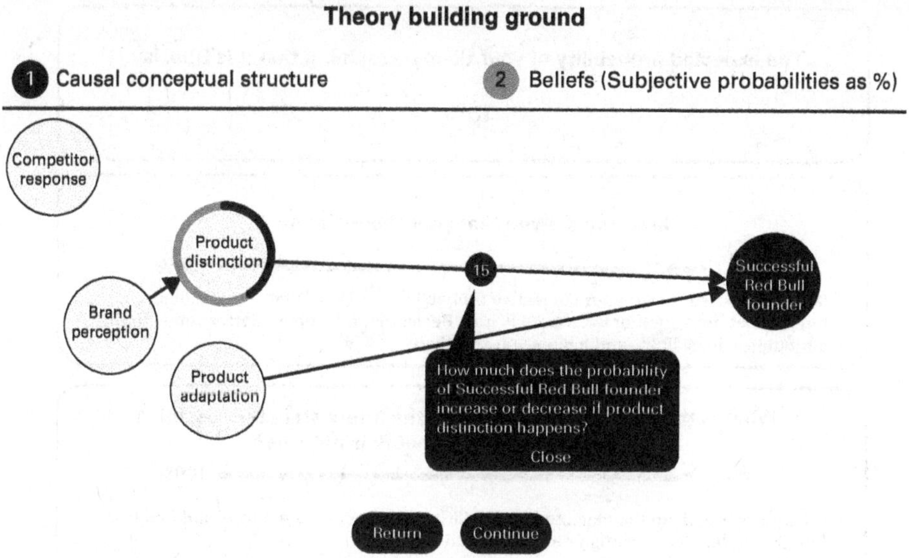


**Figure 14.3**
The theory building ground: defining prior probabilities.

priors" (Agrawal et al., 2025). These values allow the simulation to compute the unconditional expected probability of the future envisioned space.

Following the development of the first theory, the player is shown the second envisioned future space. In the Red Bull case, the alternative vision was for Mateschitz to continue his role as a marketing director, potentially reaching top executive or even CEO positions. This theory is less uncertain, given Mateschitz's extensive experience in the corporate world, but still hinges on several uncertain attributes. Given this different vision, the player is provided with different attributes and must repeat the steps undertaken for the previous theory, including the theory-building process described above.

Up to this point, the player has undergone two processes of theory formation, ideally absorbing the thought process needed to develop a theory. The next step is to start experimenting with the two theories. As shown in figure 14.5, players are prompted to choose the theory on which they want to experiment. This simulates a real-world situation in which running experiments is costly, and decision-makers have to decide what to devote their attention to.

Given the chosen experiment, the simulation platform provides players with additional information that can be used to update priors, as displayed in figure 14.6. Players are brought back to the theory building ground, where they can change the causal structure, their beliefs, and/or their prior on priors

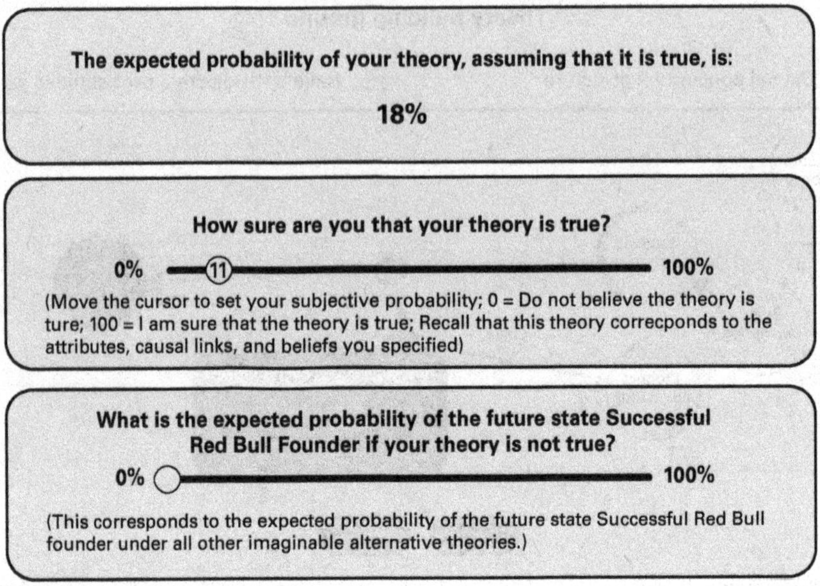


**Figure 14.4**
Expected probability and prior on priors.

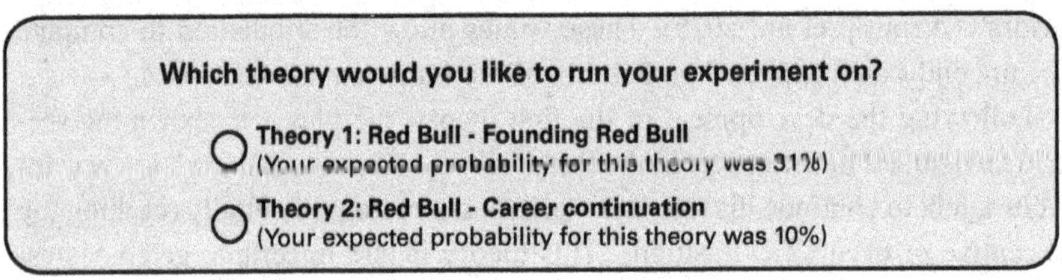


**Figure 14.5**
Experiments.

based on the experimental outcomes. As in the typical Bayesian framework, the subjective priors initially defined by the player are updated using actual observations and result in updated posteriors. Players can also choose to keep experimenting until they reach a point where they are able to decide on which theory to settle.

As explained, the probability outcomes displayed by the simulation engine depend on the players' subjective probability distributions, which directly stem from their reasoning and mental models. It is important to note that, unlike traditional case studies or other simulations, there is no right or wrong theory in this context. In the Red Bull case, this means that the first theory and related state space—which we know from actual events led to the founding of a successful

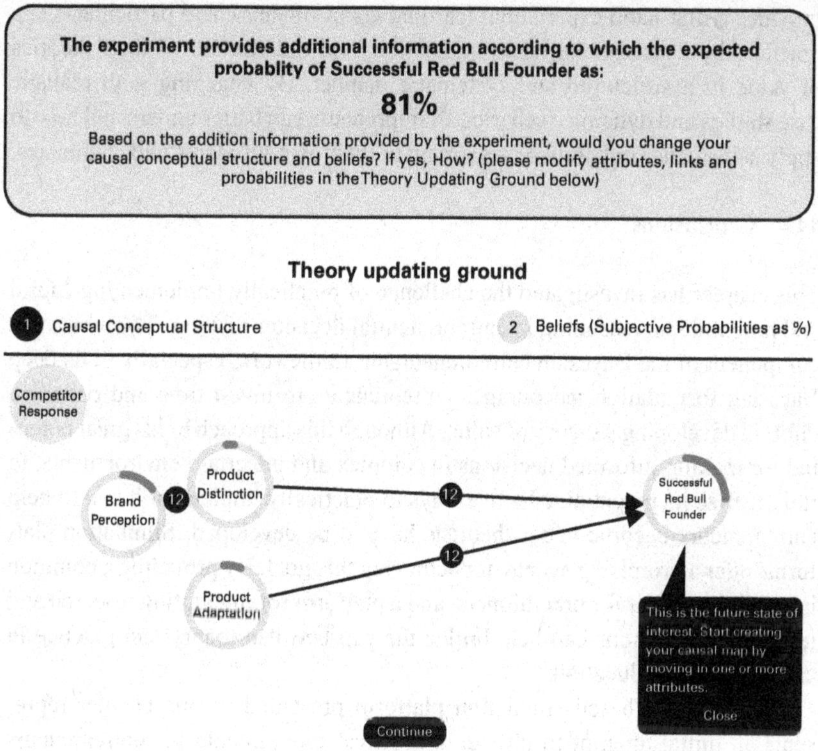


**Figure 14.6**
Experiments.

brand in the real world—might not be the best ones in the simulated world. In the backend, the simulation generates a random probability distribution that represents the likelihood of the "true" theories and future state spaces occurring. This approach allows players to receive feedback about their theorizing through the experimentation steps they take. Hence, players receive signals aligned with the unique distribution drawn by the backend, which can vary from game to game. Ultimately, this does not imply that the theory chosen by the player is right or wrong; it only indicates whether the theory the player settled on is more likely to occur, given the probability drawn from the probability distributions that populate the platform's backend. In this way, as in the real world, the player is nudged to fully acknowledge and embrace the fundamental uncertainty of strategic decision-making.

These simulations can thus serve as a powerful practical tool for entrepreneurs and decision-makers to learn how to formulate their decision-making according to the deep Bayesian framework. The prototype developed by IMSL provides a first-hand experiential learning environment where participants can practice the development, experimentation, and redefinition of their theories of value in a structured and systematic manner. By engaging with realistic case studies and dynamic scenarios, entrepreneurs can better understand how to apply causal and probabilistic reasoning to their own problems and challenges.

## 14.4 Conclusions

This chapter has investigated the challenge of practically implementing causal and probabilistic reasoning in entrepreneurial decision-making. The normative component of the Bayesian entrepreneurship framework, especially in its deep Bayesian formulation, encourages entrepreneurs to invest time and cognitive effort in developing a theory of value. Although this approach holds great potential for making informed decisions in complex and uncertain environments, to fully realize its potential, effective ways to practically implement it and to help entrepreneurs become better theorists have to be developed. Simulation platforms offer a promising avenue for achieving this goal. By providing a common language, a toolkit for practitioners, and a platform for integrating research and teaching, simulations can help bridge the gap between theory and practice in entrepreneurial education.

The computer-based simulation platform presented in this chapter represents an initial attempt to provide a practical tool suitable for entrepreneurs and decision-makers as well as to enhance the teaching of this approach in classrooms. This platform offers significant advancements in making the complex Bayesian framework more accessible and actionable. By allowing users to visualize and experiment with causal models through real-world case studies, it bridges the gap between theoretical concepts and practical application, fostering a deeper understanding and more informed decision-making. Additionally, the platform's interactive nature encourages active learning and engagement, making it an invaluable resource for both practitioners and educators.

Clearly, there is room for improvement, especially since research (and application) in these areas is still in its early stages. First, future developments of the platform could address the potential drawbacks of representing human causal reasoning with causal graphs if some assumptions are not met. Indeed, although causal graphs are arguably the best and most intuitive representation of causal relationships that can be used, scholars have noted how human causal reasoning often violates some of the assumptions behind the validity of causal graphs. For instance, causal graphs rely on the Markov independence condition—namely, that a variable is conditionally independent of its non-descendants, given its parents (direct causes). In other words, once one knows the value of a variable's direct causes, knowing the value of any other non-descendant variable does not provide any additional information about the variable in question. However, studies have shown how people often violate this condition, since it is more common to employ associative reasoning strategies rather than purely normative causal reasoning ones (Rehder, 2014). This implies that people may make wrong inferences in terms of causal structures and that causal graphs might reflect this bias. Thus, future versions of simulation platforms, other than making interfaces more user-friendly and intuitive, should also prompt players to deeply think about the parsimony of their models (Jannace, Kazemi, et al., 2025).

Second, expanding the simulation's capabilities to incorporate real-time data and generative artificial intelligence will further enhance its utility, making it a more comprehensive tool for strategic decision-making. A great advancement in the field would be the creation of tools that allow managers to develop causal models starting from their own narratives, eventually testing them in a simulated environment that draws information and data in real time from the web or other resources. Such a tool, which is easier to envision than to effectively produce, could be a game changer in how strategies are formulated and might fully popularize the Bayesian entrepreneurship approach by drastically reducing the cognitive burden associated with developing causal structures and defining prior probabilities. It would also address one of the remaining open questions: how to translate experience-based decision-making into such a causal framework. Such a tool would automatically enable entrepreneurs to visualize causal links that might have been obscured by experience-based reasoning and evaluate their correctness after deliberately seeing them mapped out and comparing them with real-world data and simulated scenarios.

Developing these tools presents a tremendous opportunity for academics to bridge the gap between academia and practice, creating a product that meets the growing demand for rigorous and practical guidance in entrepreneurial decision-making. As these tools evolve, they hold the potential to transform entrepreneurial education and practice, equipping decision-makers with the skills and insights necessary to thrive in a dynamic business environment. Ultimately, the continued development and refinement of these tools will contribute significantly to the field of entrepreneurship, offering innovative solutions that align theoretical rigor with practical application.

## 14.5 Appendix: A Short Note on Bayesian Networks and Causal Reasoning

This appendix provides an introduction to the topic of causal structures. Specifically, it covers elements to help better understand how to build causal structures, highlight their properties, and represent them through Bayesian networks. Understanding how causal structures work and can be represented is foundational to defining beliefs or priors (the subjective probability distributions on the future state space).

Since attributes are random variables and causal links define conditional probabilities among these variables, strategists can assign subjective probabilities to causal chains of attributes. One way of doing this is by means of Bayesian networks, which are defined by:

1. A directed acyclic graph (DAG) $G=(V,A)$ in which each node $v_i \in V$ corresponds to a random variable $X_i$
2. A global probability distribution $X$ of $G$ with parameters $\Theta$, which can be factorized into smaller probability distributions contingent on the edges $a_{ij} \in A$ present in the graph

Bayesian networks allow for the visualization of theories as DAGs, representing causal structures, as well as the beliefs associated with them. They also enable the updating of beliefs based on experiments conducted on the graph's elements. In these networks, the attributes of the theory are represented as nodes in the DAG, whereas the causal links are depicted as directed edges between the nodes. Consequently, a node depends on another only if there is a causal link; if there is no causal link, the nodes are assumed to be independent.

The network structure underlying the DAG articulates the (in)dependence relationships among nodes $v_i \in V$, thus specifying the potential factorization of the global probability distribution:

$$P(X)=\prod_{i=1}^{N} P(x_i | \prod X_i; \Theta_i)$$

$$\text{where} \prod X_i = \text{parents of } X_i$$

Formally, a DAG is an independence map of the probability distribution of $X$, with graphical separation among nodes implying probabilistic independence among them (Nagarajan et al., 2013). DAGs contain only *directed* edges, do *not* contain any loop (edges from a node to itself), and do *not* contain any cycle (edges that start and end in the same node). Strategists' theories are Bayesian networks characterized by:

1. One and only one node/attribute that does not cause/imply any other node (this node represents the "end state" or attribute of interest to the strategist)
2. At least one node/attribute that is not caused/implied by any other node (this node represents an assumption, a hypothesized action or a "micro-theory")

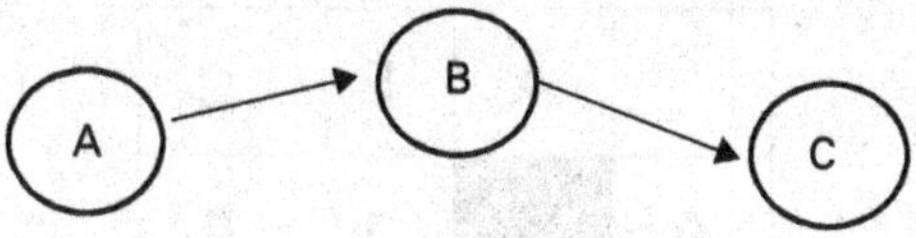


**Figure 14.7**
Sample DAG.

Therefore, once strategists select attributes, these represent the nodes of a Bayesian network. Strategists connect the attributes *only* through directed edges. With this respect, three aspects of theories as Bayesian networks are critical:

1. *How many* nodes/attributes (and which) strategists include in their theory
2. Which of the selected attributes/nodes strategists do *not* connect (i.e., are considered independent)
3. Which of the selected attributes strategists *connect* through causal links (i.e. using directed edges)

Using only directed edges, strategists can obtain a compact, factorized representation of the joint probability distribution of the attributes and causal links they consider by taking advantage of their conditional *independence*.

The methods used to represent directional dependencies among elements fall under the category of design structure matrices (DSM; Eppinger and Browning, 2012), widely implied in engineering design. DSMs represent dependencies in DAGs. The DSM is a square asymmetric matrix in which:

- nodes/attributes are on the columns and rows
- cells $a_{ij} = 1$ if node/attribute $i$ depends on node/attribute $j$ and $a_{ij} = 0$ otherwise

The whole matrix then represents the direct directional dependencies (causal links between attributes/nodes).

Consider the DAG (causal structure) represented in figure 14.7. It comprises three attributes and two causal links (chosen by the strategist). The *direct* causal dependencies in the graph are:

- *C* depends on *B* (i.e., *B* causes *C*),
- *B* depends on *A* (i.e., *A* causes *B*).

These direct dependencies (causal links) in the above DAG can be represented in the DSM shown in figure 14.8.

| | A | B | C |
|---|---|---|---|
| A | | 1 | 0 |
| B | | | 1 |
| C | | | |

**Figure 14.8**
Direct DSM.

| | A | B | C |
|---|---|---|---|
| A | | 1 | 1 |
| B | | | 1 |
| C | | | |

**Figure 14.9**
Indirect DSM.

Note that the use of directed edges (causal links) with no loops and no cycles imply that there are no links in the matrix below the diagonal. However, since we are interested in the whole causal chains, we need to map also the *indirect* dependencies. The indirect causal dependencies in the graph are:

- *C* depends on *B* which depends on *A* (i.e., *A* causes *B* which causes *C*).

This indirect dependency (causal *chain*) can be represented in the DSM represented in figure 14.9:

The indirect dependency between *A* and *C* occurs through *A* causing *B* and *B* causing *C*. If the direct dependencies are known, the first order (DSM) matrix can be raised to successive powers to reveal the causal chains among the attributes/nodes. The results display the direct and indirect dependencies that exist for successive path lengths. Summing these matrices yields the visibility matrix, which shows the dependencies that exist for all possible path lengths (causal chains among multiple nodes; MacCormack et al., 2006).

## Notes

1. For instance, the popular lean start-up framework (Ries, 2011) encourages entrepreneurs to immediately "get out of the building" after developing hypotheses on the key elements of their start-ups or innovative projects, prioritizing action and real-world testing over theoretical planning and extensive thinking (Felin et al., 2024).

2. Scholars are widely debating whether entrepreneurs should follow a rational and predictive approach such as the one advocated by this book or whether these decision-makers are better off with heuristics-based approaches, arguing that entrepreneurship is not necessarily an area where a theory-based approach is truly beneficial (see Dimov, 2024, for a critical perspective on the matter). However, to us, if a critique has to be moved toward the research program on theory-based approaches to entrepreneurship, this should have to do with the absence (at the time of this writing) of more in-depth analyses of the thought and cognitive processes put in place by entrepreneurs following such approach, rather than on the economic effectiveness of its application, which has been extensively proved by different groups of researchers. Although it is out of the scope of this chapter and book to empirically investigate this matter, this represents a promising research avenue to improve our understanding of the phenomenon.

3. The platform is freely accessible (upon registering) at https://imsl.unibocconi.it/index.php.

4. The platform includes more than fifty distinct case studies at the time of writing. Each case is based on real-world decisions made by leading companies over the past decades. Importantly, the cases are designed as the basis for simulation and discussion rather than to illustrate effective or ineffective strategizing, and they are based on public secondary sources. Therefore, they do not necessarily or accurately reflect the personal opinions or intentions of the individuals involved in the actual decisions.

## References

Agarwal, R., Bacco, F., Camuffo, A., Coali, A., Gambardella, A., Msangi, H., Sonka, S., Temu, A., Waized, B., and Wormald, A., (2025). Does a theory-of-value add value? Evidence from a randomized control trial with tanzanian entrepreneurs. *Organization Science*, 36(2), 601–625.

Agrawal, A., Camuffo, A., Gambardella, A., Gans, J. S., Scott, E. L., and Stern, S., (2025). The foundations of Bayesian entrepreneurship. *Bayesian entrepreneurship*. MIT Press.

Aken, J. E. V. van (2004). Management research based on the paradigm of the design sciences: The quest for field-tested and grounded technological rules. *Journal of Management Studies*, 41(2), pp. 219–246.

Bonabeau, E., (2003, May). Don't Trust Your Gut. *Harvard Business Review*.

Camuffo, A., Cordova, A., Gambardella, A., and Spina, C., (2020). A scientific approach to entrepreneurial decision-making: Evidence from a randomized control trial. *Management Science*, 66(2), pp. 564–586.

Camuffo, A., Gambardella, A., Messinese, D., Novelli, E., Paolucci, E., and Spina, C., (2024). A scientific approach to entrepreneurial decision-making: Large-scale replication and extension. *Strategic Management Journal*, 45(6), pp. 1209–1237.

Camuffo, A., Gambardella, A., and Pignataro, A., (2024). Theory-driven strategic management decisions. *Strategy Science*, 9(4), pp. 382–396.

Castrillon, G., Epp, S., Bose, A., Fraticelli, L., Hechler, A., Belenya, R., Ranft, A., Yakushev, I., Utz, L., Sundar, L., Rauschecker, J. P., Preibisch, C., Kurcyus, K., and Riedl, V., (2023). An energy costly architecture of neuromodulators for human brain evolution and cognition. *Science Advances*, 9(50) eadi7632.

Coali, A., Gambardella, A., and Novelli, E., (2024). Scientific decision-making, project selection and longer-term outcomes. *Research Policy*, 53(6), p. 105022.

Coali, A., Novelli, E., Sirigiri, A., and Spina, C., (2024). A scientific approach to decision-making: Key tools and design principles. *Journal of Business Venturing Design*, 3, p. 100023.

Davenport, T. H., and Bean, R., (2018, February 15). Big companies are embracing analytics, but most still don't have a data-driven culture. *Harvard Business Review*.

Dimov, D., (2024). The future in the mirror and behind it: Scientists and more. *Journal of Business Venturing Insights*, 21, e00464.

Eppinger, S. D., and Browning, T., (2012). *Design structure matrix methods and applications*. MIT Press.

European Business Review. (2021, November 14). Evolution of decision making: How leading organizations are adopting a data-driven culture. *The European Business Review*. https://www.europeanbusinessreview.com/evolution-of-decision-making-how-leading-organizations-are-adopting-a-data-driven-culture/.

Felin, T., Gambardella, A., Novelli, E., and Zenger, T., (2024). A scientific method for startups. *Journal of Management*, 50(8), pp. 3080–3104.

Felin, T., and Zenger, T., (2009). Entrepreneurs as theorists: On the origins of collective beliefs and novel strategies. *Strategic Entrepreneurship Journal*, 3(2), pp. 127–146.

Garcia-Retamero, R., and Hoffrage, U., (2006). How causal knowledge simplifies decision-making. *Minds and Machines*, 16(3), pp. 365–380.

Gareth Shepherd, N., Lou, B., and Maynard Rudd, J., (2024). Going with the gut: Exploring top management team intuition in strategic decision-making. *Journal of Business Research*, 181, p. 114740.

Hagmayer, Y., and Witteman, C., (2017). Chapter four—Causal knowledge and reasoning in decision making. In B. H. Ross (Ed.), Psychology of learning and motivation (Vol. 67, pp. 95–134). Academic Press.

Jannace, D., Kazemi, S., and Camuffo, A., (2025). Parsimonious theorizing.

Kahneman, D., (2013). *Thinking, fast and slow* (1st ed.). Straus and Giroux, Farrar.

Kahneman, D., Lovallo, D., and Sibony, O., (2019, March 4). A structured approach to strategic decisions. *MIT Sloan Management Review*.

Keys, B., and Wolfe, J., (1990). The role of management games and simulations in education and research. *Journal of Management*, 16(2), pp. 307–336.

MacCormack, A., Rusnak, J., and Baldwin, C. Y., (2006). Exploring the structure of complex software designs: An empirical study of open source and proprietary code. *Management Science*, 52(7), pp. 1015–1030.

Miller, C. C., and Ireland, R. D., (2005). Intuition in strategic decision making: Friend or foe in the fast-paced 21st century? *Academy of Management Perspectives*, 19(1), pp. 19–30.

Nagarajan, R., Scutari, G., and Lebre, S., (2013). *Bayesian networks in R*. Springer.

Novelli, E., (2025). A scientific approach to entrepreneurial decision-making as Bayesian learning. *Bayesian entrepreneurship*. MIT Press.

Novelli, E., and Spina, C., (2024). Making business model decisions like scientists: Strategic commitment, uncertainty, and economic performance. *Strategic Management Journal*, 45(13), pp. 2642–2695.

Oaksford, M., and Chater, N., (2020). New paradigms in the psychology of reasoning. *Annual Review of Psychology*, 71(2020), p. 71.

Over, D. E., (2009). New paradigm psychology of reasoning. *Thinking & Reasoning*, 15(4), pp. 431–438.

Pearl, J., (1985). Bayesian networks: A model cf self-activated memory for evidential reasoning. *Proceedings of the 7th Conference of the Cognitive Science Society*, University of California, Irvine, CA, USA, pp. 15–17.

Pearl, J. (1995). Causal diagrams for empirical research. *Biometrika*, 82(4).

Rehder, B., (2014). Independence and dependence in human causal reasoning. *Cognitive Psychology*, 72, pp. 54–107.

Ries, E., (2011). *The lean startup: How today's entrepreneurs use continuous innovation to create radically successful businesses*. Currency.

Salas, E., Wildman, J. L., and Piccolo, R. F., (2009). Using simulation-based training to enhance management education. *Academy of Management Learning & Education*, 8(4), pp. 559–573.

Santamaria, S., Abolfathi, N., and Mahmood, I. P., (2024). Demand pull versus resource push training approaches to entrepreneurship: A field experiment. *Strategic Management Journal*, 45(3), pp. 564–587.

Shepherd, D., and Gruber, M., (2021). The lean startup framework: Closing the academic–practitioner divide. *Entrepreneurship Theory and Practice*, 45(5), pp. 967–998.

Smolka, K. M., Verheul, I., Burmeister-Lamp, K., and Heugens, P. P. M. A. R., (2018). Get it together! Synergistic effects of causal and effectual decision–making logics on venture performance. *Entrepreneurship Theory and Practice*, 42(4), pp. 571–604.

Stanovich, K. E., and West, R. F., (2000). Individual differences in reasoning: Implications for the rationality debate? *Behavioral and Brain Sciences*, 23(5), pp. 645–665.

Wittrock, M. C., (1985). Teaching learners generative strategies for enhancing reading comprehension. *Theory Into Practice*, 24(2), pp. 123–126.

Zantow, K., Knowlton, D. S., and Sharp, D. C., (2005). More than fun and games: Reconsidering the virtues of strategic management simulations. *Academy of Management Learning & Education*, 4(4), pp. 451–458.

Zellweger, T. M., and Zenger, T. R., (2023). Entrepreneurs as scientists: A pragmatist approach to producing value out of uncertainty. *Academy of Management Review*, 48(3), pp. 379–408.

## Contributors

**Ajay Agrawal,** University of Toronto and NBER

**Ashish Arora,** Fuqua School of Business, Duke University and NBER

**Thomas Åstebro,** HEC Paris

**Arnaldo Camuffo,** Bocconi University

**Andrea Coali,** ION Management Science Lab, SDA Bocconi, and Cerved Group

**Susan Cohen,** Terry College of Business, University of Georgia

**Timo Ehrig.** Solvay Brussels School of Economics and Management, Universite libre de Bruxelles

**Teppo Felin,** Department of Strategy and Entrepreneurship, Huntsman School of Business, Utah State University, and Department of Strategy, Innovation and Management, Saïd Business School, University of Oxford

**Frank M. Fossen,** Department of Economics, University of Nevada-Reno

**Claudia Frosi,** ION Management Science Lab, SDA Bocconi

**Alfonso Gambardella,** Bocconi University

**Joshua S. Gans,** Rotman School of Management, University of Toronto and NBER

**Luca Gius,** MIT Sloan

**Cédric Gutierrez,** Department of Management and Technology and ICRIOS, Bocconi University

**Diego Jannace,** ION Management Science Lab, SDA Bocconi

**Saeid Kazemi,** ION Management Science Lab, SDA Bocconi

**Hyunjin Kim,** INSEAD

**Rembrand Koning,** Harvard Business School

**Joshua L. Krieger,** Harvard Business School

**Ramana Nanda,** Imperial and Harvard Business School.

**Elena Novelli,** Bayes Business School, City, University of London

**Abhinav Pandey,** ION Management Science Lab, SDA Bocconi

**Erin L. Scott,** MIT Sloan

**Chiara Spina,** Entrepreneurship and Family Enterprise Area, INSEAD

**Scott Stern,** MIT Sloan and NBER

**Todd Zenger,** David Eccles School of Business, University of Utah

# Index

Page numbers followed by an *f* denote a figure, a *t* denotes a table.